Lecture Notes in Electrical Engineering

Volume 29

Shan-Ben Chen · Jing Wu

Intelligentized Methodology for Arc Welding Dynamical Processes

Visual Information Acquiring, Knowledge Modeling and Intelligent Control

Springer

Shan-Ben Chen
Institute of Welding Engineering
Shanghai Jiao Tong University
Dongchuan Road, 800
Shangahi, 200240
P R China
sbchen@sjtu.edu.cn

Jing Wu
Institute of Welding Engineering
Shanghai Jiao Tong University
Dongchuan Road, 800
Shangahi, 200240
P R China
wujing065@gmail.com

ISBN: 978-3-540-85641-2 e-ISBN: 978-3-540-85642-9

Library of Congress Control Number: 2008935359

Cover design: eStudio Calamar S.L.

Printed on acid-free paper

9 8 7 6 5 4 3 2 1

springer.com

Preface

Welding handicraft is one of the most primordial and traditional technics, mainly by manpower and human experiences. Weld quality and efficiency are, therefore, straitly limited by the welder's skill. In the modern manufacturing, automatic and robotic welding is becoming an inevitable trend. However, it is difficult for automatic and robotic welding to reach high quality due to the complexity, uncertainty and disturbance during welding process, especially for arc welding dynamics. The information acquirement and real-time control of arc weld pool dynamical process during automatic or robotic welding always are perplexing problems to both technologist in weld field and scientists in automation.

This book presents some application researches on intelligentized methodology in arc welding process, such as machine vision, image processing, fuzzy logical, neural networks, rough set, intelligent control and other artificial intelligence methods for sensing, modeling and intelligent control of arc welding dynamical process.

The studies in the book indicate that the designed vision sensing and control systems are able to partially emulate a skilled welder's intelligent behaviors: observing, estimating, decision-making and operating, and show a great potential and promising prospect of artificial intelligent technologies in the welding manufacturing.

The book is divided into six chapters: Chap. 1 gives an introduction on development of welding handicraft and manufacturing technology; Chap. 2 mainly addresses visual sensing systems for weld pool during pulsed Gas Tungsten Arc Welding (GTAW); Chap. 3 mainly address information acquirement of arc welding process by image processing methods, including acquiring two dimensional and three dimensional characteristics from monocular image of GTAW weld pool; Chap. 4 mainly addresses modeling methods of weld pool dynamics during pulsed GTAW, including identification of linear models and nonlinear transfer function models of weld pool dynamical process; artificial neural network models and knowledge models for predicting and control of weld pool dynamical characteristics; Chap. 5 mainly addresses intelligent control strategies for arc welding process, including self-regulating PID, fuzzy, PSD controllers, neural network self-learning controllers, model-free controller and composited intelligent controllers for dynamical weld pool during pulsed GTAW; Chap. 6 mainly addresses real-time control of weld pool dynamics during robotic welding process, including intelligentized welding robot systems with real-time monitoring and control of weld pool dynamics; and an application of intelligentized welding robot systems.

The ordinal reading of this book has two outlines: one reading line is compiled by current outline, i.e., sensing, modeling and control methodology for welding process; the other by classifying of welding technics and conditions, or welding materials, e.g., bead-on-plate, welding with wire filler, gaps variation conditions; steel and aluminium alloy welding workpiece. Bead-on plate welding is addressed in Sects. 3.1.2.2, 4.1, 5.3.2. Welding with wire filler is mainly addressed in Sects. 2.2.2, 3.1.2.3, 3.2, 4.3.3, 5.6.4, 6.3. Gap variation condition is mainly addressed in Sects. 5.4.2, 5.6.2, 6.4.1. Aluminium alloy welding is mainly addressed in Sects. 2.3, 3.1.3, 3.2, 4.2.1, 4.4.2, 4.4.3, 5.1, 5.2.2, 5.6.3, 6.2.2, 6.3, 6.4. Steel welding is mainly addressed in Sects. 2.2, 3.1.2, 3.2.3, 4.2.2, 4.3, 5.2.2, 5.3.2, 5.6, 6.1.

The research results in this book were mainly implemented in the Intelligentized Robotic Welding Technology Laboratory (IRWTL), Shanghai Jiao Tong University, P R China. The content in the book involves the following doctoral dissertations: Dr. Yajun Lou, Dr. Dongbin Zhao, Dr. Guangjun Zhang, Dr. Jianjun Wang, Dr. Bing Wang, Dr. Laiping Li, Dr. Wenjie Chen, Dr. Quanying Du, Dr. Wemhang Li, Dr. Xixia Huang, Dr. Hongyuan Shen, Dr. Chongjian Fan, Dr. Fenglin Lv and Dr. Huabin Chen's works, etc. As a supervisor of their doctoral dissertations, Professor Shan-Ben Chen would like to thank their contributions to this book.

We wish to give expression on acknowledgements for the researched works in this book supported by the National Natural Science Foundation of China under Grant No. 50575144 and No. 60474036; and supported by the Key Foundation Program of Shanghai Sciences & Technology Committee under Grant No. 06JC14036 and No.021111116.

We would like to thank Professor Tzyh Jong Tarn and Professor Lin Wu for their directions on the research works included in the book. And last but not least thank to Dr. Thomas Ditznger for his advice and help during the production phases of the book.

Shanghai, China *Shan-Ben Chen*
Jing Wu

Contents

List of Figures

List of Abbreviations

ANN	artificial Neural Network
BP	Backpropagation
BRDF	bi-directional reflectance distribution function
CCD	Charge Coupled Device
δ	pulse duty ratio
D_l	depression on length
D_w	depression on width
DWP	direct weld parameters
FA	function approximator
GTAW	Gas Tungsten Arc Welding
GMAW	Gas Metal Arc Welding
H_t	topside height
I_p	peak value current
I_p	peak value welding current
MIG	metal inert gas
MFC	model-free adaptive controller
MAG	metal active gas
MIMO	Multiple-Input Multiple-Out-put
OIPWPVCS	Obtaining and Image Processing of Weld Pool Vision Characteristics System
RS	Rough sets
SFS	Shape From Shading
SISO	Single-Input-Single-Output
V_f	wire feeding velocity
VPRS	variable precision rough set
V_w	welding velocity
W_t	topside width

Chapter 1
Introduction

Abstract In this chapter, an introduction is given on the development of welding handicraft, manufacturing technology and key technologies of welding automation and intelligentization. Recent twenty years have seen great development of welding robot in modern manufacturing industry, where arc welding is one of the mainstream technology. A large number of researches show that automatic control of the welding process requires not only good performance of the equipment, but also technologies, namely sensing, modeling and controlling of the welding process. None of the technologies is neglectable for welding process control, in which sensing is to monitor the process and extract characteristic information of the welding process; modeling is to identify the process based on acquired information; and controlling is to regulate the welding process based on the established models. The main part in controlling is to design controller for multi-variables coupled, nonlinear and time-varying situations.

1.1 Development of Welding and Manufacturing Technology

Welding handicraft was invented more than 3,000 years ago, and the traditional welding was implemented mainly by welder handwork and experiences [1–4]. Welding operation by hand is a burdensome and tedious labor for welder; moreover, reliability and consistency of welding quality depends on welder's ability and experiences. Of course, efficiency of welding production is also limited straitly.

Recent twenty years have seen great development of welding robot in modern manufacturing industry, where arc welding is the mainstream technology with wide application [2, 3, 5, 6]. And weld seam recognition, tracking and weld quality control are remaining the hotspot problems in welding automation and robot with the technology of weld pool information extracting and dynamic penetration control [7–11]. Dynamic control of welding process is based on the precise obtaining of weld pool geometry information, which is difficult due to the disturbance from arc, movement of weld pool and the complex nature of welding itself. Welding is a MIMO (Multiple-Input Multiple-Out-put), nonlinear, time-varying and strongly coupled process of metallurgy, physical chemistry, material property and heating

S.-B. Chen, J. Wu, *Intelligentized Methodology for Arc Welding Dynamical Processes*, Lecture Notes in Electrical Engineering 29,

with multiple affecting variables, such as welding current, welding velocity, wire feeding velocity, shielding gas, welding torch pose, and with multiple disturbances such as deformation, variation in thermal conduction and seam gap size and mismatching of the workpiece. It is, therefore, difficult to model and control this process during automatic and robotic welding.

The "teach and playback" robot and automatic equipment with only off-line parameter regulation are not able to overcome the disturbances and fluctuation during welding practice, thus they cannot meet the requirements of high qualified welding. It is more urgent to realize the control of welding dynamics and welding seam quality [9–12] for high quality welding products with intelligentized welding robot systems.

A large number of researches show that automatic control of the welding process requires not only good performance of the equipment, but also the key technologies [13,14], namely sensing, modeling and controlling of the welding process, as shown in Fig. 1.1.

None of the technologies in Fig. 1.1 is neglectable for welding process control, in which sensing is to monitor the process and extract characteristic information of the welding process; modeling is to identify the process based on the acquired information; and controlling is to regulate the welding process based on the established models. The main part in controlling is the design of controller so that it can deal with multi-variables coupled, nonlinear and time-varying situations.

To study the intelligence of welder is significant for the development of welding automation and intelligentized robot. Much better than any welding robot, a professional welder is highly adaptive to practical situations through observing the position of welding joint, dynamics of weld pool, shape of arc and appearance of the welding seam to identify the welding status; and regulating the parameters to produce high qualified welding seam. To realize automation in welding, the first step is to develop the welding sensor similar to human sensing systems. Welding sensor is a detector that can get the interior and exterior conditions of welding. Next step is the identification of the welding process to describe the time-varying welding status, i.e. modeling the welding process. The third step is the human-brain-like controller to reason the controlling strategy.

A brief description of the sensing, modeling and controlling of the welding process will be made in the following parts.

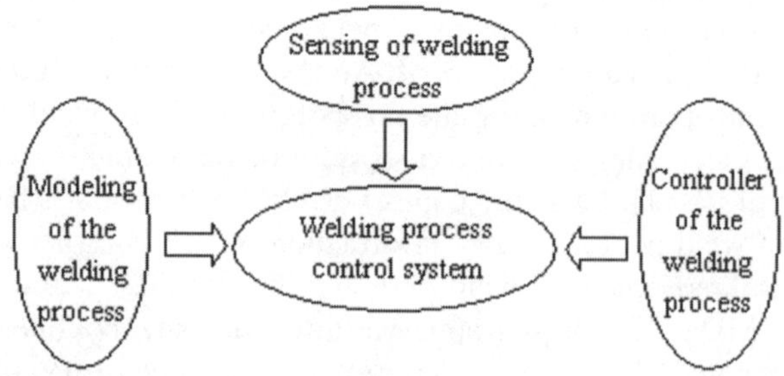

Fig. 1.1 Key technologies in the control system of the welding process

1.2 Sensing Technology for Arc Welding Process

Arc welding is roughly categorized into Gas Tungsten Arc Welding (GTAW) and Gas Metal Arc Welding (GMAW), and in this book, GTAW will be mainly discussed.

Chiefly used to monitor the state of welding process and to extract information of welding process, arc welding sensor is important for modeling and controlling the welding process [15]. Different sensing methods for welding process have been used in consideration of the disturbance from arc, high temperature, vibration, electromagnetic fields and the features of the process. Theses methods includes ultrasonic method to sense penetration [16–18], arc pressure method and arc light method to sense vibration information of the weld pool [19–30], infrared thermo scope to sense the welding temperature field [31–36], X ray method to sensor the shape of the welding pool [37, 38], acoustic sensor [39–42] and visual sensor [43–73].

Acoustic sensor is mainly used to detect the metal transfer in GMAW and keyhole plasma welding, etc [39–42]. With its main use in penetration sensing and defect detection, ultrasonic method in acoustic sensor has a limited application because its signal processing devices are complex and the coupling problem between probe and workpiece in movement is difficult of solving.

Force sensing is the recently developed weld pool oscillation method [19–25] to detect the oscillation in weld pool by arc pressure or arc light signal. This method is only applicable to step welding, and penetration information is difficult to be identified among disordered arc pressure signals due to the disturbance from arc pressure ripples of ordinary welding power.

Compared with other sensing methods, visual sensor is the most prospective sensing technology because it is not in tough with the welding circuit, thus its signal detection does not affect the welding process and it can provide with sufficient information, such as type of joint, welding edges, type of arc, position of wire and the shape of solidified welding seam. Visual sensor can be categorized into short wave (X ray), visible light and infrared by the wave length of the devices. X ray can detect the shape of weld pool because it decays with different amount in according to the thickness of the weld pool, but with the disadvantage of health-damaging, large size, complexity and high price. Infrared is used to sense temperature field and build the direct relationship between surface temperature and penetration of workpiece in avoid of the disturbance of arc light, but with the disadvantage of high price and non-appliable to practical welding manufacturing [26–30].

1.3 Visual Sensing Technology for Arc Welding Process

Visual sensing technology is well developed with the progress in electronic industry and image processing methods. It will be widely used in welding practices when visual devices are decreasing in price, increasing in reliability and protective measure and improving in image processing hardware and software. As the most studied welding sensor, it is more suitable for quality control of welding process than other means of sensing devices because it can obtain two dimensional and three dimensional information of weld pool surface, which directly reflect the welding dynamics of molten metal.

As the most frequently used visual sensor, CCD (Charge Coupled Device) obtains welding images of visible light, with the feature of non-affect to the welding process, in no touch with welding system providing rich information, such as type of joint, seld pool shape, arc state, etc. Therefore, it has become an important field of studies in direct observation of weld pool by machine vision and extraction of weld pool geometric information by image processing. Visual sensing can be divided into two categories, respectively active and passive sensing [43–75].

1.3.1 Active Visual Sensing

Active visual sensing uses laser or structural light as its light sources for the welding area, so as to obtain clear image by avoiding the effect from arc light. Laser features high intensity, directionality, monochromaticity and coherence.

Active direct visual sensor, composed of laser diode and CCD camera, is mostly used in two dimensional laser scan welding seam tracking and arc welding robot guiding system. For the application of welding quality sensing, Refs. [45–47] designed the structural light three dimensional visual sensor by light truncation to measure topside height of the weld pool surface. Point light source from laser producer is turned into line light through cylindrical lens, then into laser stripe after intersecting with workpiece. The geometric information of weld pool, such as average topside height of welding seam, is extracted from the root of weld pool.

References [48–50] proposed a stroboscopic vision method composed of a high energy density pulse laser and an electric shutter camera. Figure 1.2(a) is the schematic diagram of stroboscopic vision. Secondary light source is the pulse laser or Xe flash light source and visual sensor is CCD camera. Stroboscopic vision method can obtain clear weld pool image in GTAW and plasma arc welding. Figure 1.2(b) shows the weld pool image during GTAW by stroboscopic vision. The image is clear with strong contrast, thus it is easy to extract geometry information of the weld pool.

To obtain three dimensional information of the weld pool, Refs. [51–55] designed a weld pool visual sensing system composed of a grid-shaped structured light stripe high power laser and electronic shutter, with the schematic diagram shown in Fig. 1.3. Average power of laser pulse is 7 mW; its duration time in a period is 3 ns, its power is 50 kW and wave length is above 337 nm. Clear image of weld pool is obtained because energy density is much large than that of the disturbance from arc light. Figure 1.4 (a) is the three dimensional weld pool image obtained during GTAW of stainless steel sheet SS304 with welding current as 118 A, arc length 3 mm. Specific image processing algorithm can be used to extract the edge of frame of grid-shaped structured light stripe as shown in Fig. 1.4 (b) and three dimensional

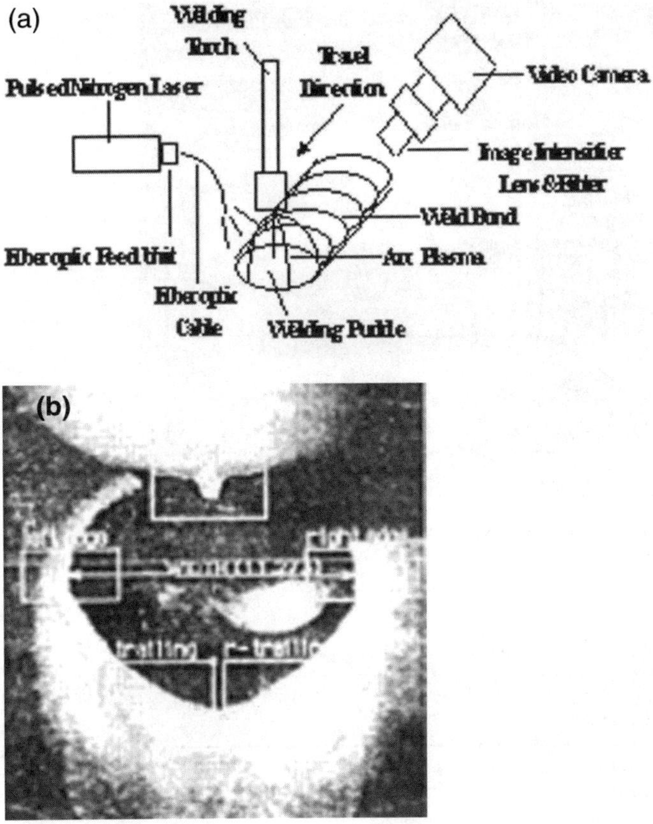

Fig. 1.2 Weld pool image with the stroboscopic vision sensing system [48] (**a**) Schematic diagram (**b**) Schematic diagram

height of the weld pool is calculated. But this method is limited due to the quality of the image and precise of calculation.

The above two methods are good among all the active visual sensing methods using secondary light source, but its application is limited due to its high cost of high energy density pulse laser and special electric shutter camera.

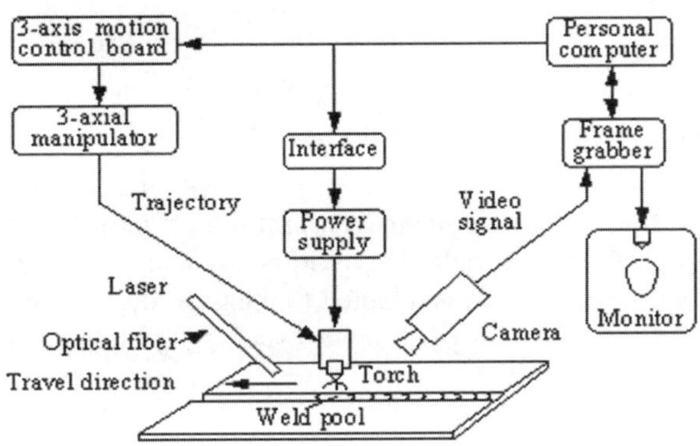

Fig. 1.3 Schematic of sensing the image of weld pool using structural light system [51]

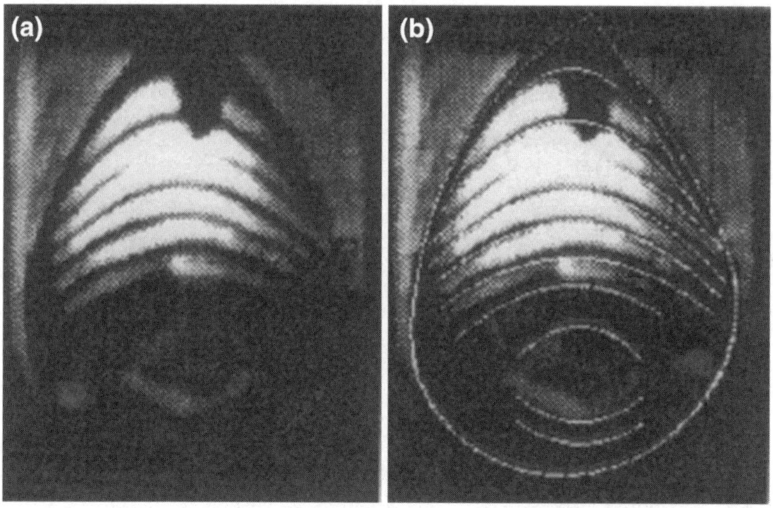

Fig. 1.4 The pool image with structural light sensing system in GTAW [51] (**a**) Original image (**b**) Stripe skeleton and boundary

Table 1.1 Research status of active vision sensing in welding process

Researcher	Sensing information	Equipment	Strength	Weakness
R. Kovacevic [22–25] (USA)	3D visual sensing of weld pool	Pulse laser, Grating	High energy density	High cost
C. G.Morgan [64] (Oxford University)	Welding seam image	CCD camera, Laser diode	Little disturbance from arc light	Difficult of installation
J.E. Agapakis [76] (USA, Automatix Inc)	Stroboscopic vision sensing	High energy pulse laser, electric shutter, CCD	Clear image, for general purpose	Hight cost

Table 1.1 shows the research status of active vision sensing technology in welding process.

1.3.2 Passive Direct Visual Sensing

Without additional secondary light source, passive direct visual sensing use the light from black body radiation of liquid metal, metallicl vapor and arc. Plasma spectrum diagnostic method is adopted to measure the spectrum intensity and width and the affect from welding parameters such as welding current, arc length, material of the workpiece, shield gas volume and welding velocity to the distribution of spectrum. Composed of molecules and atoms of shield gas, plasma of shield gas and vapor from metal, arc emits characteristic spectral line during welding. The spectrum line is mainly nonmetallic in the area of arc column but mainly metallic in the area of weld pool surface. Arc spectrum is composed of continuous part of electron transition and of discrete part of peak spectrum. The spectrums of arc column is quite different from that of weld pool surface [26–30], [56–59].

One method of passive visual sensing is to observe the image near arc area by the light from weld pool and its reflect of arc light in the window in the interval of arc spectrum lines to avoid disturbance from the strong spectrum lines of arc.

References [57] selected the optimum imaging widow at 4064±20 Å through analysis of the experimental spectrum data of low carbon steel during GTAW. The center wave length is 4044 Å and the sprctrum area is with half width of 40 Å. Figure 1.5(a) shows the distribution of spectrum line of different elements and Fig. 1.5(b) shows the image of weld pool by spectrum sensing method. Closed control is based on the data obtained from this method [50, 51].

The other method of passive visual sensing is to use arc radiation as light source and to select a window in the interval of spectrums. The spectrum is mainly continuous with few peak values but mostly nonmetal spectrums to produce steady and high intensity light source.

References [60–62] proposed a visual sensing system which is placed inside the welding torch and in the same axis of electrode. It can observe the full picture of welding area clearly because the electrode and nozzle block off the brightest part of arc and avoid excessive expose to the arc light. Primary research in weld pool observation and welding seam tracking was carried out. Figure 1.6 (a) shows the system set and Fig. 1.6(b) shows image of weld pool. Guass filtering and edge enhancing algorithm are used to obtain the edge of weld pool with the processing time of 6.5 s on VAX11/785 PC for only off-line use.

References [66] analyzed the spectrum line of arc radiation and weld pool metal together with its integral intensity under the base welding current of 60A during GTAW. It is proposed to light the weld pool by contrinuous spectrum of arc in the imaging window so as to enhance image contrast by mirror reflection from liquid metal on the surface of weld pool and diffuse reflection from solid workpiece surface, thus turning the disadvantage factors into advantages ones. The center wave length of imaging window of narrow-band filter is 661 nm, its half width is 10 nm, and transmittance is 28.8%. And double-side visual sensing system is designed. References [63–65] successfully obtain the topside weld pool image with CCD camera at base welding current because the current is low in base period and arc light is relatively weak to remove the effect of arc light in some delay after the welding current decreases during pulsed GTAW. The weld pool image is clear and features high contrast.

Table 1.2 shows the passive vision sensing method in the welding process.

After analyzing the spectrum features of aluminium alloy weld pool area, Refs. [66–68] proposed a wide band filtering method with reflected arc light as light

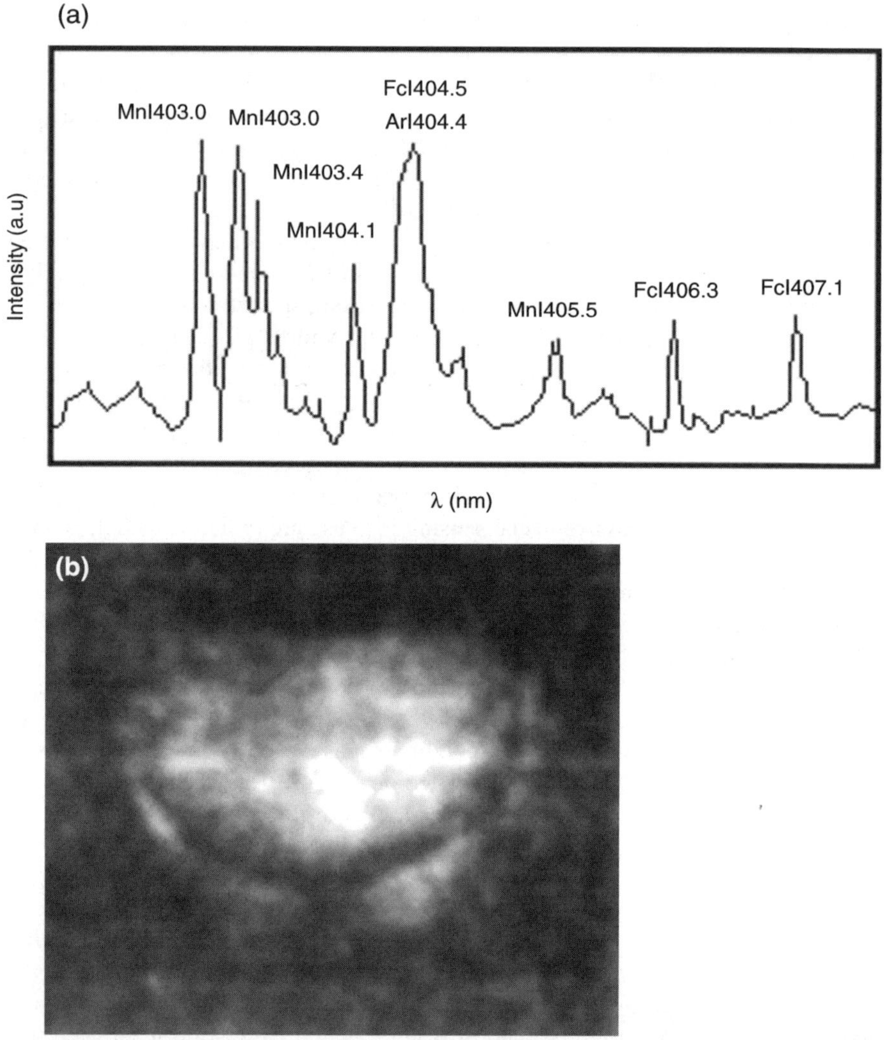

Fig. 1.5 The method of spectral censoring [57] (**a**) Intensity distribution of the spectral lines (**b**) Image of weld pool

source; light in the range of 590–710 nm is allowed to pass and peak transmittance is 25% and suitable imaging current, time and position are adopted to obtain clear double-side aluminium alloy weld pool image by increasing image contrast.

In the above mentioned cases, images quality of passive visual sensing can be improved by composite filtering technology in the specific radiation range, though it is not as good as that of active visual sensing. Moreover, passive visual sensing is less cost in equipment with only CCD camera to obtain the weld pool images and easier in equipment structure so that it is more suitable for welding manufacturing.

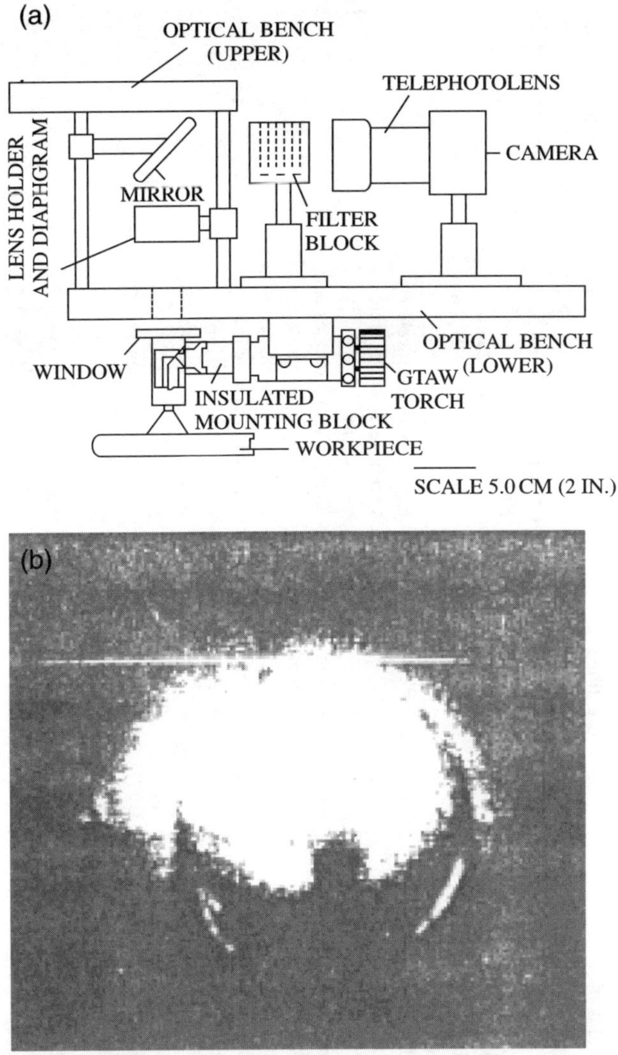

Fig. 1.6 Method of coaxial weld pool viewing in GTAW [60] (**a**) System set (**b**) Image of weld pool

1.3.3 Image Processing Methods

Image processing is aimed to extract characteristics of weld pool based on which the relationship between topside and backside weld pool geometry parameters is built and real time penetration control is realized [70–75, 90–105]. Therefore, the precise of image processing algorithm is important for control of welding process [65–75, 90–105]. Generally, original image cannot be directly used in control

Table 1.2 Passive vision sensing method in the welding process

Researcher	Sensing Information	Equipment	Strength	Weakness
R.W. Richardson (OSU) [77]	GTAW weld pool geometry, MIG welding seam tracking	Visual sensor with welding torch in the same axis of electrode	Decrease arc light to produce clear image	Large amount of calculation, off-line analysis
K. Oshima [78] (Japan, Saitama University)	MIG/MAG weld pool dynamics	CCD Camera	Imaging under last arc, primary control of weld pool	–
Hezhi Li [79] (Gansu Institute of technology)	Pulsed MAG weld pool image	CCD Camera	Decrease welding current when imaging	–
Kezhen Wang [80] (Tsinghua University)	MIG weld pool image	High speed CCD Camera	Extract weld pool width and control penetration	–
Wuzhu Chen [81] (Tsinghua University)	Online measurement of weld pool	CCD camera with composite filtering system	Detect welding process on line	Huge disturbance, high processing speed required
Shishen Huang [82] (South China university of technology)	Topside weld pool image	Area array CCD Camera	Shooting in the overlook direction of 50°	Processing time: 200 ms
Pengjiu Li [83] (Harbing Institute of technology)	Affect of welding parameters to GTAW arc spectrum	–	Strongest area arc light in 320–440 nm, 700–800 nm	–
Yuchi Liu, bin Huang (Harbing Institute of technoloty) [84, 85]	Weld pool image of low carbon steel	Camera	Imaging under base welding current and control weld pool width	–
Yajun Lou, Dongbin Zhao [86, 87] (Harbing Institute of technoloty)	Topside and backside weld pool shape	CCD camera, with composite filtering system	Both sides visual sensing and modeling between topside and backside weld pool size	–
GuangjunZhang [88] (Harbing Institute of technology)	Weld pool shape and gas size of butt welding	CCD camera with composite filtering system	Three sides visual sensing respectively from the direction of front, rear and back	–
Jianjun Wang [89] (Shanghai Jiao Tong Univ.)	Weld pool shape of aluminium alloy	CCD camera with composite filtering system	Both sides visual sensing , wide band filter	–
Congjian Fan [69] (Shanghai Jiao Tong University)	Weld pool shape of aluminium alloy	CCD camera with composite filtering system	Three sides visual sensing from the direction of front, rear and back, wide band filter	–

algorithm due to the disturbance and limitations of welding equipment, thus, specific image processing is necessary. Moreover, fluctuation in welding current and arc light also lead to image degrading [66–68] in pulsed GTAW. All the above factors add difficulties to the image processing, and the image processing algorithms are required to be adaptive to different conditions. Based on the analysis of weld pool image feature, several algorithms such as degrading image recovery, integral edge detection, projection, neural network edge identification and curve fitting are developed to extract the geometry parameters of the weld pool [76, 77, 105–119].

Weld pool shape changes in length, width, rear angle and topside height [65–75, 90–102], can be observed from the image and professional welder uses this information to regulate the welding parameters to stabilize the welding seam formation. It is the key in penetration control of welding process to build the relationship between topside shape and backside shape. Arc force depresses the weld pool surface and the workpiece turns from partly penetrated to fully penetrated accordingly. Experiments show that weld pool surface height is linear to the backside width of the weld pool.

In pulsed GTAW process, topside weld pool shapes are quite different between partly penetrated and fully penetrated conditions. Typical topside weld pool images are obtained by setting different welding peak current and wire feeding velocity in experiments. References [65], [103], [84–87, 120, 121] show that weld pool is ellipse when partly penetrated with convex image; while it is peach-shaped when fully penetrated with concave image. The change in weld pool shape is most evident in the rear angle.

1.3.3.1 Image Processing of Topside Weld Pool

Generally, weld pool images under different experimental conditions require different image processing algorithms because welding current, direction of arc light, etc. will change the contrast between solid metal and weld pool edge and the difference is greater with different materials such as low carbon steel and aluminium alloy [65–68].

The purpose of image processing is to extract the weld pool edge and calculate the weld pool shape parameters. Threshold method that use the feature of double-peak or multi-peak of gray level histogram is used to extract the weld pool edge [65]. Other methods such as edge detection algorithm combined with smoothing method and multinomial fitting method to remove noise in the image.

Due to the difference in topside height, gray value varies greatly between convex and concave weld pool. References [65, 68, 102] determined the type of weld pool according to the rear shape of the weld pool. Convex weld pool has small area of ellipse-shape with clear arc in the front side of the weld pool and smooth shape in the rear side. Concave weld pool has large area of peach-shape with front side blocked by the torch and sharp shape in rear side.

References [76–83, 106–116, 118, 119, 122–126] show general steps of weld pool image processing methods. First is filtering. Then, different methods are used for convex and concave images. For convex image, gray level histogram shows typical feature of doublet. And the binary image can be obtained by finding the dale between two peaks. Edge points can be extracted by edge tracking method. For concave image, gray level histogram also shows obvious feature of doublet, but direct threshold method will lead to miss-processing. Two dimensional edge detection will greatly increase the processing time. However, one dimensional edge detection will be used to detect in some specific direction to reduce the processing time. Finally, coordination method is used to calculate the actual shape parameters of the weld pool in work piece plane.

1.3.3.2 Image Processing of Backside Weld Pool

Coordination is also necessary for backside weld pool image processing. The light source is high-temperature radiation from the melton metal and the image is typical target image, which can be processed by threshold method after Guass filtering. The edge points can, thus, be determined [86].

1.3.3.3 Calculation of Three Dimensional Characteristics of Weld Pool

References [102] studied a simple method to extract the height of the topside weld pool according to the reflection of arc in the weld pool. The method is to extract the distance between the torch and reflection, and then indirectly calculate the height of the weld pool.

The image processing method can be divided into following steps

1. Guass filtering
2. Threshold method
3. Determine the position of electrode
4. Calculate the topside height

References [84–89, 120, 121, 127–130] calculated the topside height of the weld pool based on monocular vision method. Theoretically, three dimensional information of the image cannot be extracted from monocular image. However, some additional information, such as image geometric model, surface features and physical features, will help to extract the three dimensional information. This method is called Shape From Shading (SFS), which uses some prior knowledge as constrained conditions to remove the morbidity of the reflectance map equation.

The SFS method is to obtain the single image from experiment. And then to calculate the topside height according to the reflectance map equation that relate gray level to shape of the image. The key problem of shape from shading is to construct imaging reflection map equation and to solve the equation. To construct the reflection map equation of real weld pool surface, the characteristics of light source, camera, and object surface are the prior conditions. Then, the reflection map equation relating surface gradient to grayness of image is constructed under ideal imaging condition, and iteration method for calculating the surface height from the equation is proposed, and the validation is verified by synthetic image and real image of stationary weld point.

References [120, 121] introduced SFS method to topside height calculation. Reflection map equation based on actual conditions is established. And equation resolving algorithm was proposed with real imaging conditions of weld pool considered, such as correlation between work piece coordinates and camera coordinates, setting the intensity of light source, and determining the reflection coefficient of weld pool surface by comparison.

References [131, 132] established the general reflection map equation for GTAW weld pool images based on the analysis of arc spectrum, welding parameters, camera parameters and different image characteristics for low carbon steel, stainless steel and aluminium alloy respectively. Solution of reflection map equation was to resolve large sparse linear equations. Linear table method can reduce storage memory because of more zeros. Preprocessing conjugate gradation method was designed to assure convergence and the convergent rate. The results showed the calculated height was in accordance with the actual surface height characteristics of concave and convex molten pool.

1.4 Modeling Methods for Arc Welding Process

There are three methods to build model for weld pool dynamics [10, 11, 133–135]:

(1) Analytical model based on mechanism of the system. The analytical model is to build theoretic equation in analysis of physical or chemical dynamics of the system.
(2) Identification model of differential, integrals or difference equation based on input or output signals of the system. Single processing methods, such as least square method, maximum likelihood and parameter estimation method based on pulsed, step or stochastic signals, is used.
(3) Intelligentized model based on mass input and output signals from complex and uncertain object or environment. This is an emerging filed of study in the recent ten years.

The three types of models including analytical, identification and intelligentized models [131–188] will be briefly described in the following part.

1.4.1 Analytical Model

Also called as white box model, analytical model is developed by analyzing the motion laws of the system based on known principles and theorem.

Temperature field analytical model [136–139] is used in the early period of mechanism model building in welding. Many hypothetic conditions that are inconsistent with the facts are used in the model. In the hypothetic conditions, welding heat source is considered as point heat source; no phase transition, latent heat and temperature-related material properties change are taken into consideration; heat transfer only in metal; workpiece is infinite in length. References [75] decreased the amount of hypothetic conditions. However these hypothetic conditions still lower the accuracy of the model.

With the development of computer technology, finite element method are applied in the force computation of welding process by analyzing the flow field, temperature field and force inside the weld pool to establish dynamical model between flow, temperature field and surface deformation [140–143]. The model is more precise because many practical conditions are taken into consideration, such as size and distribution of the heat source, latent heat of the material and temperature-related physical property. These models can tell the rules such as influence of welding parameters to the shape of the weld pool, but they are only appliable to off-line model establishing due to their large amount of computation.

1.4.2 Identification, Fuzzy Logic and Neural Network Models

Owing to the uncertainties of phenomena such as metallurgy, heat transfer, chemical reaction, arc physics and magnetization, arc welding process is inherently variable, nonlinear, time-varying and strong coupling in its input/output relationships. Many factors influence the welding process, such as welding current, welding voltage, welding velocity, wire feeding velocity and even environment conditions. As a result, it is very difficult to obtain a practical and controllable model of the arc welding process by classical modeling methodology. Identification and intelligentized models are more applicable for welding dynamics [10, 11].

Identification method is frequently used in practice to develop black-box model of the system, because it is of high precise, robust and practicality with its input and output as experiment data, based on which the structure and parameters of the model is identified [131, 132, 144–163].

Table 1.3 shows the process model in open literatures.

From Table 1.3, we can see that the welding process model developed from simple to complex. Experiments show that using single information (temperature in one point, topside height, topside width and area) to predict penetration on the backside has its limitation. More information about the topside size of the weld pool makes more precise prediction of the backside width of the weld pool. In this way, multi-input-multi-output model are adopted with information in different aspects such as width, length, area, rear width and rear angle of the weld pool. Further, welding parameters are also used in the process model to predict the backside penetration of the weld pool.

Table 1.3 Models used for weld shape control

Researcher	Input	Output	Structure	Application
J. B. Song [35]	Tempture on the backside of several points	penetration	Tradition	GTAW
Y. Kozono	Tempture on the rear part of the weld pool	penetration	Tradition	GTAW
Nagarajanetc	Topside area of the weld pool	penetration	Tradition	GTAW
Chunli Yang [24]	Resonance frequency of the weld pool	penetration	Tradition	GTAW
Heqi Li [93]	Topside area of the weld pool	penetration	Tradition	GTAW
Yuming Zhang [46]	Topside height of the weld pool	Backside width of the weld pool	Tradition	GTAW
Billy Chan [158]	Welding parameters (4)	1	ANN	GMA on-plane welding
George E [159]	Welding parameters (4)	2	ANN	VPPA
J. Y. Jeng [160]	Welding parameters (3)	3	ANN	Laser welding
Yasuo Suga [161]	Welding parameters and topside size of the weld pool (6)	7	ANN	GTAW
Y. M. Zhang [147]	Welding parameters and topside size of the weld pool	1	ANN	On-plane GTAW
Di Li [189]	Welding parameters and topside size of the weld pool	1	ANN	GTAW
Yajun Lou [65]	Welding parameters and topside size of the weld pool (48)	3	ANN	Butt GTAW
Dongbin Zhao [102]	Welding parameters and topside size of the weld pool (21)	2	ANN	GTAW with wire feeder
Guangjun Zhang [103]	Welding parameters, gap size and topside weld pool size with its historical values (21)	3	ANN	Gap-variation GTAW with wire feeder
Jianjun Wang [68]	Topside width of the weld pool	Backside width of the weld pool	Stochastic model	Aluminium alloy GTAW

In recent years, fuzzy logic and neural network methods are used with the development of artificial intelligent. Since the 1990's, many significant researches have been carried out in fuzzy logic and neural network modeling for arc welding process.

References [147] developed Sugeno fuzzy model mapping topside geometry size and backside width of the weld pool. Because the parameters of the fuzzy model are obtained by neural network method, the model is called neurofuzzy model, which shows high precise in experiments. Researches were also carried in fuzzy logic model in Refs. [11, 17, 174].

Model becomes coupling when its inputs and outputs increase, which adds difficulty for traditional model building. Backpropagation (BP) artificial Neural Network (ANN) is used with a deviation, a hidden layer of S-shpae function and a linear output layer to approximate any equation. Therefore, system can be considered as a black-box, whose external dynamics can be simulated by BP ANN. During model building, inputs and outputs are sent to BP ANN to learn the node value between each layer so as to make a black-box model of the system. ANN is effective for complex process as a modeling method.

In Refs. [151, 152, 159], BP model between geometric size of the welding seam and direct weld parameters(DWP) was developed for Gas Tungsten Arc Welding (GTAW), and the model was validated by experimental data that its precise is no less than the traditional model. With the same method, Ref. [152] developed a 3-layer BP model between backside width and topside width of the weld pool during aluminium alloy variable polarity plasma arc welding for Marshall Space Flight Center of NASA. In Refs. [153, 158], 4-layer BP model was developed during Gas Metal Arc Welding (GMAW) to predict topside height after welding, topside width, penetrated depth and penetrated area of the welding seam.

References [189] primarily proved the fault tolerance, anti-interference and universality of BP network model by modeling between topside width and backside width of the weld pool. References [10, 11, 65, 102, 171–173] developed both dynamic and static BP models between welding current and topside width of the weld pool under several welding parameters, both on-line and off-line. Reference [164] developed model with Counter Propagation Network (CPN), in which hidden layer is competitive layer (also called Kohonen layer) with unsupervised learning; output layer is Grossberg layer to be multipoint interconnected to hidden with Grossberg or Widrow-Hoff learning.

Reference [158] developed two neural network model: one is to use welding parameters (welding current, welding voltage, welding velocity and workpiece thickness) to predict weld pool shape (topside width, height and penetrated depth), the other to use weld pool shape to predict welding parameters, as shown in Fig. 1.7.

Reference [102] researched on the welding formation during butt GTAW with wire feeder. Control variables of the system were backside width and topside height of the weld pool. Because the backside width was not visible in practice and image processing algorithm for topside height was not suitable for on-line calculation, Backside and Height Dynamic Model (BHDM) is developed to predict the backside width and topside height by welding parameters such as welding current, pulse duty ratio, welding velocity, wire feeding velocity and historical value of topside size of the weld pool.

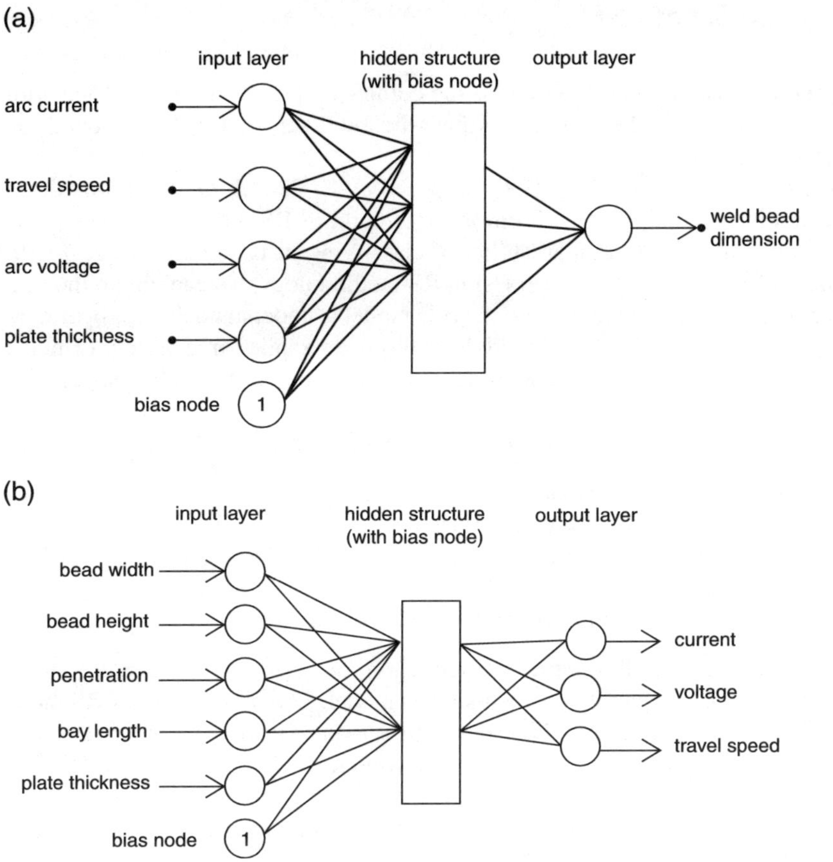

Fig. 1.7 The neural network models of welding process [158] (**a**) The forward model (**b**) The reverse model

Though the above-mentioned intelligentized modeling method is widely used, there are still some difficulties left to be overcome. For fuzzy logic modeling, a lot of subjective factors, such as prior knowledge, are used and they cannot be overcome by the method of neural network. Another drawback lies in rule explosion when the size of fuzzy model becomes uncontrollable with the increase of variable and its value. However, welding is a multi-variable coupling process, which leads to the limitation of fuzzy logic model for this process.

As a black box model, neural network model has not clear physical meaning, which will is difficult of maintenance. Furthermore, the convergence rate is not acceptable in some situation and its structure and parameters depend too much on experience. Finally, training of the neural network model needs example data.

In short, both neural network and fuzzy logic modeling depends too much on empirical knowledge.

1.4.3 Rough Set Model

With a mass of uncertainties, welding is too complex a process to be modeled with classic methods. Researcher begin to imitate the intelligence of welder to build the model [165–185].

Rough sets (RS) theory [165–169], proposed by Z. Pawlak in 1982, provides a new method of modeling for us. From the viewpoint of RS theory, knowledge has essential relationship with human ability of classifying. It is powerful to deal with uncertainty of the controlled object. Using RS methodology, we can obtain the rule model of complex process; moreover, the rule model is understandable for operators and easy to revise directly. RS methodology has been applied in a variety of fields such as data mining, pattern recognition, decision support, fault analysis and so on [174–185].

In general, main steps of RS modeling methodology are as follows:

Step 1: Preprocessing of raw data;
Step 2: Discretization of continuous attributes;
Step 3: Condition attribute reduction;
Step 4: Condition attribute value reduction;
Step 5: Rule reduction (Optimization of the set of rules).

Among all steps, condition attribute reduction, or attribute reduction for short, mainly decides the complexity of the final rule model. The first feature of RS modeling method is that it depends only on data, rather than on prior experience. The second feature is that it can efficiently extract knowledge to obtain concise model. The last feature is that the model is in the form of rules, which is easy to be maintained and intelligible. Therefore, it is very suitable for welding process where knowledge is difficult to be obtained. Usage of using human experience makes the model easy to be understood and possible to be revised by engineers.

In recent years, RS modeling method has been used in welding.

Reference [174] applied RS theory to fuzzy logic system modeling with a new identification method for GTAW modeling. It discussed the method of developing fuzzy model based on the rough set theory, presents concepts of adjustment and expansion, and the dynamic model of pulse TIG welding process was developed by the proposed method. The validations of models were done and the results demonstrated the effectiveness of the method.

References [175, 176] used RS model to primarily extract knowledge and neural network model to optimize the knowledge for lack of weld in Buick Car subframe welding assembly. References [177, 178] obtain the rule model of aluminum alloy pulsed gas tungsten arc welding (GTAW) process by rough set. A novel attribute reduction algorithm is proposed. The algorithm makes full use of human experience by means of defining a partial ordering on the set of condition attributes. It is proved that the algorithm is complete for the definition of attribute reduction. A value reduction algorithm and a rule reduction algorithm are generalized from the algorithm. It uses classical RS modeling method. However, it does not take into consideration the features of welding and it did not study discretization.

Reference [179] studied discretization method of RS modeling. A new method called secondly discretization is brought forward, which means that on the base of equal width intervals discretization or the discretization based on Kohonen net, intervals combination discretization based on importance of attributes is used. Reference [180] introduced a new reasoning method based on attributes significance. This method takes the significance of different attributes into account, correctly utilizing the heuristic knowledge in the sustaining strategy; put forward the generalized matching degree based on the attributes significance, which totally overcome the limitation of the used reasoning method. In addition, the new method inherited the means to resolve conflict matching problems in the known reasoning method, to give the fit output according to the confidence of the relative rule.

References [184, 185] introduced so-called variable precision rough set (VPRS) [181–183] to welding dynamics modeling. The VPRS can effectively obtain knowledge from mass data in little dependence on experience knowledge with high comprehensibility and easy maintenance compared with classic rough set method. Especially, the reduction algorithms in VPRS were well described. For discretization method, condition attributes' discretization and decision attributes' discretization were distinguished. Experiment showed that decision attribute's discretization was very important, which was seldom noticed in former researches. For condition attribute's discretization, some common discretization algorithms were compared and the entropy based algorithm performed better. Furthermore, a modified global algorithm grounded on entropy based method was proposed to overcome the latter's disadvantages, and the validation result was satisfying. A minority prior inference strategy that longer rules had higher priority was put forward to meet the complexity of the system. VPRS model was compared with common rough set model and BP neural network model respectively. It showed that VPRS model is more stable and can better predict the unseen data than common RS model. What is more, VPRS model has close precision to neural network model, but was much simpler than the latter.

References [186, 187] studied a new support vector machine-based (SVM) system with high comprehensibility for welding dynamics.

1.5 Intelligent Control Strategies for Arc Welding Process

Many stochastic factors such as variation of seam gap and thermal deformation exist in welding process; therefore, stable weld seam formation is not ensured under fixed welding parameters. References [45, 46, 48] adopted an adaptive method to control penetration of welding seam. With laser as light source, topside width and depressed depth are extracted and penetration model of welding process was established. However, this method has its limitation in welding practice because the model is established without the consideration of wire feeding, which greatly affects depressed depth.

Classical and modern control theory is not effective in weld pool dynamics due to the complexity of welding process.

To improve the performance of controller, intelligentized control method was introduced by researchers in the field of welding since 1980s' [188, 190–192]. Especially in the past ten years, a large number of papers on intelligentized control method for welding process have been published. And intelligentized controller developed from single fuzzy controller to expert system-fuzzy-ANN composite controller in various purposes [188–224], such as droplet transfer, welding seam tracking and welding seam formation.

Intelligentized control can imitate human behavior of dealing with uncertainty and complexity to make decisions with experience, knowledge and reasoning independent of mathematical model of the process. Fuzzy controller, artificial neural network controller, expert system controller and learning controller are suitable welding process [10, 11].

Fuzzy logic controller is developed by imitating human being's intelligent behavior of nature language. It is aimed to blur the accurate quantity of controlling error, obtain fuzzy reasoning rules based on natural language rules, and finally defuzzify the fuzzy quantity into accurate quantity.

Reference [198] studied fuzzy control on robot MIG welding penetration depth. Both the topside width and gap were extracted by two CCD cameras and penetration depth was predicted by ANN model. The FNNC, as shown in Fig. 1.8, is composed of a feedforward fuzzy controller for gap variation and a feedbackward neural network controller for welding current.

Andersen [162] designed the fuzzy controller and ANN controller combined with PID controller, as shown in Fig. 1.9. Some basic concepts relating to neural networks and how they could be used to model weld-bead geometry in terms of the equipment parameters selected to produce the weld were explained. Approaches to utilizing neural networks in process control were discussed. The need for modeling transient as well as static characteristics of physical systems for closed-loop control was pointed out, and an approach to achieving this was presented.

References [189] used ANN to build dynamic model and inverse dynamic model of welding process. Inverse dynamic model was to relate backside weld pool width to welding current to predict output controlling variable; dynamic model was to

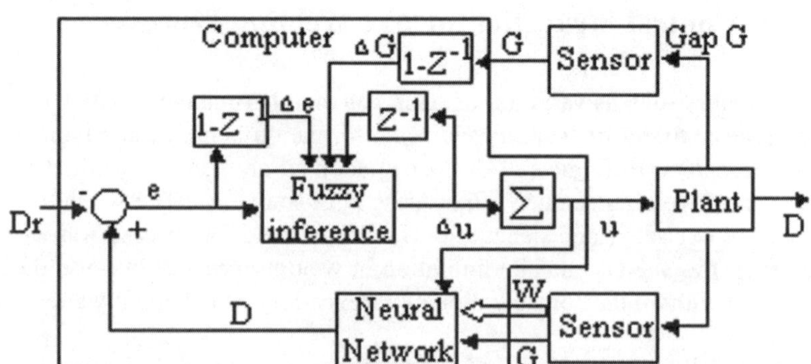

Fig. 1.8 Fuzzy neural network control system to control the penetration depth [198]

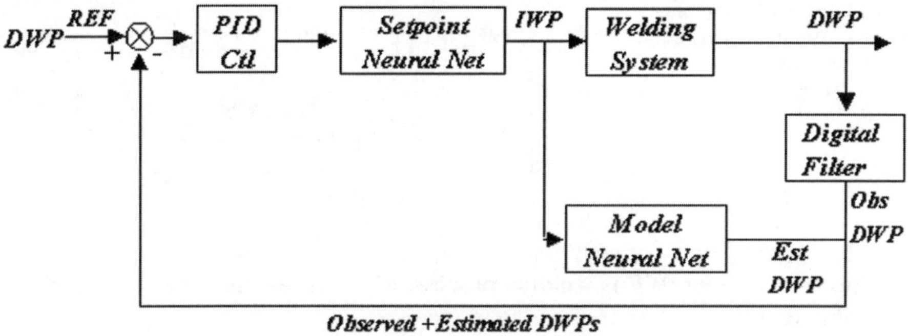

Fig. 1.9 Close-loop control system of neural network during GTAW [162]

relate welding current to backside weld pool width to predict the output of sensor. The ANN model is shown in Fig. 1.10.

References [10, 65] establised SSNNM model relating DWP to backside weld pool parameters based on visual sensing for pulsed Bead-on-Plate GTAW, and a neural network model of the dynamic process was established for predicting the backside width with the welding parameters and topside size parameters.

Reference [215] applied intelligent control methodology to improve weld quality. Based on fuzzy logic and artificial neural network theory, a self-learning fuzzy and neural network control scheme was developed for real-time control of pulsed GTAW. Using an industrial CCD camera as the sensor, the weld face width of the weld pool, i.e., the feedback signal in the closed loop system, was obtained by computer image processing techniques. The computer vision providing process status information in real-time was an integral part of a self-learning fuzzy neural control system. Such a system enables adaptive altering of welding parameters to compensate for changing environments. The control system is shown in Fig. 1.11, in which

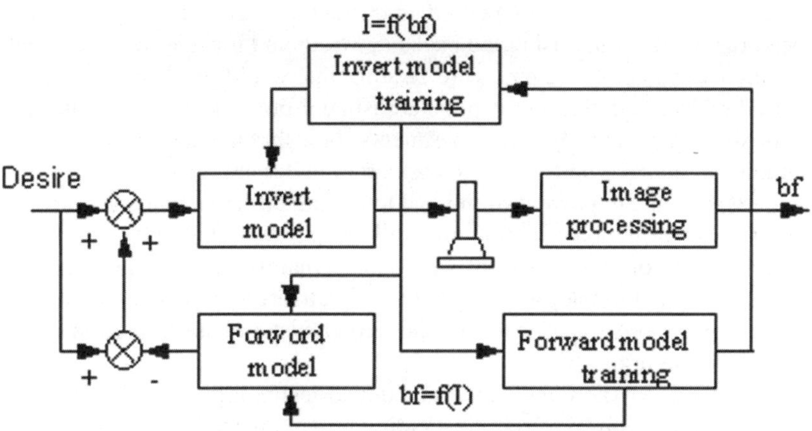

Fig. 1.10 Closed-loop control system of neural network during GTAW [189]

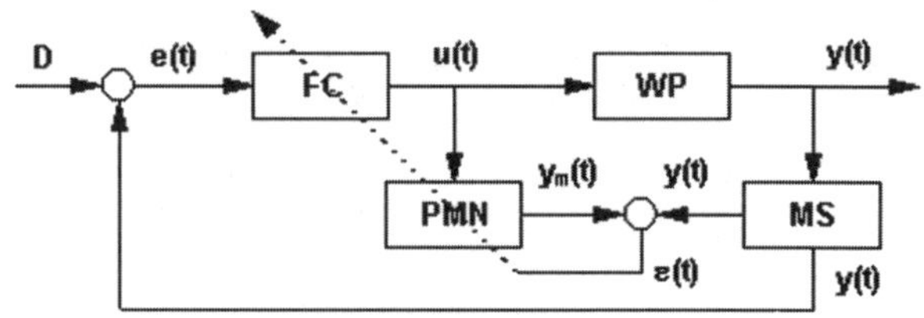

Fig. 1.11 Principle diagram for self-learning fuzzy neural control for GTAW process [215]

FC is fuzzy controller, WP is welding process, MS is measuring system and PMN is nerual network.

In Refs. [11, 65], both the size and shape neural network model (SSNNM) and single size neural network model (SNNM) were established for predicting the backside width, and the more accuracy could be attained from SSNNM rather than from SNNM. Furthermore, a double inputs and double outputs (DIDO) controller, incorporated with fuzzy neural network and expert system, was designed for control backside width and weld pool length by adjusting pulse duty ratio and travel speed.

In Refs. [102, 214], intelligent controller was designed for realizing dynamic intelligent control of weld pool shape during pulsed GTAW with wire filler. For the single requirement of stabilizing backside width or topside height, a single variable self-learning PID controller based on single neuron was proposed. And for the requirement of stabilizing backside width and topside height simultaneously, a double variable self-adaptive fuzzy controller based on single layer neural network was designed, and the output of the controller is constant, which can be adjusted on-line with the characteristic of real process.

Because of the particularity of aluminum alloy, Refs. [66, 67] designed single-variable adaptive controller. The welding current controller could regulate welding current to form uniform welding seam. However, both of them cannot completely avoid cutting. The controller of wire feeding velocity can be used to avoid cutting, but with pool welding seam formation. To overcome the shortage of single-variable adaptive controller, Ref. [68] designed the intelligent controller with fuzzy supervising and adaptive regulation (ICFA) based on current adaptive adjusting and wire feeding speed fuzzy adjusting, using the backside width and topside reinforcement as the controlled variable. The fuzzy wire feeding rate could accelerate the convergence speed of welding current adaptive adjustment process and avoid cutting. The results of simulation and welding experiments show that the double parameters control method can achieve uniform weld formation under abrupt varied heat sink conditions, moreover, it can prevent from weld to cutting effectively and succeed in controlling the weld formation of aluminum alloy during the pulsed TIG welding.

Existing researches on the control of weld pool dynamics demonstrate that intelligentized control methodology can effectively realize the real-time control of welding quality by adaptive, self-learning and expert system based on passive visual sensing and modeling.

References [210–231] show the above mentioned technology in welding robot system, thus close-loop control of robotic welding process is realized. This is a key technology of intelligentized welding robot to operate under complex situations.

1.6 The Organized Framework of the Book

The written program of the book is organized as following:

Firstly, the Chap. 1 gives an introduction on development of welding handicraft and manufacturing technology; the key technologies of welding automation; intelligentized technologies for arc welding process.

The Chap. 2 mainly addresses visual sensor and systems for arc welding process, which includes the visual sensing system and images of weld pool during pulsed GTAW for both low carbon steel and aluminium alloy.

The Chap. 3 mainly addresses information acquirement of arc welding process, which includes acquiring two dimensional characteristics from monocular image of weld pool during pulsed GTAW; computing of three dimensional characteristics from monocular image of weld pool during pulsed GTAW; the algorithm software of Image processing and characteristic extracting of weld pool during pulsed GTAW for various kind of weld pool images.

The Chap. 4 mainly addresses modeling methods of weld pool dynamics during pulsed GTAW, which includes identification models; artificial neural network models; fuzzy rules models and rough set models.

The Chap. 5 mainly addresses various control strategies for arc welding dynamics including self-regulating PID controller; fuzzy control strategies; PSD controller; neural network self-learning controller and composite intelligent controller.

The Chap. 6 mainly addresses real-time control of weld pool dynamics during robotic welding process including intelligentized welding robot systems with monitoring weld pool dynamics and applications of intelligentized welding robot system on real-time control of weld pool dynamics during robotic welding process.

The Chap. 7 mainly addresses conclusion of the whole book.

References

1. E. Kannatey-Asibu. Milestone developments in welding and joining processes. ASME Journal of Manufacturing Science and Engineering. 1997, 119(11):801–810
2. P. Drews, G. Starke. Welding in the century of information technology. Welding in the World. 1994, 34(1):1–20
3. J.L. Pan. A survey on 21th century welding science researches, Beijing :Mechanical Industry Press, Proceedings on 9th Chinese Conference on Welding (in Chinese), 1999, (1): D-001-D-017
4. T.H. Song, M.X. Li. Development of advanced manufacturing technologies and welding in the future, Beijing :Mechanical Industry Press, Proceedings on 8th Chinese Conference on Welding (in Chinese), 1997, (1):17–27
5. T.J. Tarn, S.B. Chen, C.J. Chou. Robotic Welding, Intelligence and Automation, Springer Verlag Berlin Heidelberg, Mar., 2004
6. T.J. Tarn, S.B. Chen, C.J. Chou. Robotic Welding, Intelligence and Automation, Springer Verlag Berlin Heidelberg, Mar., 2007
7. S.B. Chen, L. Wu. A survey on intelligentized welding technologies, Harbin: Heilongjiang People Press, Proceedings on 10th Chinese Conference on Welding (in Chinese), 2001, (1):84–96

8. C. Shanben, Q. Tao, L. Tao, et al. On intelligentized technologies for modern welding manufacturing, Chinese Journal of Mechanical Engineering, 2003, 16(4):367–370
9. L. Wu, S.B. Chen. Intelligentized Welding Technology, National Defence Industry Press, Beijing, 2000
10. S.B. Chen. Intelligent methodology for sensing, modeling and control of pulsed GTAW: Part1 – Band-on-plate Welding. Welding Journal, 2000, 79(6):151s–163s
11. S.B. Chen. Intelligent methodology for sensing, modeling and control of pulsed GTAW: Part2 – Butt Joint Welding. Welding Journal. 2000, 79(6):164s–174s
12. P. Ward. Improving productivity with robotic welding. Welding Journal. 2001, 80(8): 52–54
13. M. Ushio, W. Mao, Sensors for arc welding: Advantages and limitations. Transactions of JWRI. 1994, 23(2):135–141
14. H. Nomura. Sensor and Control System in Arc Welding. Chapman & Hall. 1994, 3–5
15. L. Masumoto, T. Araya, A. Iochi, H. Normura. Development and application of sensors and sensor system for arc welding. Journal of Japan Welding Society. 1983, 52(4):39–47
16. D.E. Hardt. Ultrasonic measurement of weld penetration. Welding Journal. 1984, 63(9): 273s–285s
17. R. Fenn. Ultrasonic monitoring and control during arc welding. Welding Journal. 1985, 64(9):18–24
18. N.M. Carlson, J.A. Johnson. Ultrasonic sensing of weld pool penetration. Welding Journal. 1988, 67(11):239s–246s
19. L. Wu, DX. Dong, DH. Chen. Detection and analysis of welding penetration signal in stationary TIG arc welding-An preliminary study of detection of welding penetration. Transactions of The China Welding Institution. 1986, 7(2):64–71
20. Q.L. Wang, J.H. Zhang, C.L. Yang. Detection of weld full penetration signals of a traveling arc in GTAW process. Transactions of the China Welding Institution. 1992, 11(3):175–179
21. R.J. Renwick, R.W. Richardson. Experimental investigation of GTA weld pool oscillations. Welding Journal. 1983, 62(2):29s–35s
22. S.J. Salter, R.T. Deam. A practical front face penetration control system for TIG welding. Proceedings of an International Conference on Trends in Welding Research. Gatlinburg, Tennessee, USA. 1986, 381–392
23. H. Maruo, Y. Hirata. Natural frequency and oscillation modes of weld pools. 1st Report: Weld pool oscillation in full penetration welding of thin plate. Welding International. 1993, 7(8):614–619
24. L.C. Yang. Real-Time Control of full penetration of steel sheet TIG weld pool with arc sensor based on welding pool resonance, [1.Doctorial dissertation], Harbin Institute of Technology, 1991
25. S.I. Rokhlin, A.C. Guu. A study of arc force, pool depression and weld penetration during tungsten arc welding. Welding Journal. 1993,72(8):381s–390s
26. P.J. Li, Q.L. Wang. The development of the arc light sensing method in GTAW process. Proceedings of the 78th American Welding society Annual Meeting. Los Angeles, USA, 1997:112–120
27. J.W. Kin and S.J. Na. A Study on an arc sensor for gas metal arc welding of horizonal fillets. Welding Journal. 1991,70(8):216s–221s
28. Y.L. Song. Some features of welding arc radiation. China Welding. 1996, 5(2):96–101
29. Q.L. Wang, P.J. Li, M. Naka. The welding arc spectrum and the improved arc light information for welding process control. High Temperature Physicals. 1996, 22(6):313–320
30. J.P. Li. Fundamental research on arc light sensing technology during welding process, [1.Doctorial dissertation], Harbin Institute of Technology, 1997
31. H.C. Wikle. Infrared sensing techniques for penetration depth control of the submerged arc welding process. Journal of Materials Processing Technology. 2001, 113(6):228–233
32. S. Nagarajan, W.H. Chen, B.A. Chin. Infrared sensing for adaptive arc welding. Welding Journal. 1989, 68(11):462s–466s
33. S. Nagarajan, B.A. Chin, W. Chen. Control of the welding process using infrared sensors. IEEE Transactions on Robotics and Automation. 1992, 8(1):86–93

34. Y. Koiono. Control System of Reverse Side Bead Width Surface Temperature Monitoring of TIG Weld. Quarterly Journal of the Japan Welding Society. 1986,4(2):58∼67
35. J.B. Song, D.E. Hardt. Closed-loop control of weld pool depth using a thermally based depth estimate. Welding Journal. 1993, 72(10):471s–478s
36. C.C. Doumanidis, D.E. Hardt. Simultaneous in-process control of heat affected zone and cooling rate during arc welding. Welding Journal. 1990, 69(3):186s–195s
37. A.C. Guu, S.I. Rokhlin. Computerized radiographic weld penetration control with feedback on weld pool depression. Materials Evaluation. 1989, 47(10):1204–1210
38. A.C. Guu, S.I. Rokhlin. Arc weld process control using radiographic sensing[J]. Materials Evaluation. 1992, 50(11):1344–1348
39. M. Futamata. Application of arc sound for detection of welding process. Quarterly Journal of the Japan Welding Society. 1983, 1(1):11–16
40. J.A. Johnson, N.M. Carlson, D.E. Clark. Process control of gas metal arc weld: Sensing of metal transfer mode. Welding Journal. 1991, 70(4): 91s–96s
41. M. Kutsua, M. Abraham. Metal transfer and spatter loss in double gas shielded metal arc welding using flux-cored wire. Transactions of the Japan Welding Society. 1989, 20(1):2–15
42. J.A. Johnson, N.M. Carlson, H.B. Smartt. Detection of metal transfer mode in GMAW. Proceedings of the 2nd International Conference on Trends in Welding Research. Gatlinburg, Tennessee, USA. 1989, 37–38
43. R.J. Beatlie, S.K. Cheng, P.S. Logue. The use of vision sensors in multi-pass welding applications. Welding Journal. 1988, 67(11):28–33
44. C.G. Morgan, J.S. Bromley, P.G. Davey. Visual guidance techniques for Robot arc welding. Proceeding of 3rd International Conference on Robot Vision and Sensory Controls, 1983, 615–619
45. Y.M. Zhang. Modern control analysis and design of welding process, Harbin Institute of Technology Press, Harbin, 1990, 72–76
46. Y.M. Zhang, L. Wu. Determination of top supervision parameters for state of full penetration in TIG welding, Transactions of The China Welding Institution. 1991, 12(1):39–45
47. Y.M. Zhang, R. Kovacevic, L. Wu. Closed-loop control of weld penetration using front-face vision sensing. Journal of Systems and Control Engineering. 1993, 207(3):27–34
48. J.E. Agapakis, J. Bolstad. Vision sensing and processing system for monitoring and control of welding and other high luminosity processes. International Robots & Vision Automation Conference, 1991, 23–29
49. S. Nakata, J. Huang, Y. Tsuruha. Visual sensing system for in-process control of arc welding process. Welding International. 1988, 2(12):1086–1090
50. T. Hoffman. Real-time imaging for process control. Advanced Materials & Processes. 1991, 140(9):37–43
51. S.B. Chen, J. Wu, "A Survey on Intelligentized Technologies for Visual Information Acquirement, modeling and Control of Arc Welding Pool Dynamics", Proceedings on the 33rd Annual Conference of the IEEE Industrial Electronics Society (IECON' 07), Taipei, Taiwan, R.O.C., Nov. 5–8, 2007, IEEE IECON07, 1-4244-0783-4/07, pp. 781–785
52. S.B. Chen. "The Status and Development of Intelligent Technology for Welding in China," (Invited report) The First International Conference on Mechanical Engineering (ICME'2000), Shanghai, China, Nov., 2000. (The conference on Materials, plenary report)
53. Y.M. Zhang, L. Wu, B. Walcott, D.H. Chen. Determining joint penetration in GTAW with vision sensing of weld face geometry. Welding Journal, 1993, 72(10):463–469
54. Y.M. Zhang, R. Kovacevic, L. Wu. Sensitivity of front-face weld geometry in representing the full penetration. Proc. Instn. Mech. Engrs, Part B: Journal of Engineering Manufacture. 1992, 206:191–197
55. Y.M. Zhang, R. Kovacevic, L. Wu. Dynamic analysis and identification of gas tungsten arc welding process for weld penetration control. ASME Journal of Engineering for Industry. 1996, 118:123–136
56. E.W. Kim, C. Allemand, T.W. Eagar. Visible light emissions during gas tungsten arc welding and its application to weld image improvement. Welding Journal. 1987, 66(12):369s–377s

57. Y.L. Song. Sensing technology of spectral image of welding zone in arc welding. Transactions of the China Welding Institution. 1992, 13(2):127–132
58. Y.L. Song. Some phenomena and features of the component distributions in welding arc. China Welding. 1995, 4(2):65–69
59. J.E. Agapakis, J. Bolstad. Vision sensing and processing system for monitoring and control of welding and other high luminosity processes. International Robots & Vision Automation Conference, 1991, 23–29
60. R.W. Richardson, D.A. Gutow. Coaxial arc weld pool viewing for process monitoring and Control. Welding Journal. 1984, 63(3):43–50
61. D. Brzakovic, D.T. Khani. Weld pool edge detection for automated control of welding. IEEE Transactions on Robotics and Automation. 1991, 7(3):397–403
62. K.A. Pietrzak, S.M. Packer. Vision-based weld pool width control. ASME Journal of Engineering for Industry. 1994, 116(2):86–92
63. Y.C. Liu. Experimental measurement of the change process for impulse TIG welding pool width. [1.Master Thesis], Harbin Institute of Technology, 1994
64. Y.J. Lou, M. Dai, S.B. Chen. Capturing simultaneous images of weld pool in a frame from two directions. Robot. 1997, 19, Sup.(1): 59–63
65. Y.J. Lou. Intelligent control of dynamics in pulsed GTAW process by weld pool image sensing, [1.1.1.1.Phd. dissertation] (in Chinese). Harbin: Harbin Institute of Technology, 1998
66. J.J. Wang, T. Lin, S.B. Chen. Obtaining of weld pool vision information during aluminum alloy TIG welding. International Journal of Advanced manufacturing technology. 2005, 26:219–227
67. J.J. Wang, T. Lin, S.B. Chen, W. Wang, Research for the method of image acquisition of the molten pool in the TIG welding of aluminum alloy, China Welding (English Edition), Vol. 13, n 2, November, 2004, pp. 115–118
68. J.J. Wang. Visual information acquisition and adaptive control of weld pool dynamics of Aluminum alloy during pulsed TIG welding. PhD dissertation, Shanghai Jiao Tong University, 2003
69. C.J. Fan, S.B. Chen, T. Lin. Visual sensing and image processing in aluminum alloy welding, Lecture Notes in Control and Information Sciences, LNCIS 362, pp. 275–280, 2007(ISTP)
70. S.B. Chen. "Visual Information Acquirement and Real-time Control Methodologies for Weld Pool Dynamics during Pulsed GTAW," Tthe invited paper of the THERMEC'2006, Journal Materials Science Forum vols. 539–543 in March 2007
71. S.B. Chen, D.B. Zhao, Y.J. Lou, L. Wu. Computer vision sensing and intelligent control of welding pool dynamics. Series Lecture Notes in Control and Information Sciences. 2004, 299:25–55, (SCI)
72. Y.J. Lou, S.B. Chen, L. Wu. Monitoring of Weld Joint Penetration Based on Sensing Double-side Vision Image of Weld Pool. Journal of Harbin Institute of Technology. 1999, 6(3):26–31
73. R. Kovacevic, Y.M. Zhang, S. Ruan. Sensing and control of weld pool geometry for automated GTA welding. ASME Journal of Engineering for Industry. 1995, 117:210–222
74. R. Kovacevic, Y.M. Zhang, L. Li. Monitoring of weld penetration based on weld pool geometrical appearance. Welding Journal. 1996, 75(10):317s–328s
75. D.H. Chen, L. Wu, Q.H. Xu. Study on a new measuring method of temperature distribution on weld with microcomputer. 1981, 12:1–11
76. R. Kovacevic, Y.M. Zhang. Real-time image processing for monitoring of free weld pool surface. Journal of Manufacturing Science and Engineering. 1997, 119(5):161–169
77. S.K. Pal. Automatic gray level thresholding through index of fuzziness and entropy. Pattern Recognition. 1983, 1(3):141–146
78. K.P. Horn. Height and gradient from shading. International Journal of Computer Vision. 1990, 5(1):37–75
79. K.M. Lee, C.J. Kuo. Shape from shading with a linear triangular element surface model. IEEE Transactions on Pattern Analysis and Macine Intelligence. 1993, 15(8):815–822
80. K.M. Lee, C.J. Kuo. Shape from shading with a generalized reflectance map model. Computer Vision and Image Understanding. 1997, 67(2):143–160

81. K.M. Lee, C.J. Kuo. Shape from shading with perspective projection. Computer Vision Graphics Image Process. 1994, 59(2):202–212
82. R. Kimmel, A.M. Bruckstein, Global shape from shading. Computer Vision and Image Understanding. 1995, 62(3):360–369
83. J.H. Sheng. 3D image processing system based on Shape Reconstruction. Pattern Recognition and Artificial Intelligence. 1991, 4(4):46–52
84. D.B. Zhao, S.B. Chen, L. Wu. Research on the application of the 3-D shape from a single image. Optical Technique. 2000, 23(2):147–152
85. D.B. Zhao, S.B. Chen, L. Wu. Analysis and realization of the calculus of height from a single image. Chinese Journal of Computers. 2001, 27(4):367–369
86. D.B. Zhao, S.B. Chen. Shape parameter definition and image processing of the weld pool during pulsed GTAW with wire filler. Transactions of the China Welding Institution. 2001, 19(2):5–8
87. L.P. Li, S.B. Chen, T. Lin. The light intensity analysis of passive visual sensing system in GTAW. The International Journal of Advanced Manufacturing Technology. 2003, 27:106–111, 2005
88. Q.Y. Du. Extraction and intelligent control of 3D dynamic weld pool shape information of pulsed GTAW with wire filler [1.Doctor thesis] Shanghai Jiaotong University, 2005
89. Q.Y. Du. S.B. Chen, T. Lin. Research on an algorithm of shape from shading. Journal of Shanghai Jiaotong University (English Edition), 2005, E-10(1):86–90
90. Z. Gen. Real-time control of full penetration of Aluminium alloy TIG weld pool with arc sensor based on welding pool resonance, [Doctorial dissertation], Harbin Institute of Technology, 1991
91. K.Z. Wang. Visual sensing and digital control system of MIG weld pool. Journal of Tsinghua University (Science and Technology). 1990, 30(2):22–29
92. W.Z. Chen, K. Inoue. Image processing method applied to on-line detection of narrow gap weaving. MIG/MAG. 1988, 9(4):247–253
93. H.Q. Li, K. Oshima. Study on observing and controlling MAG welding pool by digital image technique. Journal of Lanzhou University of Technology. 1988, 9(1):37–43
94. Q. Chen. Computer vision sensing technology and its application in weldin. Transactions of The China Welding Institution, 2001, 22(1):83–86
95. Y.L. Song, J.Y. Li. Thermo-equilibdum in welding are plasma. Transactions of The China Welding Institution. 1995, 15(2):138–145
96. S. Nakata, J. Huang, Y. Tsuruha. Visual sensing system for in-process control of arc welding process. Welding International. 1988, (12):1086–1090
97. T. Hoffman. Real-time imaging for process control. Advanced Materials & Processes. 1991, (9):37–43
98. Y.M. Zhang, L. Li, R. Kovacevic. Monitoring of weld pool appearance for penetration control. 4th International Conference on Trends in Welding Research, Gatlinburg, Tennessee, USA, 1995:30–35
99. R. Kovacevic, Y.M. Zhang. Vision sensing of 3D weld pool surface. 4th International Conference on Trends in Welding Research, Gatlinburg, Tennessee, USA, 1995:85–91
100. K. Oshima, M. Morita. Observation and digital control of the molten pool in pulsed MIG welding. Welding International. 1988, 2(3):234–240
101. K. Oshima, M. Morita. Sensing and digital control of weld pool in pulsed MIG welding. Transactions of the Japan Welding Society. 1992, 23(4):36–42
102. D.B. Zhao. Intelligent control of weld pool dynamics in pulsed GTAW process with filler by 3-dimesion visual sensing, [Phd. dissertation] (in Chinese). Harbin: Harbin Institute of Technology, 2000
103. G.J. Zhang. Intelligent control of shaping seam in pulsed GTAW process with varied gaps and filler by visual sensing, [1.1.1.1.Phd. dissertation] (in Chinese). Harbin: Harbin Institute of Technology, 2002
104. Z. M Guo. Vision serving strategies for guiding welding robot to weld starting, [Phd. dissertation] (in Chinese). Harbin: Harbin Institute of Technology, 2002

105. Q. Chen et al. Application of computer vision in welding. Transactions on Welding (in Chinese), 2001, 22(1):83–86
106. R.C. Gonzalez, R. Woods. Digital Image Processing. 3rd ed, Addison-Wesley, 1977
107. W.K. Pratt. Digital Image Processing, John wiley and sons, New york. 1991
108. G.A. Baxes. Digital Image Processing: Principles and Applications, Whiley, New York, 1994
109. S.D. Yanowitz, A.M. Bruckstein. A new method for image segmentation, Computer vision, 1989, (46):82–95
110. J. Bernsen. Dynamic thresholding of gray-level images. 81CPR, 1986, 1251–1255
111. H. Szu, I. Kopriva, Artificial neural networks for noisy image super-resolution. Optics Communications. 2001, 198:71–81
112. K. Etemad, R. Chellappa, A neural network based edge detector. IEEE International Conference on Neural Networks. 1993, 132–137
113. T. Warren, L. Damin, L. Yueming. Extraction of welds from radiographic images using fuzzy classifiers. Information Sciences. 2000, 126:21–40
114. R. Kovacevic, Y.M. Zhang. Three-dimension measurement of weld pool surface. Proceedings of the International Conference on Modeling and Control of Joining Processeses, Oriando, Florida, USA, 1993, 600–607
115. Y.M. Zhang, M. Beardsley, R. Kovacevic. Real-time image processing for 3D measurement of weld pool surface. Manufacturing Science and Engineering. 1994, 68(1):255–262
116. D. Brzakovic, D.T. Khani. Weld pool edge detection for automated control of welding. IEEE Transactions on Robotics and Automation. 1991, 7(3):397–403
117. T.L. Huntsberger, C.L. Javabs. Iterative fuzzy image segmentation, Pattern recognition. 1985, 18(2):131–138
118. Y.L. Gao, L. Zhang, W.W. Wu. An algorithm for threshold based on arithmetic mean of gray valu. Journal of Image and Graphes. 1999, 4(6):524–528
119. K. Chen. An image segmentation algorithm based on edge detectio. Pattern Recognition and Artificial Intelligence. 1992, 5(1):25–29
120. D.B. Zhao, J.Q. Yi, S.B. Chen, L. Wu, Q. Chen. Extraction for three-dimension parameters for weld pool surface in pulsed GTAW with wire filler. Journal of Manufacturing Science and Engineering. 2003, 125(3):493–503
121. D.B. Zhao, J.Q. Yi, S.B. Chen, L.Wu. "Surface Shape Reconstruction of Weld Pool During Pulsed GTAW From Its Single Image", Series Lecture Notes in Control and Information Sciences, LNCIS. 2004, 299:56–62, (SCI)
122. K.M. Lee, C.J. Kuo. Surface reconstruction from photometric stereo images. Journal of Optical Society of America. 1993, 10(5):855–868
123. M. Clerc, S. Mallat . The texture gradient equation for recovering shape from texture. IEEE Transactions on Pattern Analysis and Machine Intelligence. 2002, 24(4):536–549
124. R. Zhang, P. Tsai, J.E. Cryer, et al. Shape from shading: A Survey. IEEE Transactions on Pattern Analysis and Machine Intelligence, 1999, 21(8):690–706
125. K.P. Horn, M.J. Brooks. The variational approach to shape from shading, Computer Vision. Graphics, and Image Processing. 1986, 33:174–208
126. Y. Liao. Analysis and evaluation of several typical SFS algorithm. Journal of Image and Graphics. 2001, 6(10):953–961
127. Q.Y. Du, S.B. Chen, T. Lin. Inspection of weld shape based on the shape from shading, International Journal of Advanced Manufacturing Technology. 27(7–8):667–671 Jan 2006
128. J.F. Wang, L. Zhou, S.B. Chen. Measurement for three dimensional surface of welding pool in GTAW welding. Lecture Notes in Control and Information Sciences, LNCIS. 362:219–226, 2007(ISTP)\
129. D. Quanying, C. Shanben, L. Tao. Reconstruction of weld pool surface based on shape from shading. Chinese Journal of Mechanical Engineering. 2006, 19(2):168–171 (Achieved the Best paper award in 2007)
130. L. Laiping, C. Shanben, L. Tao. The modeling of welding pool surface reflectance of aluminum alloy pulse GTAW. Materials Science and Engineering: A. 2005, 394:320–326

131. B.Z. Jin, W.H. Liu, K.J. Ohshima. Control of weld pool width and cooling rate in circumferential GTA welding of pipe by using neural network model. In, International IEEE/IAS conference on industrial automation and control, Taipei, Taiwan, 1995, 41–46
132. K. Ohshima, M. Yabe, K. Akita, et al. Sensor fusion using neural network in the robotic welding, in, Conference Record - IAS Annual Meeting (IEEE Industry Applications Society), Orlando, FL, USA, 1995, 1764–1769
133. S.B. Chen, W.J. Chen, L. Tao. Identification models of weld pool dynamic process in pulsed GTAW. Transactions of the China Welding Institution, 2001, 22(3):5–8
134. Y.M. Zhang, L. Li, R. Kovacevic. Dynamic estimation of full penetration using geometry of adjacent weld pools. ASME Journal of Manufacturing Science and Engineering, 1997, 119(11):631–643
135. I.S. Kim et al. A study on prediction of bead height in robotic arc welding using a neural network. Journal of Materials Processing Technology. 2002, 130–131, 229–234
136. Y. Koiono. Relation between surface temperature and reverse side bead width in TIG welding of thin mild steel plate. Journal of the Japan Welding Society. 1982, 51(3):52–55
137. D. Rosenthal. Mathematical theory of heat distribution during welding and cutting. Welding Journal. 1941, 20(3):220–234
138. N.D. Malmuth. Transient thermal phenomena and weld geometry in GTA welding. Welding Journal. 1974, 53(3):388–400
139. W.H. Kim. Heat and fluid flow in pulsed current GTA weld pool. International Journal of Heat and Mass Transfer. 1998, 41:3213–3227
140. H.G. Fan. Heat transfer and fluid flow in a partially or fully penetrated weld pool in gas tungsten arc welding. International Journal of Heat and Mass Transfer. 2001, 44:417–428
141. M.A. Wahab. The prediction of the temperature distribution and weld pool geometry in the gas metal arc welding process. Journal of Materials Processing Technology. 1998, 77:233–239
142. R.T.C. Choo, J. Szekely, S.A. David. On the calculation of the free surface temperature of gas tungsten arc weld pools from first principles: Part 2. Modeling the weld pool and comparison with experiments. Metal Transfer B. 1992, 23B:371–384
143. D. Pradip. Thermal modeling of gas tungsten arc welding process with nonaxisymmetric boundary conditions. Numerical Heat Transfer: An International Journal of Computation and Methodology; Part A: Applications. 1995, 27:499–518
144. S. Yasuo, Measurement of molten pool shape and penetration control applying neural network in TIG welding of thin steel plates. ISIJ International, 1999, 39(10):1075–1080
145. S. Nagarajan, B.A. Chin, On-line identification and control of part-preparation and fixturing errors in arc welding. Journal of Engineering for Industry. 1993, 115(11):385–389
146. L.A. Zadeh, Outline of a new approach to the analysis of complex systems and decision processes, IEEE Transactions on Systems, Man, and Cybernetics, 1973, 3, 28–44
147. R. Kovacevic, Y.M. Zhang, On-line measurement of weld fusion state using weld pool image and neurofuzzy model. IEEE International Symposium on Intelligent Control – Proceedings. 1996, 307–312
148. P. Li, M.T.C. Fang, J. Lucas. Modelling of submerged arc weld beads using self-adaptive offset neutral networks. Journal of Materials Processing Technology. 1997, 71(2):288–298
149. I.S. Kim, Y.K.D.V. Prasad, J.S. Son et al. A study on prediction of bead height in robotic arc welding using a neural network. Journal of Materials Processing Technology. 2002, 130–131:229–234
150. D.S. Nagesh, G.L. Datta, Prediction of weld bead geometry and penetration in shielded metal-arc welding using artificial neural networks. Journal of Materials Processing Technology. 2002, 123(2):303–312
151. K. Andersen, G.E. Cook, G. Karsai, et al. Artificial neural networks applied to arc welding process modelingand control. IEEE Transactions on Industry Applications. 1990, 26(5):824–830
152. G.E. Cook, R.J. Barnett, K. Andersen, et al. Weld modeling and control using artificial neural networks. IEEE Transactions on Industry Applications. 1993, 31(6):2181–2189

153. B. Chan, J. Pacey, M. Bibby. Modelling gas metal arc weld geometry using artificial neural network technology, Canadian Metallurgical Quarterly(Netherlands), 1999, 38(1):43–51
154. Y. Suga. Application of neural network to visual Sensing of weld line and automatic tracking in robot welding. Welding in the World. 1994, 34:275–282
155. K. Andersen. Artificial neural networks applied to arc welding process modeling and control. IEEE Transactions on Industry Applications. 1990, 26(5):824–830
156. D.S. Nagesh. Prediction of weld bead geometry and penetration in shielded metal-arc welding using artificial neural networks. Journal of Materials Processing Technology. 2002, 123(2):303–312
157. L. Di, T. Srikanthanb, R.S. Chandelc. Neural-network-based self-organized fuzzy logic control for arc welding. Engineering Applications of Artificial Intelligence. 2001, 14:115–124
158. B. Chan. Modeling gas metal arc weld geometry using artificial neural network technology. Canadian Metallurgical Quarterly. 1999, 38(1):43–51
159. G.E. Cook. Weld modeling and control using artificial neural networks. IEEE Transactions on Industry Applications. 1995, 31(6):1484–1491
160. J.-Y. Jeng. Prediction of laser butt joint welding parameters using back propagation and learning vector quantization networks. Journal of Materials Processing Technology. 2000, 99:207–218
161. SUGA. Yasuo. Measurement of molten pool shape and penetration control applying neural network in TIG welding of thin steel plates. ISIJ International. 1999, 39(10):1075–1080
162. K. Andersen, E. George. Artificial neural networks applied to arc welding process modeling and control. IEEE Transactions on industry applications. 1990, 26(5):824–830
163. S.Y. Li Fuzzy control, Neural Network and Intelligentized control theory (2nd edn) Harbin Institute of Technology Press, Harbin, 1996
164. Y.S. Tarng, J.L. Wu, S.S. Yen, S.C. Juang. Intelligent modeling and optimization of the gas tungsten arc welding, Journal of Intelligent Manufacturing. 1999, 10:73–79
165. Z. Pawlak. Rough sets. International Journal of Information and Computer Science. 1982, 11(5):341–356
166. Z. Pawlak, Granularity of knowledge, indiscernibility and rough sets. Proceedings of IEEE International Conference on Computational Intelligence, 1998, 2:933–938
167. Y.Y. Yao. Comparative study of fuzzy sets and rough sets. Information Sciences. 1998, 109(1–4):227–242
168. D. Coker, Fuzzy rough sets are intuitionistic L-fuzzy sets. Fuzzy Sets and Systems. 1998, 96(3):381–383
169. S.T. Wang, E. Scott, A. Gamermann. Extract rules by using rough set and knowledge-based NN. Journal of Computer Science and Technology. 1998, 13(3):279–284
170. W. Li, H. Sun, S.B. Chen. A rough set method of building fuzzy model. Control Theory & Applications. 2001, 18(1):69–75
171. G.J. Zhang, S.B. Chen, X. Liu, et al. Predicting the backside width of weld pool during pulsed gtaw process based on a neural network model [1.1.1.1.4565-30]. Proceeding- SPIE the International Society for Optical Engineering. 2001, 131–137
172. Y.J. Lou, S.B. Chen, L. Wu. Fuzzy-Neural network modeling and control of pool dynamic process in pulsed GTAW. Acta Automatica Sinica. 1997, 18, sup.12:194–198
173. Z. Dongbin, C. Shanben, W. Lin, C. Qiang. Neural network modeling for dynamic pulsed GTAW process with wire filler based on Matlab, China Welding v 10 n 2 November 2001, pp 88–93 (EI)
174. W. Li, H. Sun. Fuzzy logic identification method for TIG welding process modelin. Journal of the China Railway Society. 1998, 20(6):111–114
175. B. Wang, T. Lin. Knowledge acquiring in intelligent detecting system for lack of weld. Transactions of the China Welding Institution. 2001, 22 (003):29–32
176. B. Wang, S.B. Chen, T. Lin, Rough set based knowledge acquiring method in intelligent detecting system for lack of weld. Proceedings of the 4th World Congress on Intelligent Control and Automation. 2002, 4:2887–2891
177. B. Wang. Study on knowledge modeling method based on rough set theory for welding dynamic processed. [PhD dissertation] Shanghai Jiao Tong University, 2003

178. B. Wang, S.B. Chen, J.J. Wang. Rough set based knowledge modeling for the aluminum alloy pulsed gtaw process, The International Journal of Advanced Manufacturing Technology. 2005, 25(9):902–908
179. G.G. Zhu. Study on discretization method in rough set based modeling method for welding. [Master dissertation] Shanghai Jiao Tong University, 2003
180. S.R. Zhang Study on reasoning method in rough set based modeling method for welding. [Master dissertation]. Shanghai Jiao Tong University, 2004
181. W. Ziarko. Variable precision rough set model. Journal of Computer and System Sciences. 1993, 46(1):39–59
182. J.S. Mi, W.Z. Wu, W.X. Zhang. Approaches to knowledge reduction based on variable precision rough set model. Information Sciences. 2004, 159(3–4):255–272
183. A. Mieszkowicz-Rolka, L. Rolka, Variable precision fuzzy rough sets model in the analysis of process data. In Lecture Notes in Computer Science, Regina, Canada. 2005, 354–363
184. W.H. Li, S.B. Chen, B. Wang. A variable precision rough set based modeling method for pulsed GTAW. International Journal of Advanced Manufacturing Technology, Mar, 2007 (SCI)
185. W.H. Li, S.B. Chen, T. Lin. Discretiztion in rough set modeling method for welding process. Lecture Notes in Control and Information Sciences. LNCIS 362:325–332, 2007(ISTP)
186. X.X. Huang, F.H. Shi, S.B. Chen, A new support vector machine-based fuzzy system with high comprehensibility. Lecture Notes in Control and Information Sciences. LNCIS 362:421–427, 2007(ISTP)
187. X.X. Huang, S.B. Chen, SVM-based fuzzy modeling for the arc welding process. Materials Science and Engineering A-Structural Materials Properties Microstructure and Processing. 427(1-2):181–187 JUL 15 2006
188. A.E. Bentley, S.J. Marberger. Arc welding penetration control using quantitative feedback theory. Welding Journal. 1992, 71(11):397s–405s
189. S.S. Huang, D. Li. Study on neural network modellling and control of welding pool dynamic process. Chinese Journal of Mechanical Engineering. 1994, 30(3):24–29
190. J.L. Pan. Modern control of arc welding, Beijing. Mechanical Industry Press (in Chinese), 2000
191. J.A. Johnson, N.M. Carlson, D.E. Clark. Process control of gas metal arc weld: Sensing of metal transfer mode. Welding Journal. 1991, 70(4):91s–96s
192. D.A. Hartman. Intelligent control in arc welding. Intelligent Engineering Systems Through Artificial Neural Networks, ASME. 1999, 11:715–725
193. Y.M. Zhang. Study of adaptive top-vision control for full penetration in TIG welding. [1.Doctorial dissertation]. Harbin Institute of Technology, 1990
194. S.S. Huang, J.F. He, Y.L. Song. Fuzzy logic control system design of welding seam penetration during arc welding. Transactions of The China Welding Institution. 1996, 17(2):94–98
195. C. Doumanidis. Multivariable adaptive control of the bead profile geometry in gas metal arc welding with thermal scanning. International Journal of Pressure Vessels and Piping. 2002,79:251–262
196. K.A. Pietrzak, S.M. Packer. Vision-based weld pool width control. ASME Journal of Engineering for Industry. 1994, 116(2):86–92
197. Y.M. Zhang. Adaptive control of full penetration gas tungsten arc welding. IEEE Transactions on Control System Technology. 1996, 4(4):394–403
198. A. Hirai. Sensing and control of weld pool by Fuzzy-Neural network in robotic welding system. The 27th Annual Conference of the IEEE Industrial Electronics Society. 2001, 238–242
199. M.A. Austin. Real-time multi-processing fuzzy logic adaptive control gas tungsten arc welding system. The IEEE Proceedings of the 4th WPDRTS. 1996:139–142
200. A. Bigand. Arc welding fuzzy control using neural net supervisor. IEEE. 1996:2003–2006
201. Y.M. Zhang. Neuro-fuzzy model based control of weld fusion zone geometry. The Proceedings of the American Control Conference. 1997, 6:2483–2487
202. K.J. Hunt. Neural networks for control system: A survey. Automatica. 1992, 28(6): 1083–1112

203. B. Irving. Neural networks are paying off on the production line. Welding Journal. 1997, 76(10):59–64
204. S.S. Huang, D. Li. Weld quality control by neural network Welding in the Word. 1994, 34:359–363
205. Y. Kaneke, T. Iisaka, K. Oshima. Neuro-fuzzy control of the weld pool in pulsed MIG welding. Welding International. 1995, 9(3):191–196
206. T.G. Lim, H.S. Cho. Estimation of weld pool sizes in GMA welding process using neural networks. Journal of Systems and Control Engineering. 1993, 207(1):15–26
207. J.W. Kim, S.J. Na. A self-organizing fuzzy control approach to arc sensor for weld joint tracking in gas metal arc welding of butt joints. Welding Journal. 1993, 72(1):60s–66s
208. Y. Suga. Application of neural network to visual sensing of weld line and automatic tracking in robot welding. Welding in the World. 1994, 34:275–282
209. K. Andersen. Artificial neural networks applied to arc welding process modeling and control. IEEE Transactions on Industry Applications. 1990, 26(5):824–830
210. X. Gao. Shisheng huang. Fuzzy Neural Networks for Control of Penetration Depth during GTAW. China Welding. 2000, 1:1–8
211. S. Yamane. Application of fuzzy adaptive control to a welding robot. Proceedings of Second International Symposium on Signal Processing and its Application, Gold Coast, Australia. 1990, 8:271–274
212. C. Lin. Neural-network-based fuzzy logic control and decision system. IEEE Transaction on Computer. 1991, 40(12):1320–1336
213. S.B. Chen et al. Intelligentized welding robot technology. Harbin Institute of Technology Press, Harbin, 2001
214. D.B. Zhao, S.B. Chen, L. Wu. Intelligent control for the shape of the weld pool in pulsed GTAW with filler metal. Welding Journal. 2001, 80(11): 253s–260s
215. S.B. Chen, L. Wu, Q.L. Wang, Y.C. Liu. Self-learning fuzzy neural networks and computer vision for control of pulsed GTAW. Welding Journal, 1997, 76(5):201s–209s
216. S.B. Chen, L. Wu, Q.L. Wang. Self-learning fuzzy neural networks for control of uncertain systems with time delays. IEEE Transactions on Systems, Man, and Cybernetics, Part B: Cybernetics. 1997, 27(1):142–148
217. Y.J. Lou, S.B. Chen, L. Wu, Neuron PSD control based on sensing image of weld pool during pulsed GTAW. Proceedings of International Conference on Manufacturing Science, Wuhan, P.R.China, 10–12, June, 1998
218. Z. Guangjun, C. Shanben, W. Lin. Neuron self-learning PSD control for backside width of weld pool in pulsed GTAW with wire filler. China Welding. 2003, 12(1)
219. C. Wenjie, C. Shanben, L. Tao. Comparsion of three control methods in pulsed gas tungsten arc welding. Journal of Shanghai Jiaotong University, 2003, 8(1):63–66
220. S.B Chen, Y. Zhang, T. Qiu, T. Lin. Welding robotic systems with vision sensing and self-learning neuron control of arc weld dynamic process. Journal of Intelligent and Robotic Systems. 2003, 36(2):191–208
221. S.B Chen, Y. Zhang, T. Lin, T. Qiu, L. Wu., Welding robotic systems with vision sensing and real-time control of dynamic weld pool during pulsed GTAW. International Journal of Robotic and Automation. 2004, 19(1):28–35
222. S.B. Chen, X.Z. Chen, J.Q. Li, T. Lin. Acquisition of welding seam space position information for arc welding robot based on vision. Journal of Intelligent & Robotic Systems. 2005, 43(1):77–97
223. S.B. Chen, T. Qiu, et al., Intelligentlized technologies for robotic welding. Series Lecture Notes in Control and Information Sciences. 2004, 299:123–143
224. S.B. Chen,On the key intelligentized technologies of welding robot. Lecture Notes in Control and Information Sciences. 2007, LNCIS 362:105–116
225. H.Y. Shen, H.B. Ma, T. Lin, S.B. Chen, Research on weld pool control of welding robot with computer vision. Industrial Robot. 2007, 34(6):467–475
226. H.Y. Shen, J. Wu, T. Lin, S.B. Chen. Arc welding robot system with seam tracking and weld pool control based on passive vision. The International Journal of Advanced Manufacturing Technology. 2007. (SCI)

227. C. XiZhang, Z. Zhenyou,C. Wenjie,C. Shanben. Vision-based recognition and guiding of initial welding position for arc-welding robot. Chinese Journal of Mechanical Engineering. 2005, 18(3):382–384
228. L. Zhou, T. Lin, S.B. Chen. Autonomous acquisition of seam coordinates for arc welding robot based on visual servoing. Journal of Intelligent & Robotic Systems. 47(3):239–255 NOV 2006
229. X.Z. Chen, S.B. Chen, T. Lin, et al. Practical method to locate the initial weld position using visual technology. International Journal of Advanced Manufacturing Technology. 30(7–8):663–668 OCT 2006
230. Trailer. Manufacturer depends on robotic welding to boast production. Welding Journal. 1995, 74(7):49–51
231. S.B. Chen et al. Intelligentized welding robot technology. Mechanical Industry Press, Beijing, 2006

Chapter 2
Visual Sensing Systems for Arc Welding Process

Abstract Visual sensing technology is widely used in welding practices because visual devices are decreasing in price, increasing in reliability and improving in image processing hardware and software. As the most studied welding sensor, CCD(Charge Coupled Device) is more suitable for quality control of welding process than other means of sensing devices because it can obtain both two dimensional and three dimensional information of weld pool, which directly reflect the welding dynamics of molten metal. In this chapter, according to the analysis of arc spectrum and radiation of different materials, visual sensing systems with filters are described. Based on the filtering method, clear images of weld pool are obtained during pulsed GTAW.

The first step of intelligentized arc welding is to imitate the visual system of a welder to extract weld pool size. Passive visual sensing technology has seen great progress in the past years for the abundant information it extracts. Brzakovic et al. [1] obtained the weld pool image in two directions and extracted its geometry information. Wang et al. [2] attempted to get aluminium alloy image for the first time, but it is not clear enough. Zhao et al. [3] developed a passive three-dimension visual sensing method through monocular image which is processed by Shape from Shadow (SFS) algorithm to get the three-dimension geometry of the pool. In this chapter, passive visual sensing system for both aluminium alloy and low carbon steel will be discussed.

2.1 Description of the Real-Time Control Systems with Visual Sensing of Weld Pool for the Pulsed GTAW Process

As one of the dominant arc weld methods, pulsed GTAW is widely used in the high-quality weld manufacturing, especially for high-precise thin sheet. High-quality pulsed GTAW requires precise penetration and fine formation of the weld seam, thus real-time regulation of the welding process, i.e., regulating weld pool size, is necessary. Main influences on weld pool size involve electrical conditions, such as pulse duty ratio, peak current, base current, arc voltage, and welding speed; workpiece conditions such as the root opening or geometry of the groove, material, thickness,

S.-B. Chen, J. Wu, *Intelligentized Methodology for Arc Welding Dynamical Processes*,
Lecture Notes in Electrical Engineering 29,

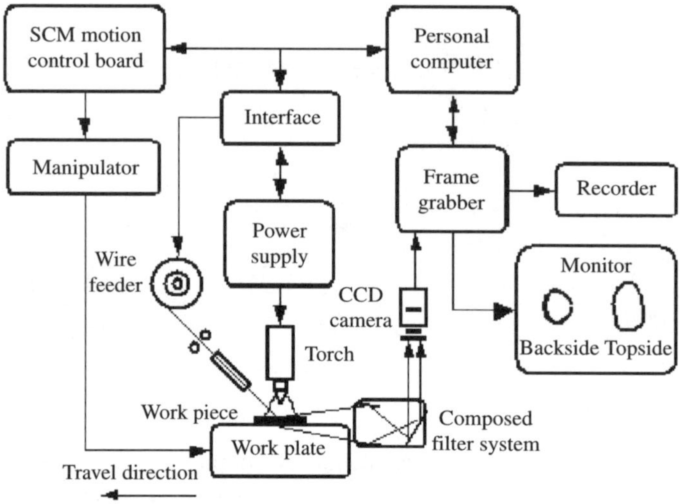

Fig. 2.1 The structure diagram of experimental system for pulsed GTAW

work piece size, electrode tip angle, and rate of shielding gas flow; welding conditions such as heat transfer condition, arc emission and so on.

The welding experiment is carried out with the monitoring system, which consists of filter system, CCD camera, recorder, frame grabber, and monitor, as shown in Fig. 2.1. The equipment is shown in Fig. 2.2.

The sensing system consists of the following parts as shown in Fig. 2.3(a):

(1) Light path of double-side imaging simultaneously in a frame. The light path is composed of topside and backside imaging light path. As shown in Fig. 2.3(b), the light from the weld pool reaches the reflector O_1 with the reverse X axis,

Fig. 2.2 The photograph of experimental equipment

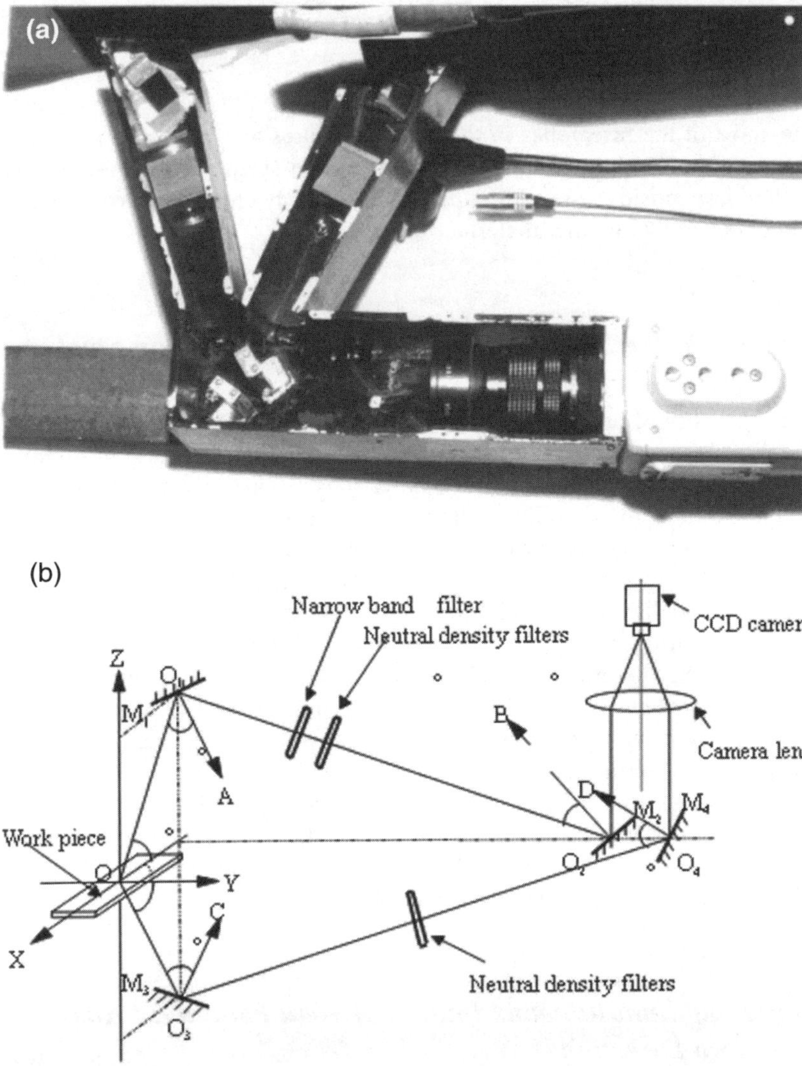

Fig. 2.3 The sensing system (**a**) the photograph of sensing system (**b**) The light path

which is reflected to pass composite filters, then reflected by O_2, finally focused on the target of the CCD camera. The principle of the backside light path is the same. The light path system is mounted behind the weld pool with a large distance from weld pool to eliminate the pollution from spatter, fume and smoke.

(2) CCD photograph system transferring the optical signal to video signal. The photograph system includes CCD camera and optical lens.
(3) Video recorder and monitor system. This subsystem includes V512B frame grabber, recorder and monitor.

2.2 The Visual Sensing System and Images of Weld Pool During Low Carbon Steel Pulsed GTAW

Up to now, most of the researches in the open literatures are on low carbon steel. This is due to good welding quality of the low carbon steel, such as proper heat dissemination, less sensitive to welding parameters, steady chemical capability and weldability compared with other materials.

2.2.1 Analysis of the Sensing Conditions for Low Carbon Steel

Due to high intensity of arc emission, the weld pool image is strongly interfered. The main task of sensor design is to eliminate the arc interference and to improve the contrast degree of the images. In this book, various visual sensing systems based on the principle of arc emission illumination on the weld pool are discussed.

The arc emission is very complex including continuous spectrum with low intensity and line spectrum with high intensity (consists of metal line, Ar atom, and Ar ion spectrum). The radiance of metal line spectrum is much weaker than that of the continuous arc spectrum, thus not suitable for image capturing. The reflection and diffusion of arc light, therefore, are suitable for image capturing because the concave weld pool serves as a good mirror to send the light into CCD camera. Using this light, the image is with strong intensity and clear weld pool edges.

The intensity of spectral distribution of GTAW with low carbon steel anode is shown in Fig. 2.4 (A). In the range of 600–700 nm there is main continuous spectrum, with few kinds of line spectrum. Figure 2.4 (B) shows the radiation flux under the same conditions. We can see the radiation flux in the range of 600–700 nm is low and flat suitable for light eliminating control.

2.2.2 Capturing Simultaneous Images of Weld Pool in a Frame from Two Directions

The backside image of weld piece is not available in many practical cases. According to the experience of skilled welder, the geometry of weld pool in both topside and backside can provide instantaneous information about welding penetration. Topside and backside images of weld pool are needed to be captured and their size characteristics extracted for modeling and controlling of the dynamic welding process. Therefore, a double-side visual sensor is discussed to detect both top and back of the work piece. The sensing system consists of topside and backside light path and composite filters. The schematic diagram of the sensing system is shown in Fig. 2.5.

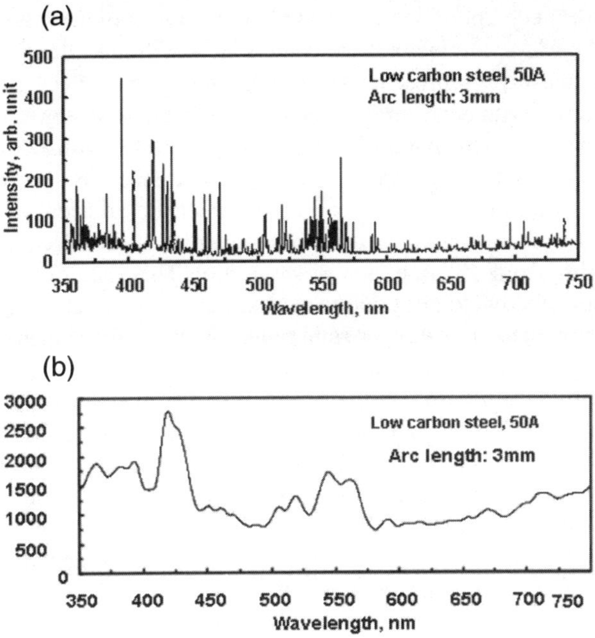

Fig. 2.4 Arc light radiation of GTAW with mild steel anode. (**a**) The spectral distribution (**b**) arc light radiation flux

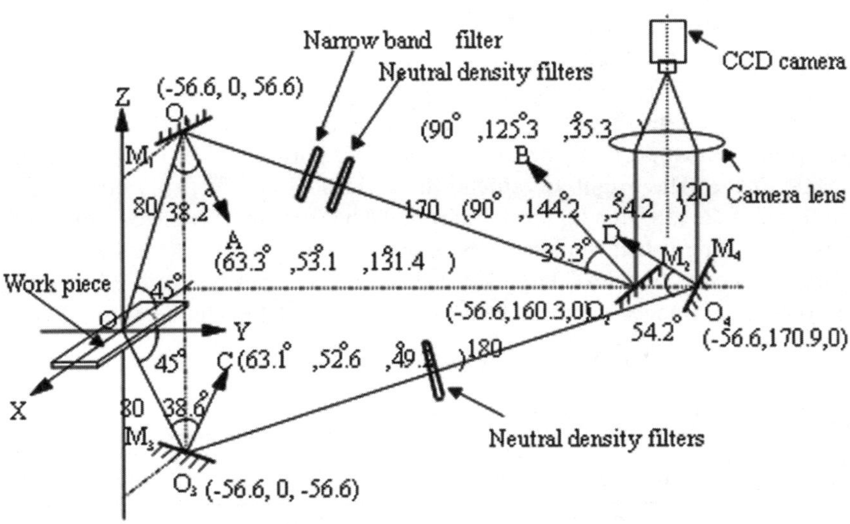

Fig. 2.5 The light path of simultaneous double-side visual image sensing system of weld pool in a frame

In Fig. 2.5, O_XYZ is work piece coordinate; point O is the center point of the weld pool image. M_1, M_2, M_3, M_4 are reflectors, the centers of which are O_1, O_2, O_3 and O_4. O_1A, O_2B, O_3C and O_4D represent the normal line of the reflectors respectively, denoted as the angles with each single axis in the coordinate system.

In the previous studies [4–6], the narrowband filtering light system is established to pitch on a center wavelength of the arc spectrum. For low carbon steel, its spectrum intensity is greater than radiant intensity of continuous spectrum adjacent to this metal spectrum line, so that the arc light of other wavelength can be filtered by the selected filter for imaging from self-radiation of weld pool. This kind of filter is feasible to the low carbon steel weld pool due to its distinct contrast between radiation spectrum of the melting metal in weld pool and radiation or reflected spectrum of solid metal on the edge of the weld pool. The composite filter system includes topside and backside light path with different filter. The topside image of weld pool is formed by the illumination from arc emission in the spectral window of 600–700 nm. Topside light path consists of a neural density filter (2 mm depth, and the speed of lens is 1%) and a narrow band filter (the center band is 661 nm, half width is 10 nm, and the peak speed of lens is 28.8%). Backside image is formed by the radiance of the backside metal with high temperature. Two neutral density filters are used in the backside light path with speed of lens of 10% and 50%. Both the topside and backside images concentrate on the same target of the CCD camera through the above double-side imaging light path system.

The system includes CCD camera and optical lens. The focal distance of the lens is 500 nm. The sensitivity of the camera is 0.4 Lux, the area of target is 5.24×6.4, and the shutter is set at 1/1000s.

(1) Welding process without wire filler

The experiment conditions for GTAW without wire filler are shown in Table 2.1. Peak current is set at 120 ampere, welding velocity is 2.5 mm/s [7]. A complete weld pool image in a frame is shown in Fig. 2.6, where the left is backside image and the right is the topside image. The image contrast is high, for the nozzle, arc center, topside molten portion, and topside solidified portion can be clearly seen in the topside image. The bright arc around the weld pool is effectively eliminated, and the shape of the tungsten tip emerged from the background. Backside weld pool image is also distinguished from the background.

Figure 2.7 shows top/backside pool serial images in a pulsed cycle. Figure 2.8 (b),(c),(d) are pool images in pulsed peak current and (e),(f),(g) in the pulsed based

Table 2.1 Experimental conditions of pulsed GTAW

DWP	Pulse frequency	Pulse duty ratio	Base current	Electrode diameter	Angle of tip	Arc length	Flow rate	Specimen size
Unit	f(Hz)	$\delta(\%)$	Ib (A)	ϕ(mm)	$\theta(°)$	l(mm)	L(l/min)	mm×mm×mm
Value	1	45	60	3.2	30	3.5	8.0	280×50×2

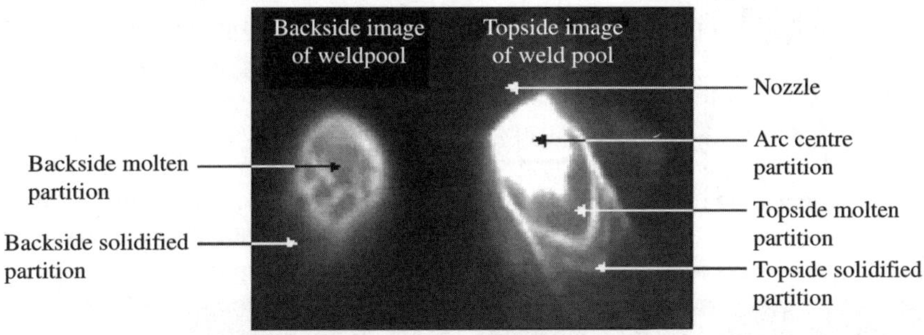

Fig. 2.6 A frame complete weld pool image of pulsed GTAW

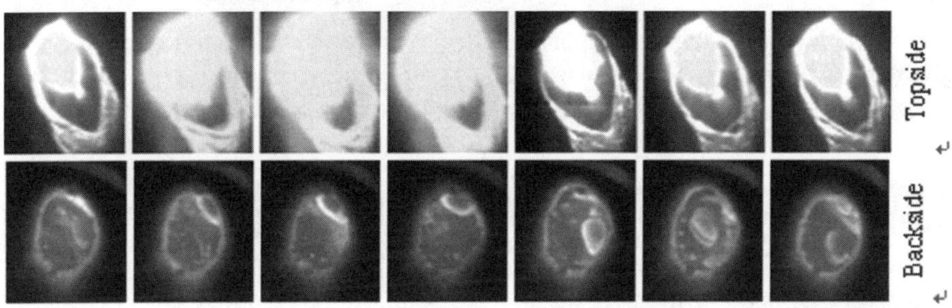

Fig. 2.7 The visual images of the weld pool in different time of a pulse cycle

current. The image is captured under the current of 60 A at 80 ms for a frame complete weld pool image.

The contrast between light reflected from the molten metal surface and that from solid metal surface is distinct; the disturbance can be turned into an advantage for taking a clear image of weld pool.

(2) Welding process with wire filler

The use of wire filler will lead to many changes in the weld pool images, including larger welding current, darker images with blurred weld pool edges. Therefore, a new design of welding experiment parameters is necessary for the welding process with wire filler [8]. Figure 2.8 shows the welding images captured in different period of a pulse. T_0 is at the peak time of the pulse, T_1 is 40 ms after the peak time, T_2 is 100 ms after peak time, T_3 is 200 ms after peak time. It can be inferred from the images that at T_0, the image is too bright to be dealt with; due to the speed limitation of current regulation, image remain bright at T_1; at T_2, the current is 30 A and the image is clear enough for weld pool edge identification.; At T_3, however, the image again get blurred. Therefore, the best time for image capturing is 100 ms after peak time.

Under the supposed experiment conditions, welding parameters are set as follows: pulse peak current 120 A, base current 60 A, pulse duty ratio 40%, and welding velocity 2.5 mm/s. According to imaging principle, image is determined by light

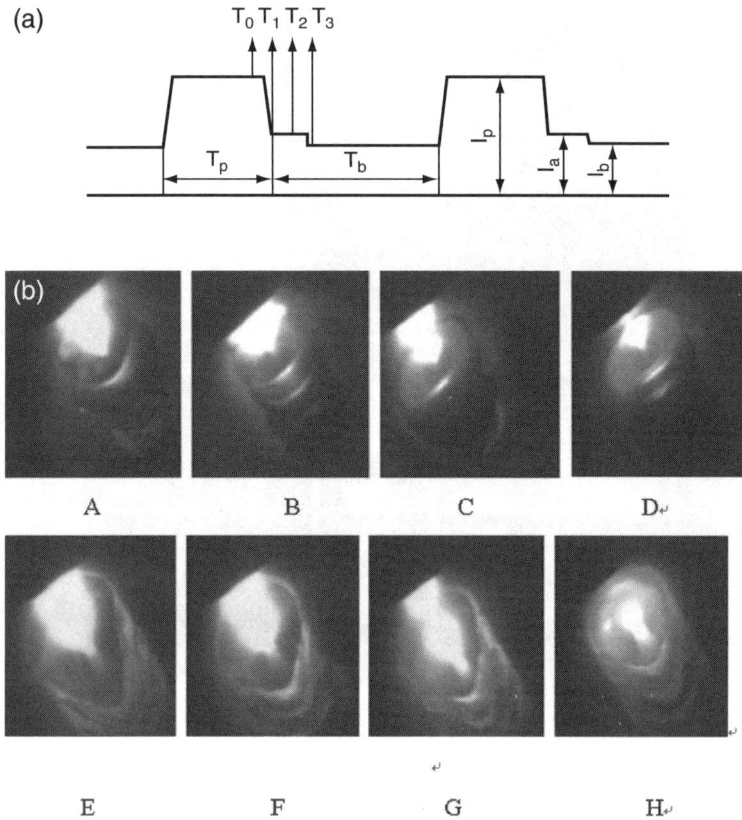

Fig. 2.8 Influence on the weld pool image during different imaging time (**a**) time sequence (**b**) weld pool images; A – 60 A, convex; B – 50 A, convex; C – 40 A, convex; D – 30 A, convex; E – 60 A, concave; F – 50 A, concave; G – 40 A, concave; H – 30 A, concave

source, camera and object shape. Here, imaging current is set as 30 A, and the light source of arc can be supposed as a point light source. The shutter of CCD camera is set as 1/1000s, and the iris diaphragm and filter ratio are fixed.

In Fig. 2.9, when the weld pool is partially penetrating, the top shape of weld pool is approximate to an ellipse, i.e., a convex model image; while on full penetrating, the pool image is similar to a peach shape, i.e. a concave model. The length and stem shape varieties of the weld pool are most obvious.

Due to the impact of arc plasma, the surface of the weld pool depressed in full penetration, while the surface of the weld pool can be convex in partial penetration or with wire filler. The rear part of the weld pool shows the different concave or convex shape distinguishably, as shown in Fig. 2.9.

With the imaging current decreasing, different images of the weld pool are shown in Fig. 2.8 (b–d) and (f–h). It shows that with the imaging current decreasing, the shape of the arc center get concave, and the weld pool gets brighter. And under

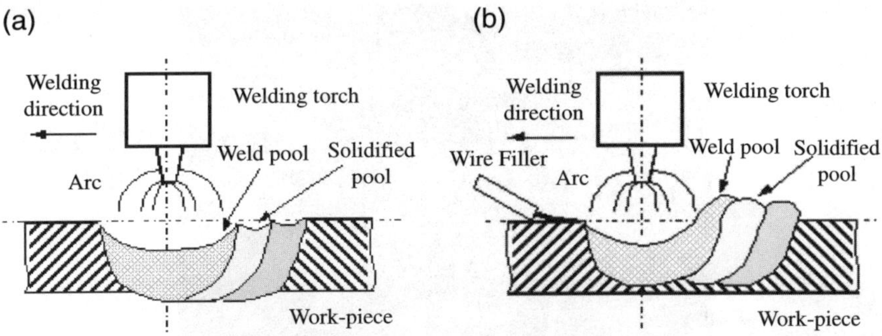

Fig. 2.9 Definition for different type of the weld pool surface **(a)** Concave type **(b)** Convex type

imaging current of 30 A, both images of the concave or convex weld pool are clear enough for image processing. The weld pool, the solidified metal, the inverted image of the arc center, and the nozzle of the torch are clearly seen.

2.2.3 Capturing Simultaneous Images of Weld Pool in a Frame from Three Directions

The addition of one light path can offer more visual information. And here the control system includes a double-side visual sensing system from 3 directions, the topside front, rear, and the backside of weld pool. The light path of the visual system is shown in Fig. 2.10.

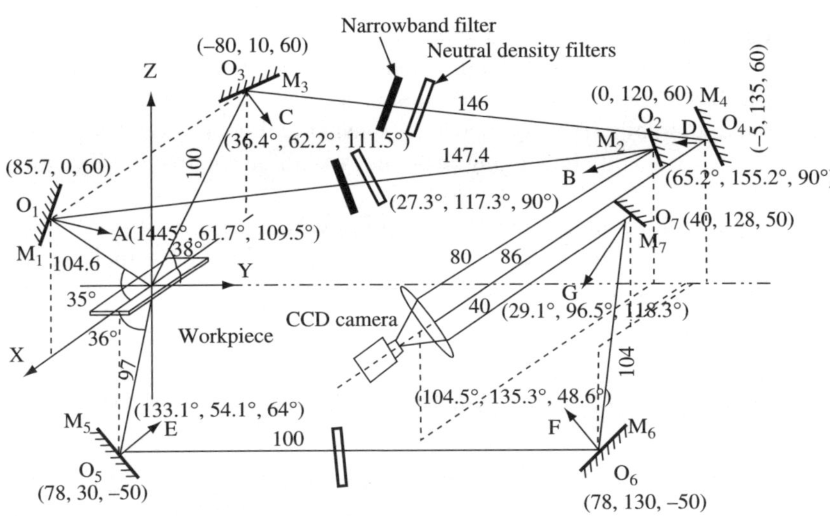

Fig. 2.10 The light path of simultaneous visual imaging system of weld pool in a frame

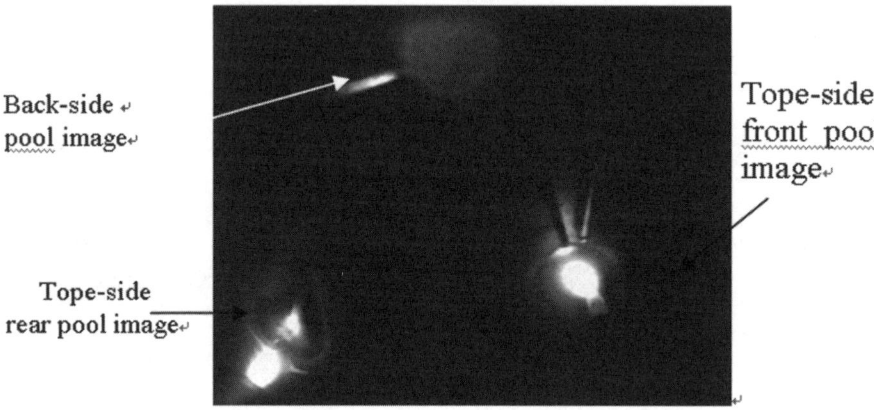

Fig. 2.11 A frame complete weld pool image of pulsed GTAW

A complete weld pool images in a frame are shown in Fig. 2.11. And Fig. 2.12 shows the image in different periods.

2.3 The Visual Sensing System and Images of Weld Pool During Aluminium Alloy Pulsed GTAW

Up to now, most of the researches on GTAW sensing systems are for steel plate. Owing to the special features of aluminium alloy welding, such as heat disseminating rapidness, strong oxygenation, evident effects of accumulating heat during welding process, weld seam cutting phenomena, non-distinctive changes in color and luster between melting weld pool and solidified metal region of aluminium alloy, etc., it gets very difficult to control stability and shaped quality of aluminium alloy welding [9, 10]. Since aluminium alloy welding is necessary techniques in aviation, spaceflight and automobile industry, real-time sensing and control of this process is becoming a pressing and challenging technology with development of welding automation.

The sensing system of aluminium alloy welding will be discussed in detail on aluminium alloy arc spectrum analysis, wideband filter design and visual sensing systems construction and image capturing for weld pool.

2.3.1 Analysis of the Sensing Conditions for Aluminium Alloy

In contrast with the low carbon steel weld pool, aluminium alloy weld pool has non-distinctive changes in color and luster between melting and solid states, which results in blurred images because radiating spectrums of not only melting pool but

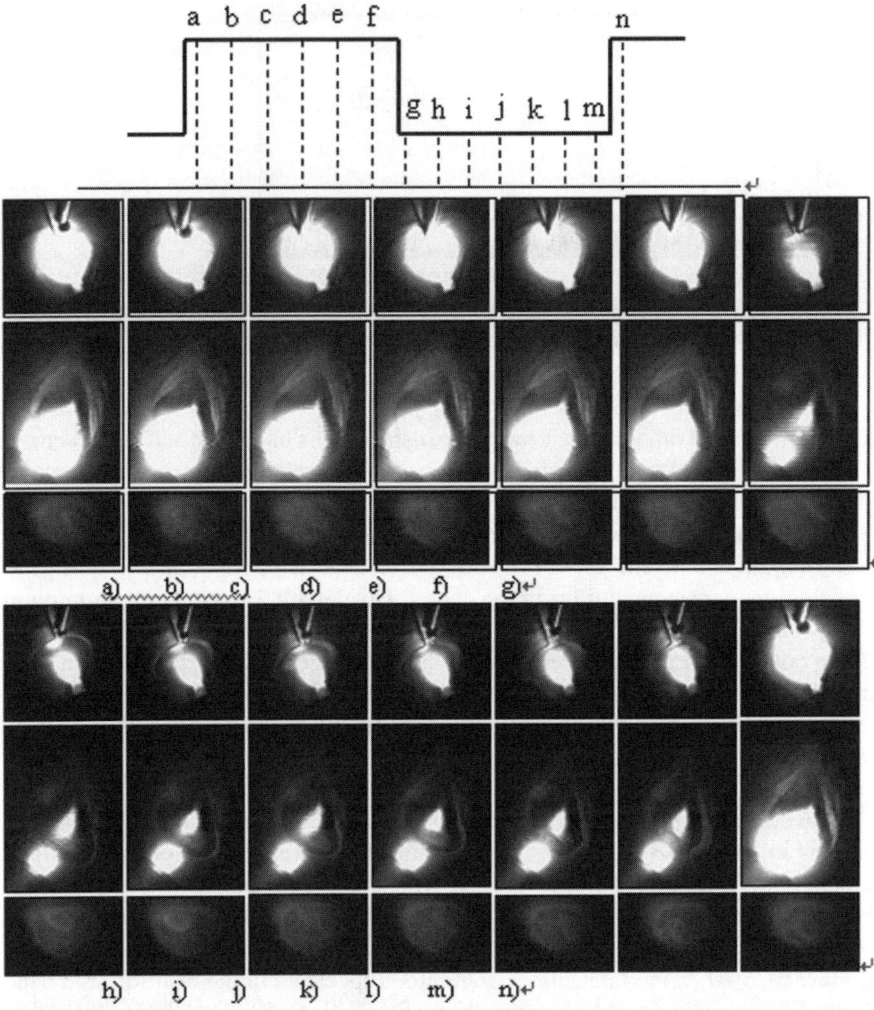

Fig. 2.12 The weld pool images of different time in a pulse

also of reflecting arc from solid metal surface around the pool exist after narrow filtering. In addition, the narrowband filter also results in a blurred image due to the metal steam on the surface of aluminium alloy weld pool.

Experiments [10] showed that arc spectrum distribution of aluminium alloy GTAW process is composed of both lower intensity continuous spectrum and other different intensity spectrum lines, therefore, it will vary with different technical parameters, e.g., welding current, voltage, materials, etc.. The arc spectrum near the surface region of aluminium alloy weld pool is mainly composed of Al atom spectrum, Al ion spectrum and continuous spectrum radiating from metal black body of the weld pool. The spectrum in the arc pole region includes the spectrum lines of

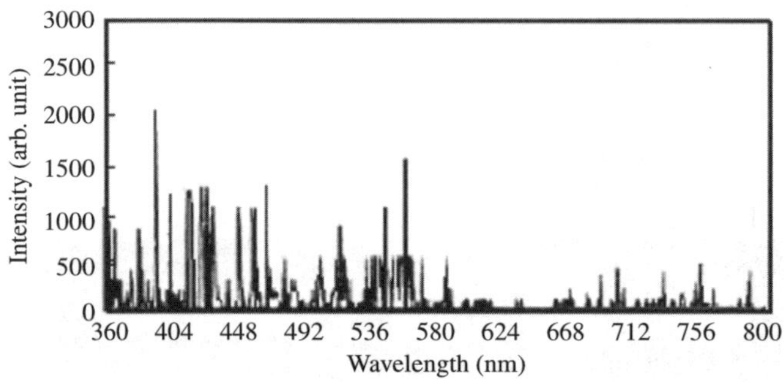

Fig. 2.13 The distribution of characteristic spectrum of A_r

argon atom, ion and other metal steam. The distributions of atomic and ion spectrum lines of argon and Al under blazing condition in arc welding process are shown as Figs. 2.13 and 2.14. The Figures indicate that within the visible light spectrum band of 380–760 nm, the density of Al spectrum is stronger than that of argon only when the wavelength is 396 nm.

If using the narrowband filter in the center wavelength 396–560 nm, aluminium alloy spectrum lines would be submerged in the spectrums of argon. Moreover, welding current variation will result in intensity changes of discrete spectrum lines and pollution of weld pool images, and common CCD camera is less sensitive to visible light. Experiments [10] also show that the narrowband filter used for low carbon weld pool is unsuitable to capture clear weld pool image during aluminium alloy pulsed GTAW process.

However, in the near infrared band 580–720 nm, according to the spectrum distributions in Figs. 2.13 and 2.14, the spectrum lines of argon and other metal and nonmetals are relatively weak compared with those of aluminium alloy, because the spectrum lines of aluminium alloy are continuous, while almost no argon spectrum line is distributed in 640–670 nm band. Moreover, in the case of welding current more than 80 A, the intensity of continuous spectrum in the near infrared band keeps stable even when welding current had a fluctuation of ±20 A. Therefore, using

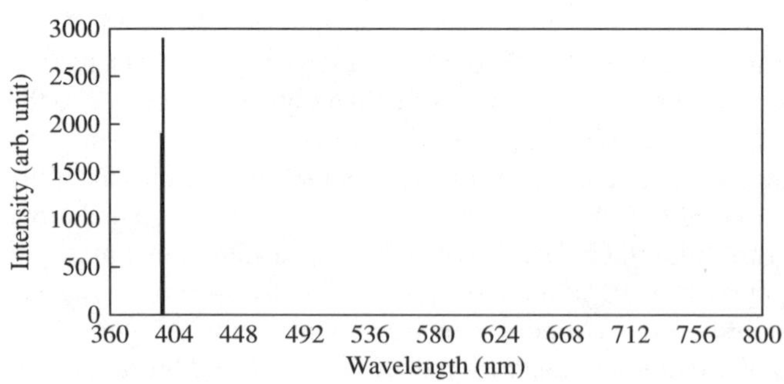

Fig. 2.14 The distribution of characteristic spectrum of aluminium alloy

the arc continuous spectrum in the near infrared band for illuminating aluminium alloy welding pool would greatly decrease interferences from other various spectrum lines.

According to the above characteristics of aluminium alloy weld pool, a wideband filtering method is presented and the filtering system is established to enlarge permeating light range of the filter and to improve anti-interference ability of the visual sensing system by illumination of continuous and discrete spectrum in the wideband and an appropriate reducing light measures. Based on a large number of experiments, the parameters of the wideband filtering system are determined as follows, permeating light range is 590—710 nm, permeating ratio of reducing light lens in the upside light path of the weld pool is 20%, and permeating ratio of reducing light lens in the backside light path of the weld pool is 90%. The response curve of the frequency spectrum of the designed wideband filter is shown as Fig. 2.15. Using the developed wideband filtering system, clear images of aluminium alloy weld pool during pulsed GTAW process can be captured.

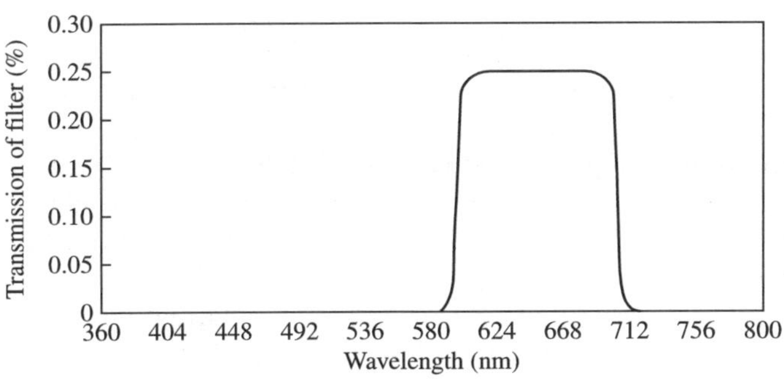

Fig. 2.15 Response curve of the frequency spectrum of the wideband filter

2.3.2 Capturing Simultaneous Images of Weld Pool in a Frame from Two Directions

Combining the analysis of arc spectrum features of aluminium alloy weld pool and the idea of wideband filter, a visual sensing system with topside and backside light paths and composite filters is developed to capture the topside and backside images of the aluminium alloy weld pool simultaneously in the same frame [10].

The system restrains some particular wavelength from passing through the primary filter and can observe the weld pool with the continuous spectrum of the arc light. Based on welding experiment and related result analysis, the parameters of the light filter system are determined as follows: the primary filter is a 560–700 nm glass filter, only the light of wavelength longer than 560 nm or shorter than 700 nm can pass through it, so it filters out the high intense noise of argon's and etc. An attenuation of the dimmer glass is 30%. Depth of field is 1/1000, and the shutter is

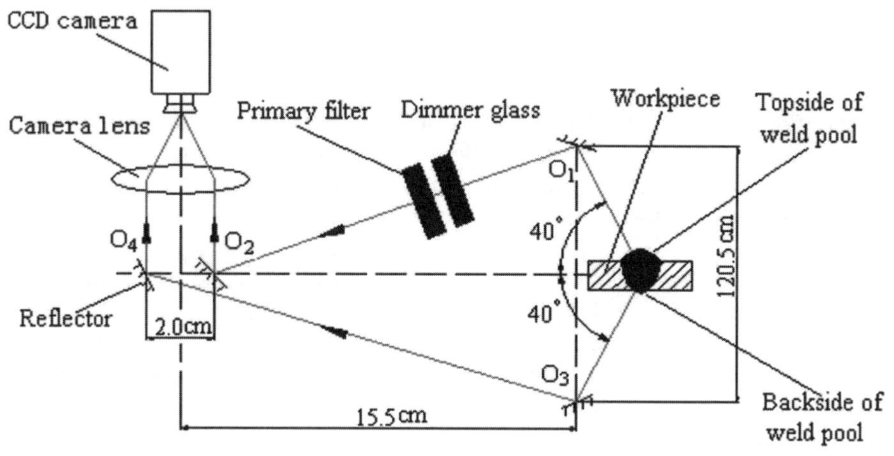

Fig. 2.16 Light path structure of double-side sensing systems for Al alloy weld pool

set at 1/125 s. The function of reducing light is realized by the dimmer glass and the CCD camera aperture is adjustable. Figure 2.16 shows a schematic diagram of the double-sided visual sensing system, which contains the topside and backside imaging light paths. The light from the topside weld pool reached the reflector O_1 at an angle of 40-deg with the X-axis, and is reflected through the dimmer glass and primary filter, then reflected O_2, and finally focused on the target of the CCD camera. The backside light path is shown in the bottom part of Fig. 2.16.

Through investigation of Al welding experiments, the pulse current pattern is designed as Fig. 2.17 for realizing a higher efficiency of welding heat input during the peak time of the pulse level and acquiring of the pool image in the based level. In practical welding, the main aim of control welding process is to ensure a stable welding with desirable appearance.

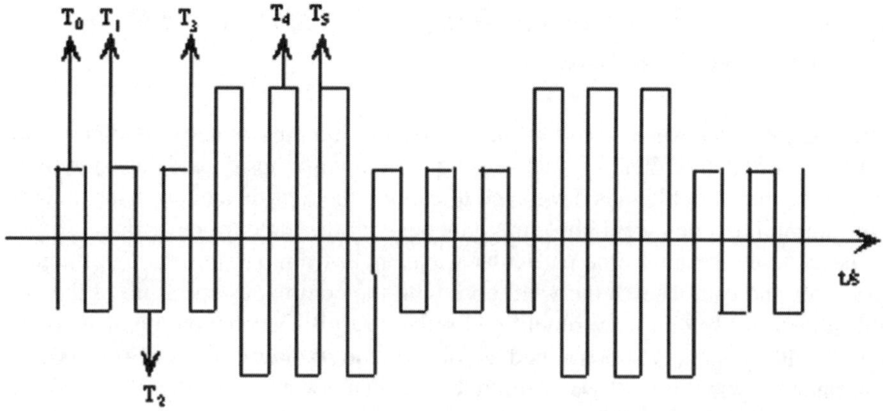

Fig. 2.17 Pulsed wave of welding current

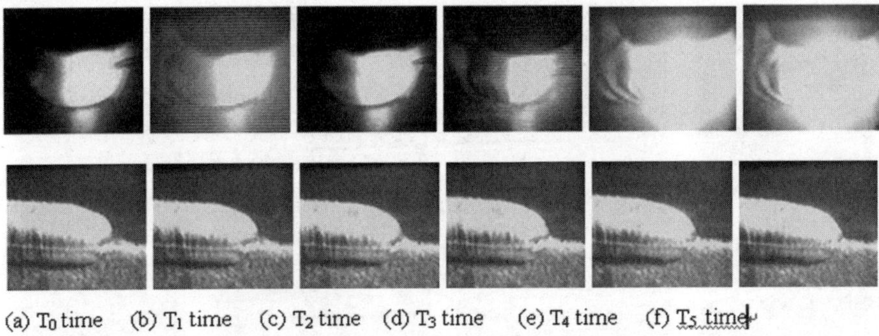

Fig. 2.18 Images of different time molten pool in a pulse cycle (**a**) T_0 time (**b**) T_1 time (**c**) T_2 time (**d**) T_3 time (**e**) T_4 time (**f**) T_5 time

Because the difference of polarity and intensity in different time pulsed current, image quality of the weld pool depends on different image capturing time. Corresponding to the different time in Fig. 2.18, T_0, T_1, ... T_5, where T_0 is the based current time at the stable positive polarity, T_1 is the transition time of the based current from negative to positive polarity, T_2 is the based current time at the stable negative polarity, T_3 is the transition time of the based current from positive to negative polarity, T_4 is the peak current time at the stable positive polarity, and T_5 is the transition time of the peak current from negative to positive polarity. The captured images of topside and backside weld pool during Aluminium alloy pulsed GTAW are showed as Fig. 2.18, (a), (b), ... (f). The images in Fig. 2.18 are continuous topside and backside images taking in a pulsed current period, the image (a) is corresponding to T_0 time, the image (b) to T_1, ... and the image (f) to T_5.

Under the welding experiment conditions: the frequency of pulse peak current is 2 Hz, the width of pulse peak current is 375 ms, the duration of pulse base current is 125 ms, the main influence on definition of weld pool images is the based current value. In Fig. 2.19, four different based current values at the time A, B, C and D are chosen for comparing the image quality. The images corresponding to base current 70 A at time A, 80 A at time B, 90 A at time C, and 100 A at time D are shown as Fig. 2.20. One can see the evident conclusion as following: if the based current value

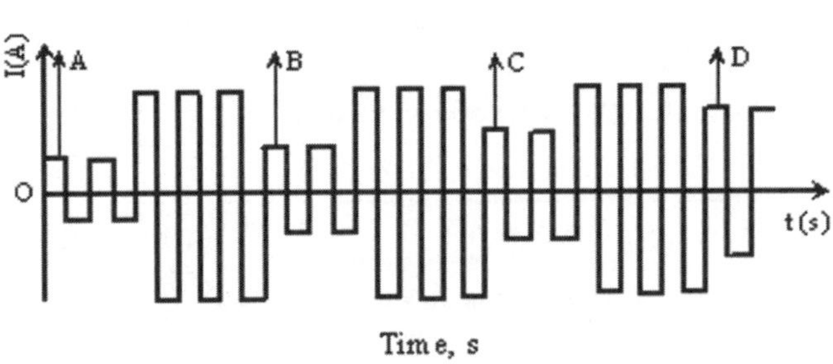

Fig. 2.19 The different based current

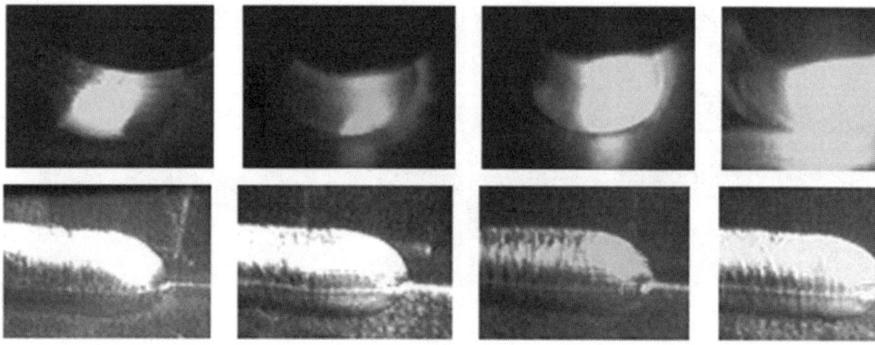

Fig. 2.20 The aluminium alloy weld pool images of different based current (**a**) 70 A (**b**) 80 A (**c**) 90 A (**d**) 100 A

is too large or small, i.e. the arc light is too strong or weak, the contrast between the pool region and the background in the image is unclear so that the boundary of weld pool and image characteristic can't be distinguished easily. Comparing with images at different based current values, the 90 A at the C time is selected the proper base current value for taking image of aluminium alloy weld pool.

Based on investigation of the above experiment results, the proper DWP for taking fine images of aluminium alloy weld pool during pulsed GTAW process are designed as Table 2.2 A typical images of the topside and backside weld pool are captured, shown as Fig. 2.21. The profile image of the weld pool is obtained from the direction paralleled to the one in which the welding gun moved. The image of the topside weld pool in Fig. 2.21 can be divided into the following parts: nozzle, deposited area of metal heap, weld brim, base metal, center of weld pool, cathode spot area and arc column etc. The nozzle reflects light least, so the gray level is low and looks black; the cathode spot area is the part whose oxidized film is removed by the arc, its gray level lies between the highest and that of the image background; arc column shines most strongly and it has a high gray level; the molten metal in the front of weld pool also reflects intensely, and is approximately like a mirror, so its gray level is the highest and it looks white. In the rear region of weld pool, the welding wire and base metal are melting and flowing backward, and the metal piles up, which produces a scattered reflectance of the arc and so only the part arc

Table 2.2 Experimental conditions of pulsed GTAW for aluminium alloy

Pulse frequency f, Hz	2	Traveling speed V, mm/s	3.3
AC frequency f, Hz	50	Arc length ι, mm	5
Peak current I_p, A	220	Electrode diameter ϕ, mm	3
Based current I_b, A	90	Argon flow rate L, l/min	8.0
Wire feed speed V_f, mm/s	15	Workpiece size, mm^3	250×50×3

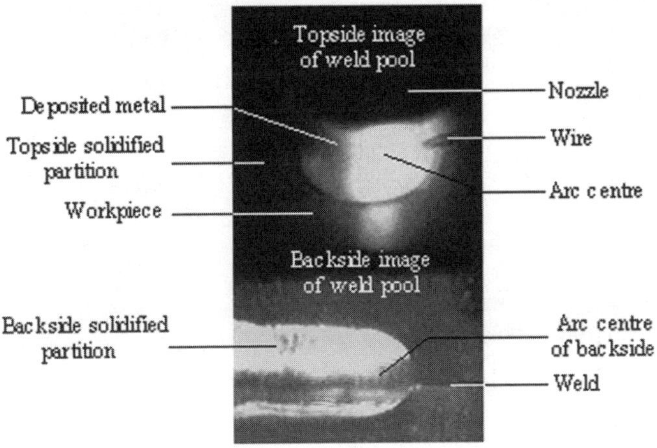

Fig. 2.21 A frame complete molten pool image of Al alloy in pulsed GTAW

is received by the CCD, and this region has a weaker light. The weld brim is clear and the border of welding seam is clearly observed. The image of the backside weld pool in Fig. 2.21 contains some information of welding direction and weld width.

2.3.3 Capturing Simultaneous Images of Weld Pool in a Frame from Three Directions

In this part, a visual sensing system of three directions, namely, frontal, rear-upside and backside direction, is presented [11, 12]. In Fig. 2.22(a), the visual sensing subsystem is composed of a CCD camera, lenses and special filters and image processing algorithms. Figure 2.22(b) shows the visual sensor for GTAW pool with three light path. The visual sensor can acquire nearly all information in a frame about the weld pool from three directions at the same time [12].

The Fig. 2.23 shows the structure diagram of visual sensing and control systems with three light paths for aluminum alloy pulse GTAW. And Fig. 2.24 is a photograph of the visual sensing and real-time control experimental systems for the aluminum alloy GTAW.

At basic current, the intensive arc light momentarily extinguishes periodically with short-circuit. The short-circuit phenomena is utilized in order to acquire an image of the weld pool and its vicinity using the vision sensor.

Figure 2.25 is a whole frame image of weld pool from top-back, top-front and back directions. And Fig. 2.26 is the top-front image of the weld pool. Beside welding pool, there are many other parts in the image such as arc, gap, groove and wire.

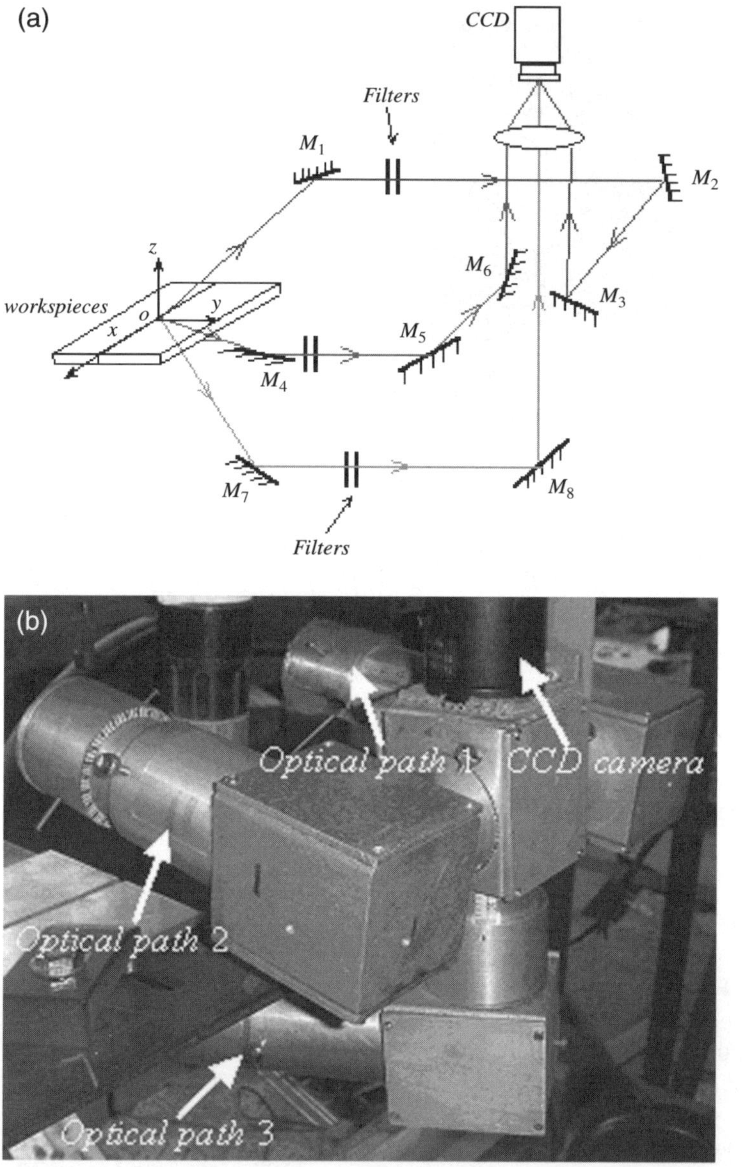

Fig. 2.22 The visual sensor subsystem (**a**) Diagram of visual sensing system (**b**) The visual sensor for GTAW pool with three light paths [12]

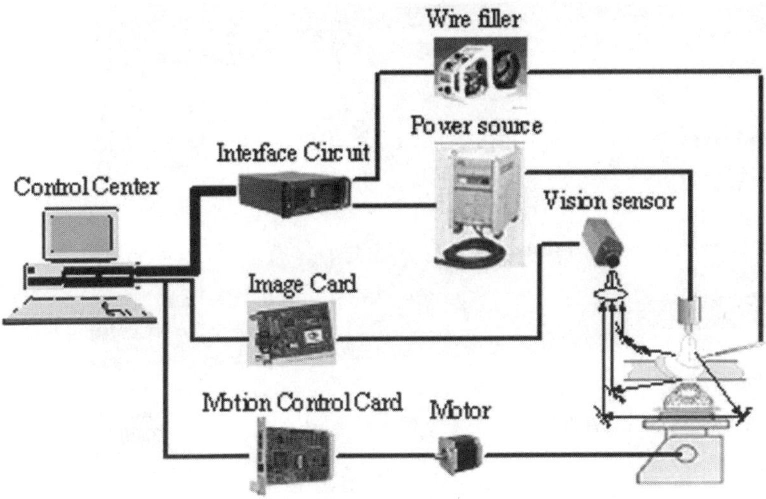

Fig. 2.23 The structure diagram of visual sensing and control systems for aluminum alloy pulse GTAW [12]

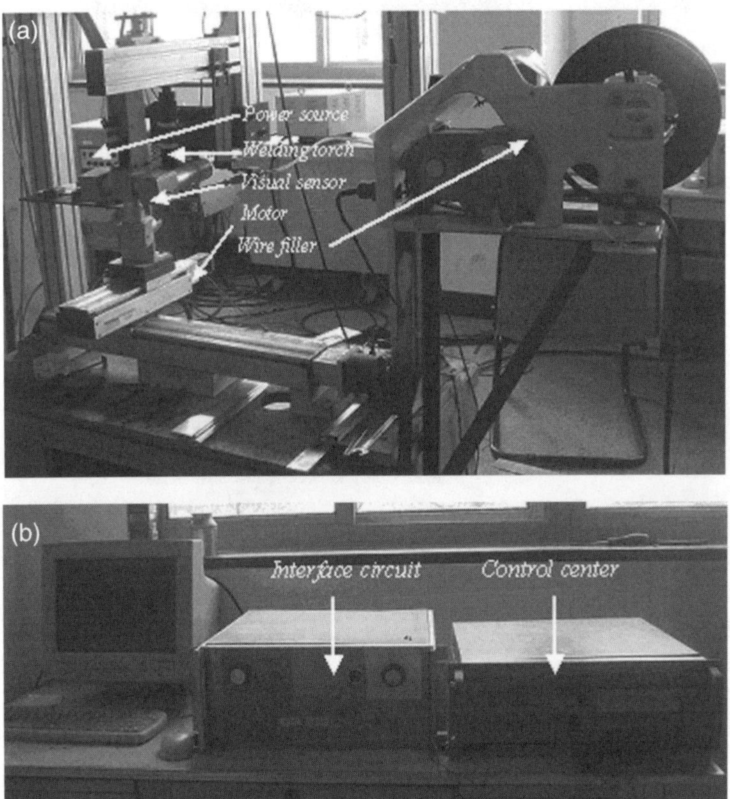

Fig. 2.24 A photograph of the experimental systems for aluminum alloy GTAW [12] (**a**) Welding unit (**b**) Control center

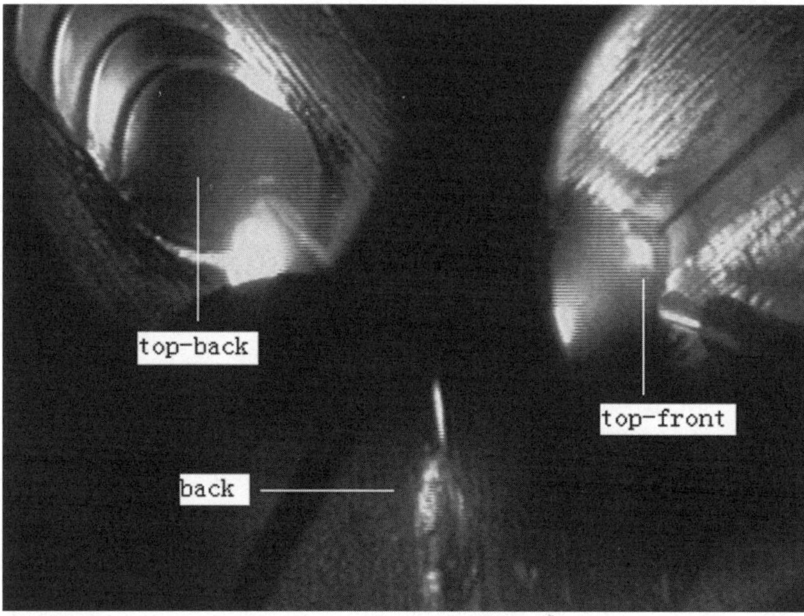

Fig. 2.25 The three-direction weld pool image

Fig. 2.26 The top-front part image

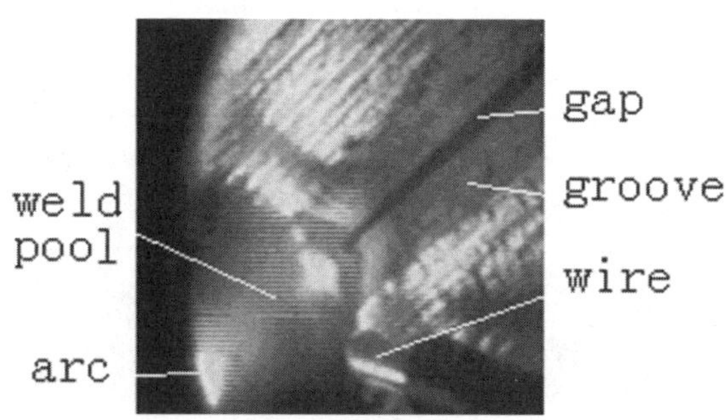

2.4 The Chapter Conclusion Remarks

According to the analysis of arc spectrum and radiation of low carbon steel and aluminium alloy, visual sensing systems with filters are described. Based on the filtering method, and proper welding parameters, clear images of weld pool are obtained during pulsed GTAW.

References

1. D. Brzakovic, D.T. Khani, Weld pool edge detection for automated control of welding. IEEE Transactions on Robotics and Automation. 1991, 7(3):397–403
2. J.J. Wang, T. Lin, S.B. Chen. Obtaining weld pool vision information during aluminium TIG welding. International Journal of Advanced Manufacture Technology, London, UK, 2005, V26:219–227
3. D.B. Zhao, Y.J. Lou, S.B. Chen, L. Wu. Surface height and geometry parameters for describing shape of weld pool during pulsed GTAW. SPIE International Symposium on Intelligent System and Advanced Manufacturing, Boston, Massachusetts, USA, 1999, V3833:91–99
4. S.B. Chen, Y.J. Lou, L. Wu, D.B. Zhao. Intelligent methodology for measuring, modeling, control of dynamic process during pulsed GTAW – Part I Bead-on-plate welding. Welding Journal. 2000, 79(6):151s–163s
5. S.B. Chen, D.B. Zhao, L. Wu, Y.J. Lou, Intelligent methodology for measuring, modeling, control of dynamic process during pulsed GTAW – Part II butt welding. Welding Journal. 2000, 79(6):164s–174s
6. D.B. Zhao, S.B. Chen, L. Wu, Q. Chen. Intelligent control for the double-sided shape of the weld pool in pulsed GTAW with wire filler. Welding Journal. 2001, 80(11):253s–260s
7. L. Yajun. "Intelligent Control for Pulsed GTAW Dynamic Process Based on Image Sensing of Weld Pool", PhD dissertation, Harbin Institute of Technology, 1998
8. D. Zhao. Dynamic Intelligent Control for Weld Pool Shape during Pulsed GTAW with Wire Filler Based on Three-Dimension Visual Sensing, [Doctorial dissertation], Harbin Institute of Technology, 2000
9. Q.L. Wang, C.L. Yang, Z. Geng. Separately excited resonance phenomenon of the weld pool and its application. Welding Journal. 1993, 72(9):455–462
10. J.J. Wang, Visual information acquisition and adaptive control of weld pool dynamics of Aluminum alloy during pulsed TIG welding. PhD dissertation, Shanghai Jiao Tong University, 2003
11. C. Fan, F. Lv, S. Chen, 5–8 Nov. 2007, "A visual sensing system for welding control and seam tracking in aluminum alloy gas tungsten arc welding", Industrial Electronics Society, 2007. IECON 2007. 33rd Annual Conference of the IEEE, Taipei: 2700–2705
12. C. Fan, Visual densing and intellignet control of varied gap AI alloy pulsed GTAW process, [2.Doctorial dissertation]. Shanghai Jiao Tong University, 2008

Chapter 3
Information Acquirement of Arc Welding Process

Abstract Precise image processing algorithm is important for welding process control. Generally, original image cannot be directly used due to the disturbance from welding equipment. Moreover, fluctuation in welding current and arc light also lead to image degrading. All the above factors add difficulties to the image processing, and the image processing algorithms are required to be adaptive to different conditions. In this chapter, both 2D and 3D image processing methods are described. The 2D image processing methods used in this chapter include degrading image recovery, integral edge detection, projection, neural network edge identification and curve fitting to extract the length and width of the weld pool. 3D image processing methods include experimental and SFS(Shape-from-Shading) method to extract topside height of the weld pool. And image processing software exclusively for weld pool images is introduced at the end of the chapter.

Real time control of weld pool dynamics is crucial for welding quality, which depends primarily on extracting and calculating geometric characteristics of the weld pool [1–4]. The weld pool contains abundant information about the welding process. Actually, in practice, a skilled welder can estimate the appearance of backside of weld pool by observing the shape, size and dynamic change of the topside of the weld pool and adjust accordingly. Image processing is aimed to obtain the relevant information by enhancing the necessary image features and suppressing undesired distortions. However, many disturbances, such as alternating magnetic field and the relative motion between CCD and weld pool, will affect the information acquirement. Therefore, image processing technology is necessary for the welding process.

3.1 Acquiring Two Dimensional Characteristics from Weld Pool Image During Pulsed GTAW

In this chapter, two frequently-used weld application are described, repectively low carbon steel and aluminium alloy.

S.-B. Chen, J. Wu, *Intelligentized Methodology for Arc Welding Dynamical Processes*, Lecture Notes in Electrical Engineering 29, © Springer-Verlag Berlin Heidelberg 2009

3.1.1 Definition of Weld Pool Shape Parameters

The problems of describing the topside shape of the weld pool accurately with some simple parameters obsessed the researchers. Reference [5] proposed a nonlinear regressive function for describing the shape of the depressed weld pool during GTAW.

$$y_r = \pm a x_r{}^b (1 - x_r),\ x_r = x/L_t,\ y_r = y/L_t \tag{3.1}$$

The definition could be seen in Fig. 3.1(a). The front corner of the weld pool was defined as the original of the orthogonal coordinate system, and the inverse welding direction was defined as the positive x-axis, and the vertical welding direction was as the y-axis. The edge point coordinate (x, y) of the weld pool was normalized with the pool length as (x_r, y_r). Parameters a and b were regressive coefficients, and $a > 0$, $0 < b \leq 1$.

Large errors occurred in describing the shape of the convex weld pool during GTAW with wire filler, especially for the weld pool with ellipse shape. Based on the shape variation of the weld pool during GTAW with wire filler, new regressive formula and shape parameters were proposed.

$$y_r = \pm a x_r{}^b (1 - x_r)^c,\ a = W_t/L_t,\ x_r = x/L_t,\ y_r = y/L_t \tag{3.2}$$

Where the coordinate system was as the same, and $a > 0$, $0 < b \leq 1$ and $0 < c \leq 1$.

The maximum width Wt should be occurred at the point $P_w(x_r, 0)$,

$$x_r = \frac{b}{b+c} \tag{3.3}$$

The pool length could be divided into two parts L_{th} and L_{tr} with the point P_w. L_{tf} denoted the front half-length of the topside weld pool from the original to P_w, and L_{tr} denoted the rear half-length from P_w to the rear corner of the weld pool. Then half-length ratio R_{hl} was introduced.

$$R_{hl} = L_{tr}/L_t = 1 - x_r = \frac{c}{b+c} \tag{3.4}$$

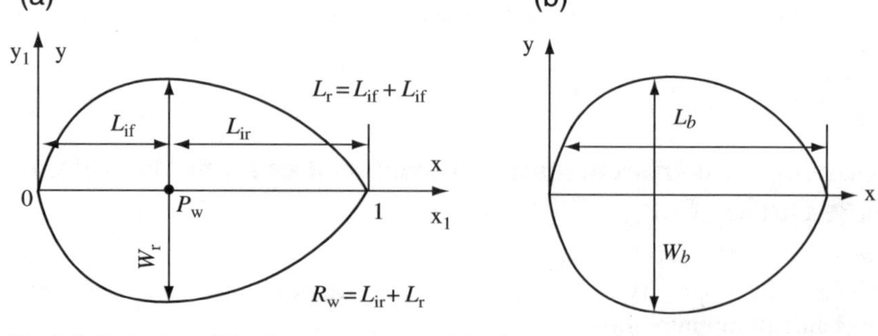

Fig. 3.1 Definition of the shape parameters of the double-sided weld pool (**a**) Topside (**b**) Backside

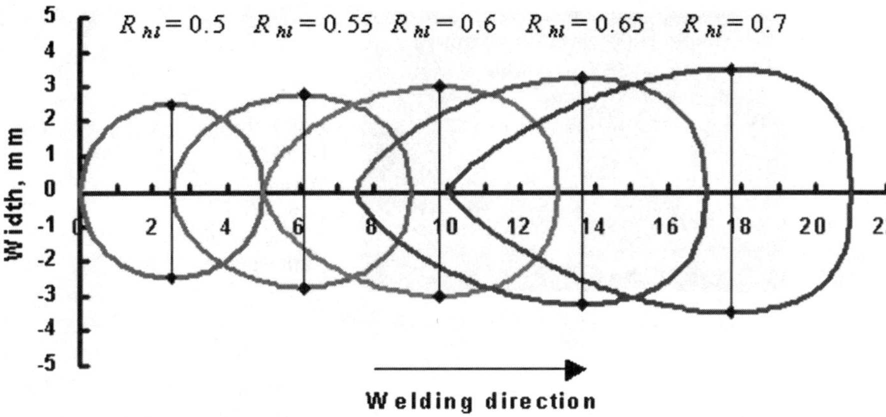

Fig. 3.2 Simulation of the weld pool shape variation during the ignition period of pulsed GTAW

Since the coefficient a is defined as the ratio of the width to length. With the Eq. (3.3), the equation (3.2) could be unfolded as follow.

$$2\left(\frac{b}{b+c}\right)^{b}\left(\frac{c}{b+c}\right)^{c}=1 \tag{3.5}$$

So, with the known topside length L_t, width W_t, and half-length ratio R_{hl}, the coefficients a, b and c, and the shape of the topside weld pool were uniquely decided. The whole definition of the topside shape parameters was shown in Fig. 3.1(a).

During the ignition of pulsed GTAW with wire filler, the shape of the weld pool changed complexly from circle-shape to ellipse-shape, and to peach-shape. The shape variation during the ignition was simulated accurately by the proposed nonlinear regressive formula with various shape parameters, shown in Fig. 3.2.

The shape of the backside weld pool was similar to ellipse; therefore, the shape was decided by the backside length L_b and width W_b, shown in Fig. 3.1 (b).

3.1.2 The Processing and Characteristic Computing of Low Carbon Steel Weld Pool Images

3.1.2.1 Analysis of the Weld Pool Images

A typical weld pool image during low carbon steel pulse GTAW is shown in Fig. 3.3, where the topside image is the topside weld pool image and the bottom one is the backside weld pool image. The structure of the topside weld pool is described in the following part.

The topside image of the pool, as shown in Fig. 3.3, can be divided into the following parts: nozzle, arc center partition, topside molten partition, topside solidified

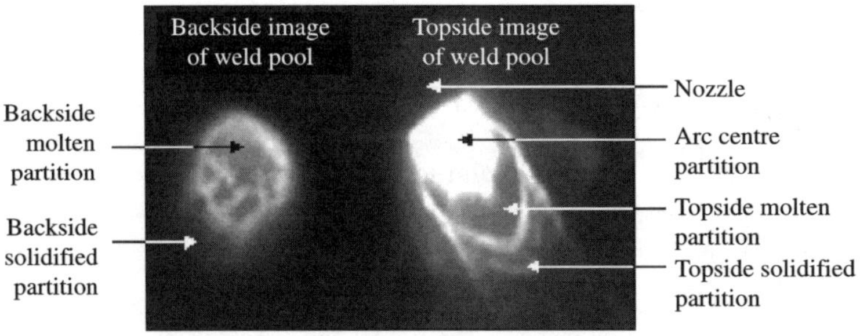

Fig. 3.3 The characteristics of the low carbon steel weld pool

partition. In the image, the nozzle gains little light from the arc so it is the darkest part in the image and looks black. Whereas the topside molten partition is relocated by the arc from the center of the pool, its gray level lies between the highest and that of the image background. In the arc center, the arc shines the most strongly and it has a high gray level, while the molten metal in the front of pool also reflects the arc intensely like a mirror, so its gray level is the highest and it looks white. In the topside solidified partition, the welding wire and base metal are melt and flow backward, where the metal piles up and produces a scattered reflectance of the arc, therefore this region possesses a low gray level. The backside image has a clear image, and the image process for this area is relatively easier.

3.1.2.2 Image Processing of the Weld Pool Image Without Wire Filler

Bead-on-plate experiment is conducted on low carbon steel during pulsed GTAW using the double-side imaging system. A complete weld pool image in a frame is shown in Fig. 3.3, in which the left is backside image and the right is the topside image. In topside weld pool, the nozzle, arc center, topside molten portion, and topside solidified portion is clearly distinguished from the topside image of weld pool.

Figure 3.4 is top/backside pool serial images in a pulsed cycle. Figure 3.4 (b), (c), (d) are pool images in pulsed peak current and (e), (f), (g) in the pulsed based current.

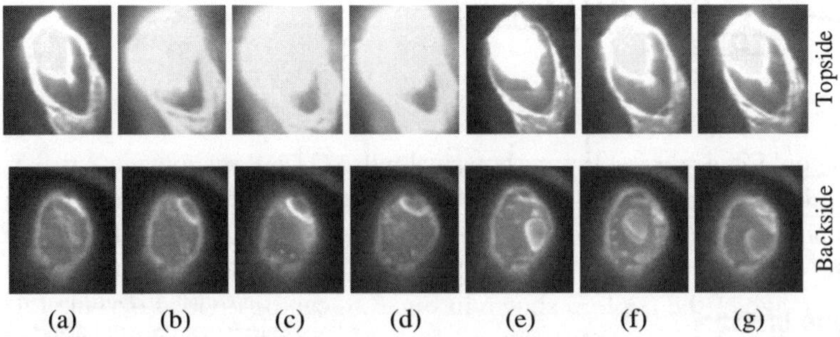

Fig. 3.4 The serial images of different time's weld pool in a pulse cycle

With the weld pool image obtained, specialized image processing algorithm is designed to get its geometric size [6].

3.1.2.3 Exponential Base Filter Processing of Weld Pool Images

The first step of weld pool image processing is filtering and wiping off noise disturbance. In order to reduce calculation, the recursion exponential base filter is used to smooth images. The response property of exponential base filter is similar to the Guass filter [7].

One Dimensional Recursion Exponential Base Filter

The unit sample response $S(n)$ of Exponential Base Smoothing (EBS) is defined as:

$$S(n) = k(\alpha|n|+1)e^{-\alpha|n|} \tag{3.6}$$

In (3.6), n is a discrete variable, α is a constant of the filter for one-dimesional space range, k is a proportion factor defined as :

$$k = \frac{(1-e^{-\alpha})}{1+2\alpha e^{-\alpha}-e^{-2\alpha}} \tag{3.7}$$

Decomposing $S(n)$ as cause-effect and non-cause-effect parts, and supposing filter input as $x(n)$, $y(n)$ is the output response of the EBS filtering $S(n)$, we have:

$$y(n) = y_c(n) + y_a(n) \tag{3.8}$$

$$y_c(n) = k\left[x(n) + e^{-\alpha}(\alpha-1)x(n-1)\right] + 2e^{-\alpha}y_c(n-1) - e^{-2\alpha}y_c(n-2)$$
$$y_a(n) = k\left[e^{-\alpha}(\alpha+1)x(n+1) - e^{-2\alpha}x(n+2)\right] + 2e^{-\alpha}y_a(n+1) - e^{-2\alpha}y_a(n+2)$$

Initial conditions as:

$$x(0) = 0,\ y_c(0) = y_c(-1) = 0,\ n = 1,2,\cdots\cdots M$$

$$x(M+1) = x(M+2) = 0\ y_a(M+1) = y_a(M+2) = 0 \qquad n = M,\cdots\cdots 2,1$$

In Eq. (3.8), $y_c(n)$ is the response for the filtering cause-effect part, $y_a(n)$ is the response for the non-cause-effect part. One dimensional recursion exponential base filtering algorithms can be realized by (3.8).

Two Dimensional Recursion Exponential Based Filtering Algorithms

Based on one-dimensional exponential based filtering function, the separable two-dimensional exponential base filter is developed. For two-dimensional input $x(m,n)$,

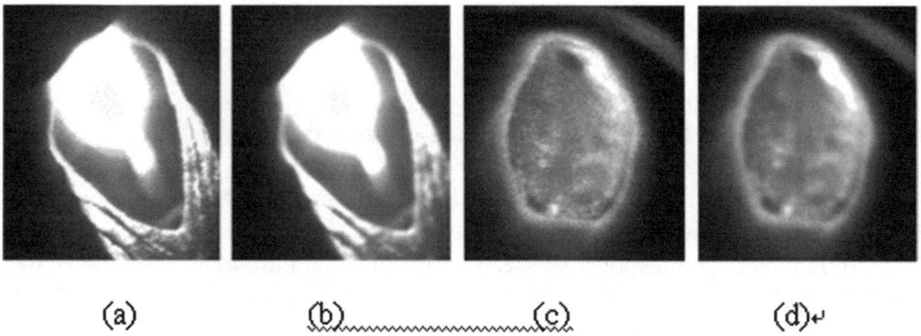

Fig. 3.5 The smoothed image of weld pool with EBS algorithm (**a**) Original topside image (**b**) Topside image smoothed with EBS algorithm (**c**) Original backside image (**d**) Backside image smoothed with EBS algorithm

one-dimensional recursion filtering algorithms along one direction can be first completed due to separability of the filter, and then taking its output as input for next one-dimensional filtering algorithms. The top/backside images of the weld pool are smoothed by the above algorithms, i.e. Fig. 3.5. The results shown that noises in the topside image is filtered and the image margin is fully maintained while $\alpha = 1.41(\sigma = 1.0)$. Taking $\alpha = 0.94(\sigma = 1.5)$, the noises in backside image is ideally sieved by the exponential base filter. Since the two-dimensional exponential base filter can be separated into two one-dimensional recursion filters in two directions, the computing time and space costs of the algorithms is greatly reduced, e.g., the algorithm time cost on a PC-486 computer with main frequency 100 MHz is not more than 8 ms for an image window with 160×190 pixel. The time cost processing the same size image by the Guass filter is about 50 ms. It is obvious that exponential base recursion filter is suitable to process images in real-time.

3.1.2.4 Contrast Enhancement Algorithms for the Weld Pool

Based on the contrast enhancement algorithms presented by Gordon [8] and Beghdadi [9], for supposed center pixel point (x, y) with greyscale G_{xy} and window W_m and the each pixel point with greyscale G_{ij}, the following contrast enhancement algorithms, named as CE, is adopted to enhance the contrast of the weld pool. The CE algorithms are as follow:

Step1: For the pixel point (x, y), calculating average greyscale $\overline{G}$ in its adjacent region:

$$\overline{G} = \sum_{(i,\ j) \in W_m} G_{ij}/(m \times m) \tag{3.9}$$

Step2: For all pixel points in the window W_m, calculating the edge value Δ_{ij} of the pool:

$$\Delta_{ij} = \left| G_{ij} - \overline{G} \right| \tag{3.10}$$

Step3: For the window W_m, calculating weighted average greyscale $\overline{E}_{xy}$ for the pool edge:

$$\overline{E}_{xy} = \sum_{(i,j)\in W_m} \Delta_{ij} * G_{ij} \Big/ \sum_{(i,j)\in W_m} \Delta_{ij} \tag{3.11}$$

Step4: Calculating the contrast degree C_{xy} of the pixel point (x, y):

$$C_{xy} = \left|G_{xy} - \overline{E}_{xy}\right| / \left|G_{xy} + \overline{E}_{xy}\right| \tag{3.12}$$

Step5: Changing C_{xy} to contrast transfer function $C'_{xy} = F(C_{xy})$

$$C'_{xy} = F(C_{xy}) = (C_{xy})^{a/b},\ b = 2^p,\ a < b \tag{3.13}$$

Step6: Defining greyscale of the pixel point (x, y) as G'_{xy} and calculating functions as following:

$$\begin{aligned} R_{xy} &= (1 - C'_{xy})/(1 + C'_{xy}) && if\ G_{xy} \le \overline{E}_{xy} \\ &= (1 + C'_{xy})/(1 - C'_{xy}) && if\ G_{xy} > \overline{E}_{xy} \end{aligned} \tag{3.14}$$

$$G'_{xy} = \overline{E}_{xy} \cdot R_{xy} \tag{3.15}$$

Step7: Repeat algorithms Step1 to Step6 for each pixel point.

Using the above CE algorithm for the topside images, the contrast enhancement processing results are shown as Fig. 3.6. Let $\beta = a/b$, window size m, for $\beta = 0.5$ to $\beta = 0.25$ and $m = 5$ to $m = 13$, the edges and background of the weld pool are learly distinguished in Fig. 3.7. Trading off the enhancement effect and calculating complexity, $\beta = 0.5$ and $m = 9$ are determined in processing and calculating time is not more than 16 ms.

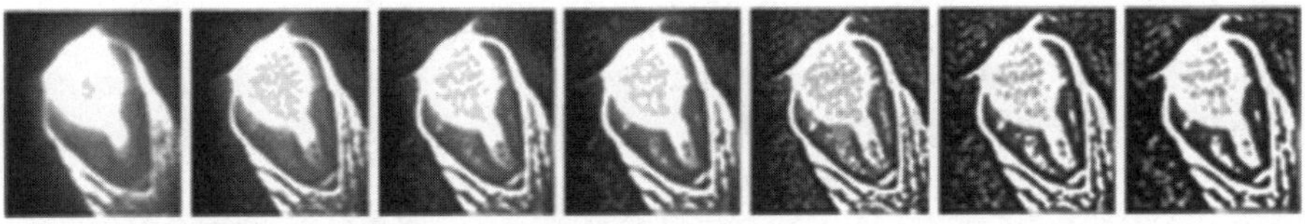

Fig. 3.6 Contrast enhancement of the topside image of weld pool **(a)** EBS smoothed image **(b)** CE ($\beta = 0.5$, m = 5) **(c)** CE ($\beta = 0.5$, m = 9), **(d)** CE ($\beta = 0.5$, m = 13) **(e)** CE ($\beta = 0.25$, m = 5) **(f)** CE ($\beta = 0.25$, m = 9) **(g)** CE ($\beta = 0.25$, m = 13)

3.1.2.5 Algorithm for Extracting the Topside Characters of the Weld Pool

The algorithm ETG for extracting top geometry of the weld pool is described as the following:

The algorithm ETG: extract-top-geometry (image window *topw*), with the result of topside characters extration shown in Fig. 3.7.

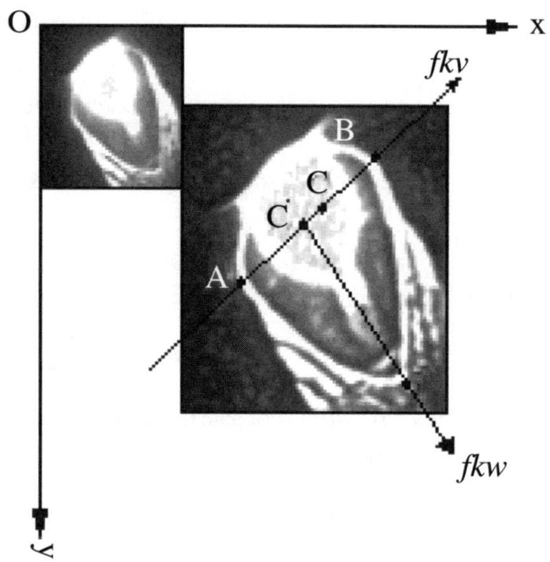

Fig. 3.7 Characteristic points of the *topside* image of weld pool

{ **Step1:** get-centre-point C (image window *topw*);
Step2: for ($i = -5; i <= 5; i = +2$)
{get-line -grey (along weld pool width direction *fkv*, through point C, line-*i*)
find-edge-point (line-*i*) (find point A and B from greyscale distribution);
calculate-width (line-*i*);
}
Step3: find-max-width ($W_{f\max}$ from 11 lines);
Step4: find-max-centre(point C');
Step5: get-line-grey (along weld pool length direction *fkw*, from C');
Step6: find-peak-point((along *fkw*);
Step7: find-max-peak(point D);
Step8: calculate-mid-length($L_{f\max}$ from C' to D);
}

3.1.2.6 Algorithm for Extracting the Backside Characters of the Weld Pool

The binaryzation of the back weld pool images, as shown in Fig. 3.8, is obtained by the EBS filter and the threshold division. From Fig. 3.8, the pool length is determined by the edge point Sp and Ep along *bkw*, and the width is determined by the edge point Lp and Rp along *bkv*. Based on weld pool vertical-square picture distribution by the EBS, the algorithm extracting back geometry (EBG) is developed as the following:

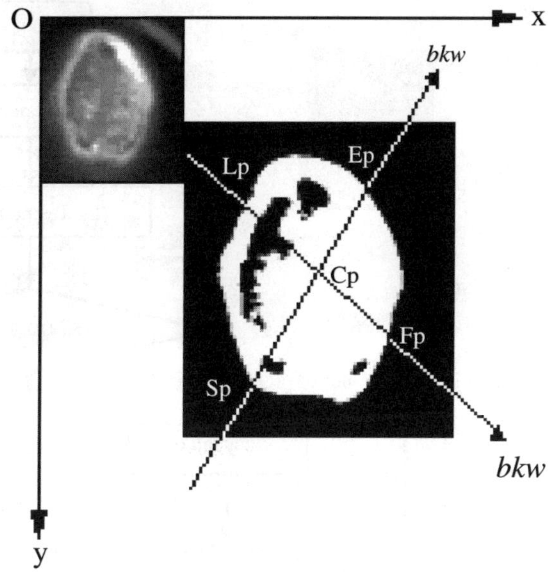

Fig. 3.8 Characteristic points of the backside image of weld pool

The algorithm EBG: extract-back-geometry(image window *backw*)

{ **Step1:** calculate-histm (*histm*-histogram, for back pool image);
Step2: histm-moving-smooth(averaging-moving algorithm for histogram);
Step3: find-threshold
Step4: binaryzation(for back pool image);
Step5: get-centre-point(Cp in Fig. 3.8);
Step6: for($i = -5$; $i <= 5$; $i = +2$)
{ get-line-grey (line-i grey from Cp, along *bkw*,)
find-edge-point(Sp and Ep, along *bkw*);
calculate-length (along line-i);
}
Step7: find-max-length ($L_{b\max}$ from 11 lines);
Step8: get-interval (determining 21 lines from $L_{b\max}$ along *bkv*);
Step9: for($i = 0$; $i <= 20$;$i++$)
{ get-line-grey (line-i along *bkv*)
find-edge-point (Lp and Rp along *bkv*);
calculate-width (along line-i);
}
Step10: find-max-width ($W_{b\max}$ from 21 lines);
Step11: calculate-area(S_b from 21 lines and $L_{b\max}$);
}

The signal flowchart of processing images of weld pool by the above algorithms is shown as Fig. 3.9, EBS, CE , ETG, EBG as the above, TD is the binary algorithm for finding threshold from the histogram distribution of the backside weld pool image.

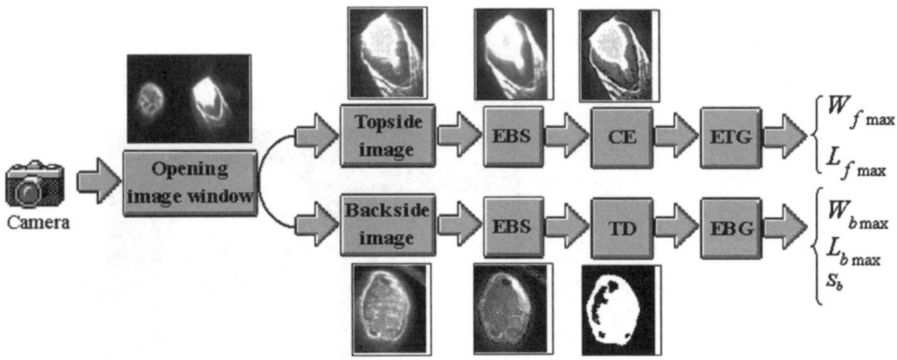

Fig. 3.9 Signal flowchart of processing images of weld pool

The above real-time processing algorithms for the top/backside weld pool images is running on the PC-486 computer, the calculating time for a frame image is not more than 50 ms for the top image and 30 ms for back image.

3.1.2.7 Image Processing of the Weld Pool Image with Wire Filler

Although many achievements of sensing, modeling and control of GTAW pool are proposed, studies on GTAW with wire filler are seldom carried out and reported. The topside image of the weld pool with wire filler is shown in Fig. 3.10. When the weld pool is partially penetrating, the top shape of weld pool is approximate to

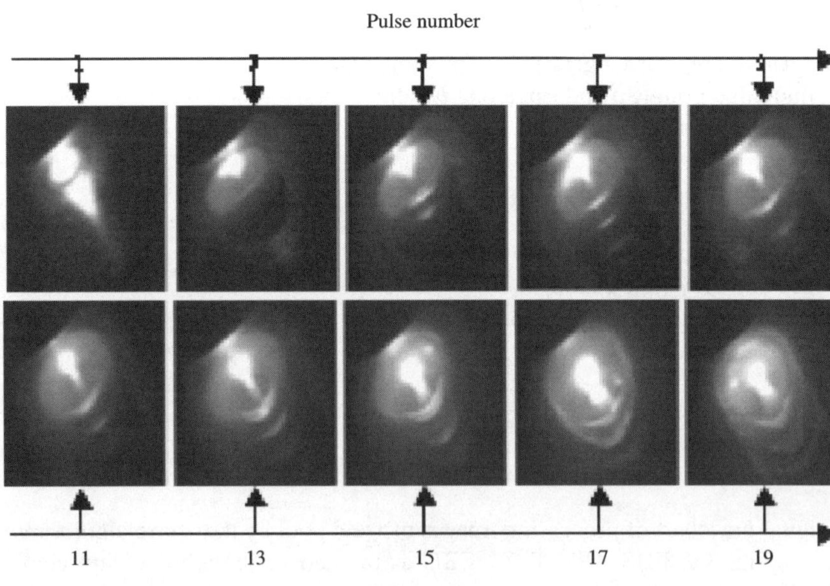

Fig. 3.10 The shape variation of topside weld pool

a ellipse, i.e., a convex model image; while on full penetrating, the pool image is similar to a peach shape, i.e. a concave model. The length and stem shape varieties of the weld pool are most obvious.

3.1.2.8 Image Processing of the Topside Weld Pool

The different surface heights of the filling wire weld pool lead to obvious difference between grey scale distributions of concave and convex type images, so that the different processing algorithms should be designed. The threshold division method is more suitable to convex type object images, and the edge extraction method is suitable to concave type images. In this section, the image processing algorithm TSE(Topside Shape Extraction) is developed to extract top characters of the filling wire weld pool. It consists TI (Type Identification) algorithm for distinguishing concave or convex type images, EE(Edges Extraction) for extracting topside edge points of the weld pool, and ER(Edges Regression) for imitating topside edge curve of the weld pool.

Type Identification of Concave or Convex Images (T1)

The Hough transformation is used to distinguish the rear pool shapes and judge concave or convex type images. The calculating results are shown in Fig. 3.11

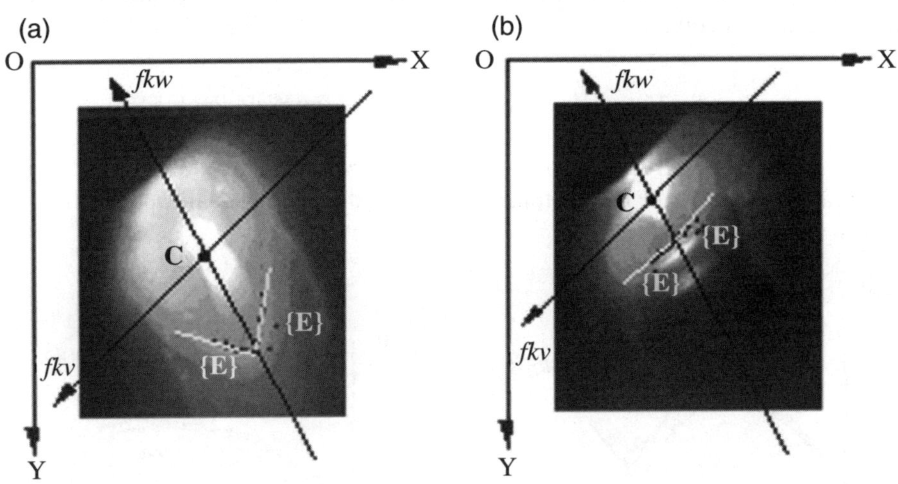

Fig. 3.11 Type identification of topside image (**a**) Convex type (**b**) concave type

Edges Extraction of Top Edge Points of the Weld Pool (EE):

For the convex type image, usually, the image histogram indicates that the image grey scale distribution has typical twin apex property, binary threshold value is determined by finding the vale point of twin apices, and the convex type binary

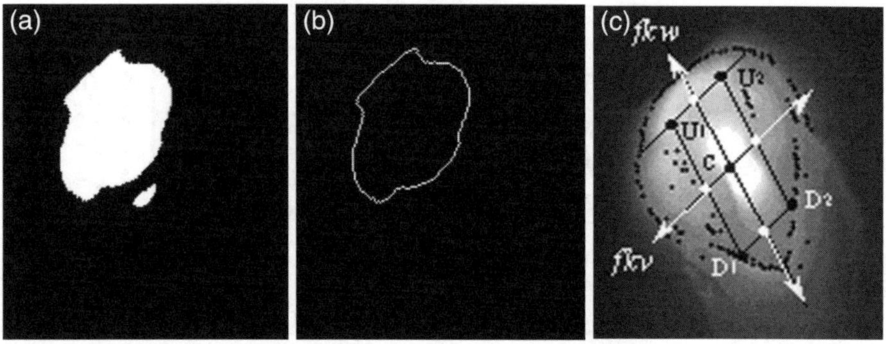

Fig. 3.12 Extracting edge points of topside image (**a**) Thresholding of convex image (**b**) Edge tracing of the thresholding image (**c**) Edge extraction of concave image

image is obtained in Fig. 3.12(a). The edge points of the weld pool can be precisely extracted by the edge tracking method, as shown in Fig. 3.12(b).

Although the grey scale histogram of the concave images also has an evident twin apices, the division algorithm of the pool rear edge easily brings some error by the threshold method. The image processing time using the two-dimensional edge checking method is too long to satisfy real-time requirement. So one-dimensional edge checking method along *fkw* direction is adopted to inspect the edge point of the weld pool and determine the initial scanning points D_1, D_2, U_1 and U_2 as Fig. 3.12(c). Here "+" denotes the checked edge points, and the arrowhead orients the scanning direction.

Edges Regression for Fitting Topside Edge Curve of the Weld Pool (ER)

Based on the checked edge points, the shape curves of the weld pool with wire filler can be fitted by the Hough transformation and the least square approximation, shown as Fig. 3.13

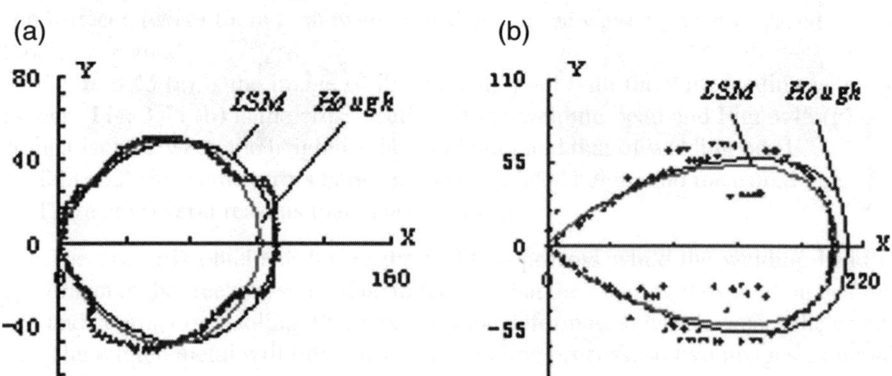

Fig. 3.13 The results of edges regression for topside pool (**a**) Convex type (**b**) Concave type

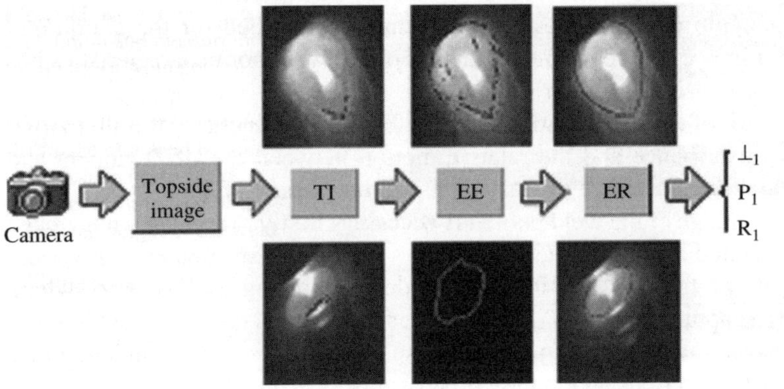

Fig. 3.14 Signal flowchart of image processing for topside pool image

The signal flowchart of processing images of weld pool by the above algorithms is shown as Fig. 3.14, the algorithms TI, EE and ER are defined as the above. The whole processing algorithms for the top/backside weld pool images is running on the PC-486 computer with main frequency 100 MHz, the calculating time for a frame image is not more than 50 ms. It is sufficient for real-time sensing and control of welding dynamic process. The backside image processing of the weld pool with filling wire is similar to that of no-filling wire pool, here is omitted.

3.1.3 The Processing and Characteristic Computing of Aluminium Alloy Weld Pool Image

Figure 3.15 shows three kinds of typical images of aluminium alloy weld pool, namely, the intact image, the fragmentary image and the degenerative image. In this

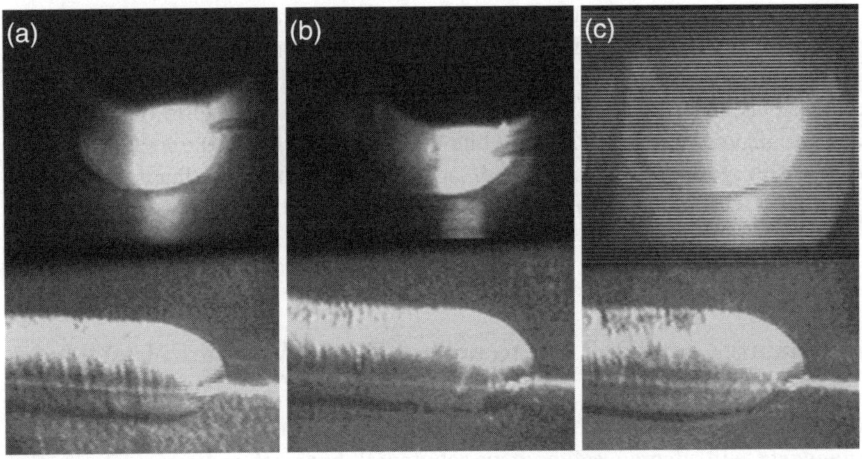

Fig. 3.15 Three kinds of image of the weld pool (**a**) Intact image (**b**) Partial image (**c**) Degenerative image

section, a systematic image processing algorithm is described for the aluminium alloy weld pool image, the real-time extraction of the weld pool characteristics has been realized for real-time monitoring.

The main factor of the degenerative image in Fig. 3.15(c) includes the alternating magnetic field disturbance and the relative motion between the CCD camera and weld pool. The common image processing algorithm cannot be directly applied to the degenerative image of the weld pool, it is necessary firstly to recover a fine image from the degenerated image by filtering noise.

The principle of filtering and imaging model is showing as Fig. 3.16, where $g(x,y)$ denotes a captured image, $f(\xi,\eta)$ denotes the original object, $n(x,y)$ denotes a noise disturbance, and $h(x,y;\xi,\eta)$ is the pulse response function of imaging system. The Fig. 3.17 describes an imaging process; it shows that the noise disturbance results in degeneration or retrogression of the actual image.

Generally, the imaging process can be described as follows:

$$g(x,y) = \int_{-\infty}^{+\infty}\int f(\xi,\eta)h(x-\xi,y-\eta)d\xi d\eta + n(x,y) \tag{3.16}$$

The recovering of image aims at obtaining an optimal estimation of $f(x,y)$ from the degenerated image $g(x,y)$, noise $n(x,y)$ and pulse response function of imaging system $h(x,y;\xi,\eta)$. In the degenerative image of aluminium alloy weld pool, the motion and noisy functions should be determined to recover the image by filtering blurred effects of welding motion and electromagnetic field disturbance.

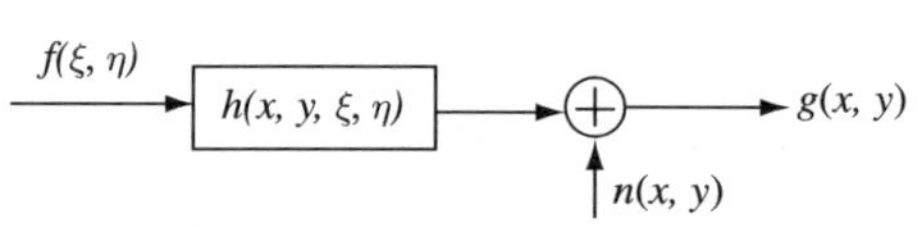

Fig. 3.16 The principle of filtering and imaging model

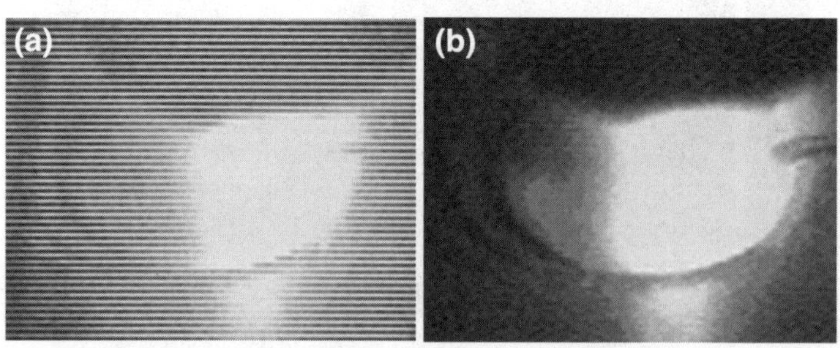

Fig. 3.17 Recovery of the degenerated image (**a**) The degenerated image (**b**) Recovered image

3.1.3.1 Image Recovering from Blurring Effect of Welding Pool Motion

Supposing that there exists a relative motion between the weld pool surface function, $f(x,y)$, and the imaging system, the movement in X direction denotes as $x'(t)$, the movement in Ydirection denotes as $y'(t)$, then the light intensity imaging in sensitization time T can be described as

$$g_1(x,y) = \int_0^T f(x - x'(t), y - y'(t))dt \tag{3.17}$$

which is also an expression of motion blurring image.

In the imaging system of aluminium alloy weld pool, supposing a constant speed movement between the weld pool and CCD camera in Xdirection and the moving distance L in sensitization time T, the movement change is:

$$x'(t) = (L/T)t \tag{3.18}$$

due to $y'(t) = 0$, and then

$$\begin{aligned} g_1(x,y) &= \int_0^T f(x - x'(t))dt \\ &= \int_0^T f(x - (a/T)t)dt \end{aligned} \tag{3.19}$$

let $t' = at/T$, we have

$$\begin{aligned} g_1(x,y) &= \int_0^a f(x - t')(T/a)dt \qquad (h(x) = T/a) \\ &= f(x)^{\times} h(x) \end{aligned} \tag{3.20}$$

Therefore, if it is possible to select a function $m(x)$, the optimal estimation of $f(x)$ can be denoted as

$$\begin{aligned} \hat{f}(x) &= g_1(x,y) * m(x) \\ &= f(x)*h(x)*m(x) \end{aligned} \tag{3.21}$$

If $h(x)*m(x)$ is considered as an ideal pulse function, the optimal estimation of the real image can be obtained as:

$$\hat{f}(x) = \int_{-\infty}^{\infty} f(\xi)\delta(x - \xi)d\xi = f(x)*\delta(x) \tag{3.22}$$

Because of no ideal pulse function to be available, using the experiential knowledge on convolving product, we selected $m(x)$ as a rectangle function for recovering blurred image of motion weld pool.

3.1.3.2 Image Recovering from the Disturbance of Alternating Magnetic Field

During aluminium alloy pulsed GTAW, the alternating welding current brings an alternating magnetic field and interfering CCD camera, which results in the degenerated image. The polarity of welding current has many time exchanges during taking a frame of the image, so that the degenerating extent in different regions in a frame of the weld pool is different.

The disturbance of the alternating magnetic field exists in the pixel of interleaved line in a frame image. The optimal estimation of the aluminium alloy weld pool image can be carried by the gray degree of the current point and the gray degree adding from the magnetic field disturbance. Because the field interfering point and non-interfering point exists in the interleaved line of the image, the field interfering point can be found by contrasting the difference of the pixel gray degree in interleaved line region. e. g., in the region D at the central point (x,y), the gray degree difference of the interleaved line is denoted as

$$d = \frac{\sum_{i=0}^{(m-1)/4}\sum_{k=0}^{(n-1)/2} g(x \pm 2i, y \pm k)}{n(m+1)/2} - \frac{\sum_{j=1}^{(m-1)/4}\sum_{k=0}^{(n-1)/2} g(x \pm (2j-1), y \pm K)}{n(m-1)/2} \tag{3.23}$$

where i and j are number of row and columns in current image, n and m are the width and height of the region D. If d is overstep the supposed region, the field disturbance can be determined at the current point, and then the pixel in the interleaved line is non-interfering point. Therefore, an estimation of ideal image can be realized by a statistical gray degree of the adjacent row pixel instead of that at the current point. A recovered gray degree at the current point pixel becomes:

$$g(x,y) = \frac{\sum_{j=1}^{(m-1)/4}\sum_{k=0}^{(n-1)/2} g(x \pm (2j-1), y \pm K)}{n(m-1)/2} \tag{3.24}$$

The Fig. 3.17 shows a recovered image from the degenerated image of Aluminium alloy weld pool in pulsed GTAW.

In real time processing of the aluminium alloy weld pool during pulsed GTAW, the computer doesn't distinguish if the image is degenerated, the recovering image algorithm is used for all typical images of the weld pool, so that the recovering algorithm for the field disturbance should be suitable to process both degenerative image and non-degenerative image, and which doesn't change the features of normal image.

The non-degenerative image includes intact image and fragmentary image of the weld pool. Image processing is mainly aimed at obtaining a complete edge of the weld pool. And processing algorithm of the degenerative image mainly includes filtering, enhancing, edge detecting, threshold, edge thinning, edge recognizing, edge curve fitting, whole edge recovering of weld pool image, etc.

In real-time control of welding process, the above algorithms should be orderly combined to realize real time processing of weld pool dynamic image. Considering the topside and backside vision images of the weld pool in the same frame simultaneously, the following real-time image processing algorithms are developed to acquire topside and backside sizes of the weld pool, the integrated algorithm procedure is designed as follows: image filtering, edge detection, binary processing, edge thinning, edge extracting, whole edge recovering of weld pool image, calculation of geometrical size of weld pool, and so on. The algorithms are described as follows.

3.1.3.3 Image Processing After Recovering

The image processing algorithms can be divided into the following five steps, namely image filtering, edge detection, binary processing, projection mode of edge thinning and network edge extraction.

Step 1. Image filtering

Filtering removes the noise. In the aluminum alloy welding, we obtained a degraded image, which is caused by the following factors:

(1) During the acquisition and quantification of the image, electromagnetic interference is introduced into the reference current, and the transmission line is too long, which causes a decreasing of the signal, and the move makes the image fuzzy.
(2) There is some high frequency noise, which is introduced in the process of acquisition and transmission of the image.

A weighted median filter was designed to deal with those effects. Its principle is to arrange the gray level of the current pixel and those of the eight adjacent ones in increasing order, supposing the arrangement is as follows:

$$\{a_{-4}, a_{-3}, a_{-2}, a_{-1}, a_0, a_1, a_2, a_3, a_4\} \tag{3.25}$$

Then, the current pixel's gray value is

$$a_c = \frac{1}{4}a_1 + \frac{1}{4}a_2 + \frac{1}{2}a_3 \tag{3.26}$$

This method effectively attenuated the systematic effect by random noise.

Step 2. Edge detection

The aim of the edge detection is to weight and obtain the information about the edge of weld pool, and enhance the profile of the image so that it can be easily recognized. The algorithm is mainly based on the gray levels between different pixels. From the characteristics of the pool image, the systematic low-frequency noise is attenuated by the difference of the gray levels of pixels in the direction of horizon; if we detect the boundary in the vertical direction, false information would be

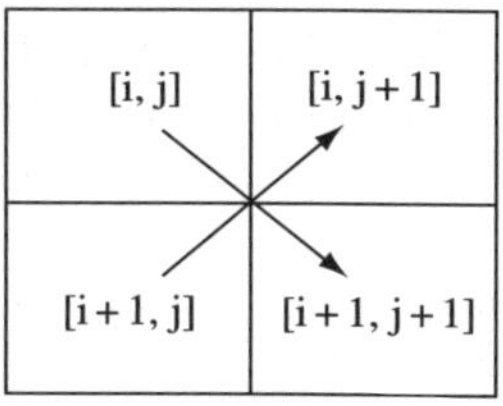

Fig. 3.18 Direction of detected edge

obtained, so in this paper ±45° is taken as the direction to detect the scale, shown as Fig. 3.18. The discrete description of the edge detecting operator is as follows:

$$G(x,y) = \left|f[i,j] - f[i+1,j+1]\right| + \left|f[i+1,j] - f[i,j+1]\right| \quad (3.27)$$

The above formula can not lead to a desirable result, because the gray value of the image background is low, so it can be almost equal to the result obtained from the formula, which is bad for detecting. The solution is to combine the threshold with statistics. Double thresholding θ_1 and θ_2, when $G(x,y) > \theta_1$, the current point is the edge-point; considering the gray levels of neighboring pixels, if the average value of eight neighborhood pixels is greater than θ_2, then multiply $G(x,y)$ with a factor k which is greater than 1. So the actual output is:

$$G\prime(x,y) = \begin{cases} G(x,y) & M \le \theta_2 \\ KG(x,y) & M > \theta_2 \end{cases} \quad (3.28)$$

Step 3. Binary processing

The target is to convert a common image into a binary image so that objects can be conveniently separated from the background. During the process it is necessary to set a threshold, however, because of the absence of the apparent hollow, which could be thought as a threshold between peaks in the histogram, the traditional method is not apposite. Based on the characteristics of the aluminum alloy weld pool image, a method of segmented expected values based on statistics was designed. Supposing the gray value in the image can be $I_0, I_1, I_2, \ldots, I_{m-1}, I_m, I_{m+1} \ldots I_n, I_m$ is the value of the current pixel according to the gray value. Let us present an arbitrary pixel using a discrete random variable X, then X can be any one value among $I_0, I_1, I_2, \ldots I_{m-1}, I_m, I_{m+1}, \ldots, I_n$. So we can describe the distribution of the gray level using a probability distribution. Supposing the probability of the gray value distribution is $P_0 = P(I_0), P_1 = P(I_1), \ldots, P_n = P(I_n)$, and $\sum_{n=1}^{N} p_n = 1$.

The mathematical expectation presents a mean, which is a kind of weighted average taking the probability as weight. In statistics, mathematical expectation shows the average level of a particular random variable, and it can be regarded as the center of the variable. In the image, the mathematical expectation of the gray value can be regarded as the desirable mean value of all gray levels, which is the ideal choice for the threshold value T:

$$T = \sum_{n=1}^{N} I_n P(I_n) \tag{3.29}$$

We think of the pixels whose gray value are greater than T as the pool pixels.

Step 4. Projection method of edge thinning

Edge thinning is to convert the boundary of the pool into a curve consisting of individual pixels which is convenient for the acquired data. The ordinary algorithms include Hilditch, Rosenfeld's and so on. The common characteristic of these algorithms is that the obtained curve lies in the center of the original rectangle, but in this study, the edge lies at the inner side instead of the center of the pool, so the above ordinary algorithms will produce errors. The algorithm for projecting edge thinning is based on the following principle: In each column of the image, if there exists a breakpoint, to project the group of pixels in the direction of the column to the breakpoint, and in this direction, leave only the breakpoint and pixels in the connected region located most nearly to the breakpoint, see Fig. 3.19. If the current pixel is a breakpoint, copy this pixel to the current position in the thinning image. This algorithm preserves most of the connectivity of the original image, shown as Fig. 3.20. It should be mentioned that in the process of thinning, the edge of the nozzle is removed using the filtering technology.

Step 5. Network edge extracting

After the above processing methods, the boundary of the image is not perfect. Since there exists some dot noise and false edge detection of the shape of line, both of which will produce bad effects on the accurate measurement. Because a neural network is excellent for dealing with problems which are difficult to describe by

Fig. 3.19 Original image

Fig. 3.20 Thinning image

accurate mathematical models or formulae, neural network have been used for a long time for image processing, for example, recovering of an image, analyzing of an image and so on. However, neural network needs a large amount of learning samples to train the network and a complex net structure to process the original image, so it can not obtain results quickly. If we extract the edge from the binary image, it needs fewer samples and the processing speed will be improved greatly, so the binary image was analysed using a BP (Back propagation) network to extract the edge.

Step 1. The structure of neural network for extracting edge

From the binary image, the noise and pool edge can be separated by experience, which is suitable for a neural network because of its ability for data clustering non-linear, optimization and pattern classification. The structure of the network of this study is shown in Fig. 3.21. The network comprises three layers: input layer, hidden layer and output layer. The input layer includes nine nodes to present a filtering window covering a 3×3 range, the output also includes nine nodes to present results. The number of nodes of the hidden layer is not definitely set. In this paper, it has seven based on experiments. The target-value of the output is 1 or 0, and the activation function of the hidden layer is the ordinary Sigmoid Function:

$$\tan sig(x_i) = \frac{e^{x_i} - e^{-x_i}}{e^{x_i} + e^{-x_i}} \tag{3.30}$$

which makes the output of the hidden layer range from −1 to 1. In the output layer, the function of the neural unit has linear form, which makes the output approach 0 or 1 more closely, and in this paper, is:

$$f(x_i) = x_i \tag{3.31}$$

$$x_i = \sum_{j=1}^{n} w_{ij} p_j + b_i \tag{3.32}$$

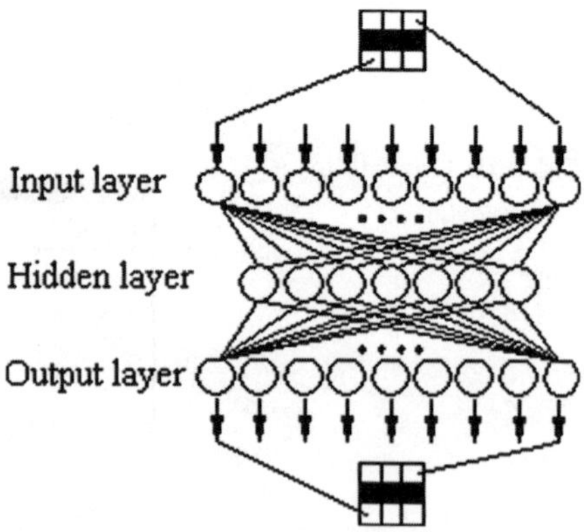

Fig. 3.21 BP network structure

In above two equations, x_i is the input of node i; w_{ij} is the connection weight between node j and node i; b_i is the error of node i.

Step 2. Learning algorithm of the network

The traditional BP algorithm always converges too slowly, and if the parameters are not apposite, there may exist a local minimum problem so that it will not converge. A modified BP algorithm is used in this section. The learning factor η is a dynamic value instead of a fixed value obtained through local optimization. After every modification of the weights, if the error decreases, multiply η with a coefficient greater than 1,or else with a coefficient less than 1. Impulse constant α is modified according to the gross error; if the gross error does not decrease, α is equal to 0, or else α is nonzero. The next modification value of weight is

$$\Delta W_{ij}(m+1) = \eta \sum_q (\delta_{qj} O_{qj}) + \alpha \Delta W_{ji}(m) \tag{3.33}$$

where m presents the iteration time; δ_{qj} and O_{qj} are separately the error signal of a hidden node and output signal; q is the number of samples.

Step 3. Training mode of the network

The input of the network is a 3×3 neighborhood window based on the current pixel, which includes nine pixels, which can detect the pixels lying in the upper, bottom, left and right side of the current pixel, or a corner pixel and curve. From theory to practice, the more input nodes, the more accurate the result is, but when the number of the input nodes rises, the structure of network becomes complex and there is also difficulty in training and learning. Moreover, the number of samples rises, required also, which is difficult in practice. When processing the image, the window should be moved over the image until item counters a black pixel, then this pixel and the other eight neighboring pixels make up of the inputs, and the outputs are the above pixel values after the noise is removed. The network adopts the supervised training mode; every input corresponds to a output to form a sample.

There exist various patterns and shapes of noise in the image, for example, strip, circle, dot, corner point and so on. Increasing the number of learning samples enhances the fault tolerance of the network, but, if an attempt is made to increase samples by getting all the noise patterns and shades, the effectiveness is not ideal, and it can even decrease the fault tolerance or make the network converge slowly or not converge at all. After repeated experiments, the author selected 7 groups of samples summing up to 42 samples, which makes the systematic error converge and can also obtain good fault tolerance. The Fig. 3.22 lists all the learning samples and edge patterns.

The first set of samples is to remove one or two noise points, and the other set is determined by the connection characteristics of curves after observing the approximate profile of the pool combining the patterns and the connection characteristics of the noise. The assessment of the teacher signal is based on the greatest probability of the four directions and corner points. The experiment is carried using MATLAB, and after iterating 16000 times, the error approachs 1.89%. To carry on experiments

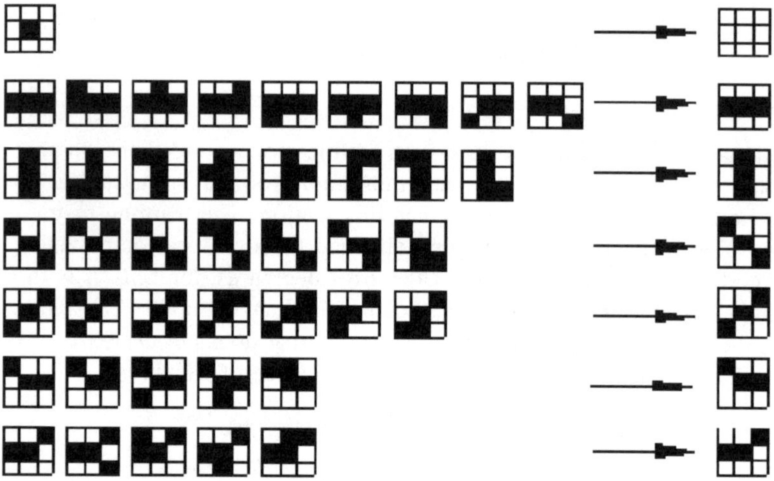

Fig. 3.22 Sets of the learning patterns

on the thinning edge image of the pool, the matrix whose size is 3×3 should be converted into a column vector whose size is 1×9 as the input of the trained network, a new matrix whose size is 3×3 is got. Then copy the new matrix to the original image.

From the result we can see that the neural network can preserve the edge characteristics of image perfectly, and can identify the edge effectively. The procedure of image processing algorithm is shown as Fig. 3.23. The whole processing algorithm is realized in a P II 350 computer and cost time is about 604 ms.

3.2 Acquiring Three Dimensional Characteristics from Monocular Image of Weld Pool During Pulsed GTAW

For GTAW with wire filler, the process becomes more complex with the feeding wire. Two-dimensional image processing extracted the weld pool edges, width and length, etc., which only reflected the plane shape variety, if there is a convex and concave changes on weld pool surface and a high precise requirements for weld seam height figuration, extracting three-dimensional characters of the weld pool is necessary.

3.2.1 Definition of Topside Weld Pool Height

The definition of the surface height parameters is shown in Fig. 3.24. A work-piece coordinate system is defined as Cartesian coordinate system. The original is the surface point of the work-piece under the welding torch. Welding direction is defined

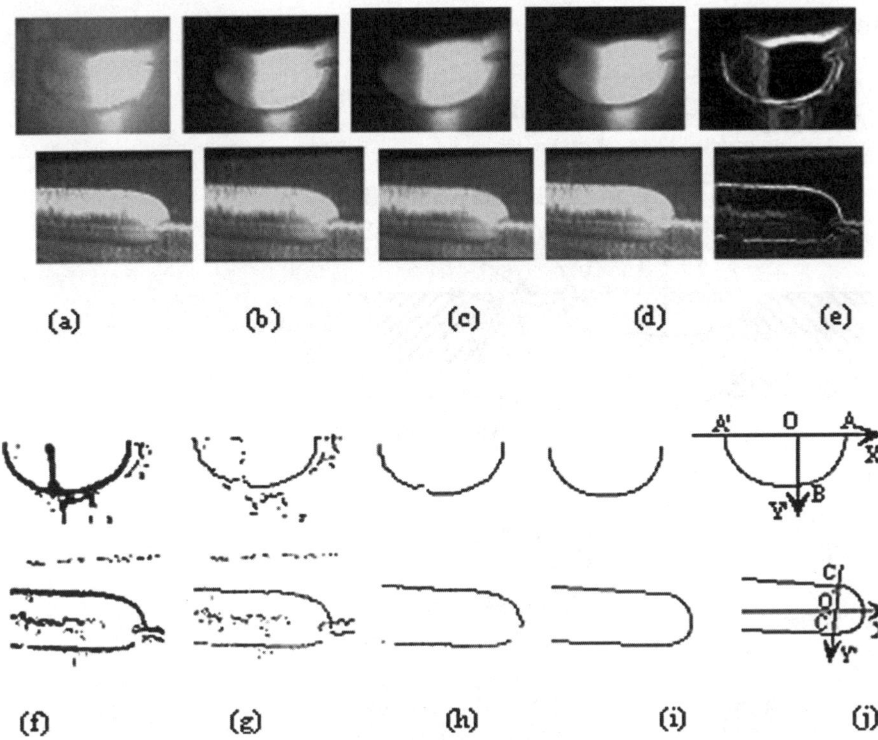

Fig. 3.23 The whole flow of image processing of Al weld pool

as x-axis (length direction), y-axis (width direction) is vertical to the welding direction, and z-axis is vertical to the work-piece surface. The average depression of the weld pool D_w along the width direction, and D_l along the length direction are defined for describing the height shape of the weld pool.

$$D_l = S_l/L_t, \quad D_w = S_w/W_t \tag{3.34}$$

Where S_w and S_l are the sum of the depression along the width and length direction, W_t and L_t are the same as the above definition.

The topside height H_t is defined as the maximum height of the weld pool along the rear center of the welding direction, also shown in Fig. 3.25. The topside height of the weld pool H_t is similar to the bead height.

3.2.2 Extracting Surface Height of the Weld Pool from Arc Reflection Position

Arc reflection is distinct in the weld pool, as shown in Fig. 3.25. Using this feature, a simple method for extracting the topside height of weld pool is presented, i.e.

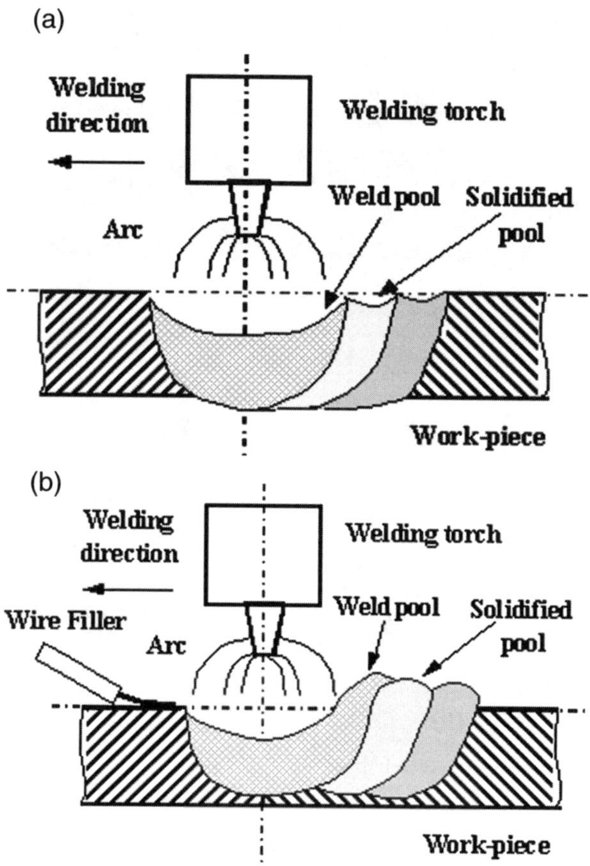

Fig. 3.24 The height parameters definition of the *topside* weld pool (**a**) Concave; (**b**) Convex

the surface height is corresponding to the position of arc reflection in the image. Therefore, the topside height is calculated as the distance between the tungsten tip and the arc inverted reflection in weld pool. The tungsten tip, arc shape in the images can be clearly distinguished in Fig. 3.25.

The algorithm calculating topside height from the weld pool images is as follows:

Step 1: Image filtering (Guass filtering);
Step 2: Binaryzation of images;
Step 3: Determining the position of tungsten tip;
Step 4: Calculating the surface height of the weld pool.

Because the tungsten pole is warded by welding torch nozzle in welding, the distance between the tungsten tip and its inverted reflection in weld pool image cannot be directly calculated. So we obtained distance between the tungsten pole and nozzle, L_0 as Fig. 3.26(e), and let arc length as L_a, then the top surface height of the weld pool, H_t is calculated by

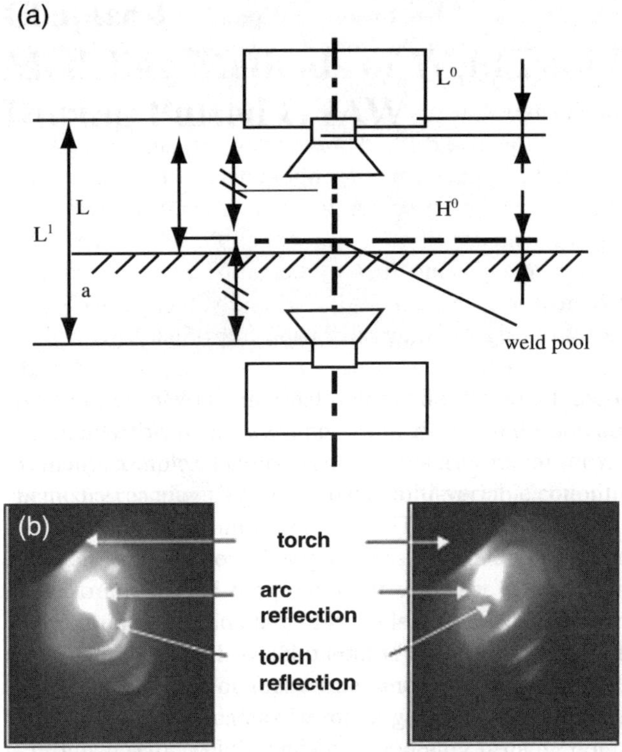

Fig. 3.25 Weld pool image with different surface height of weld pool

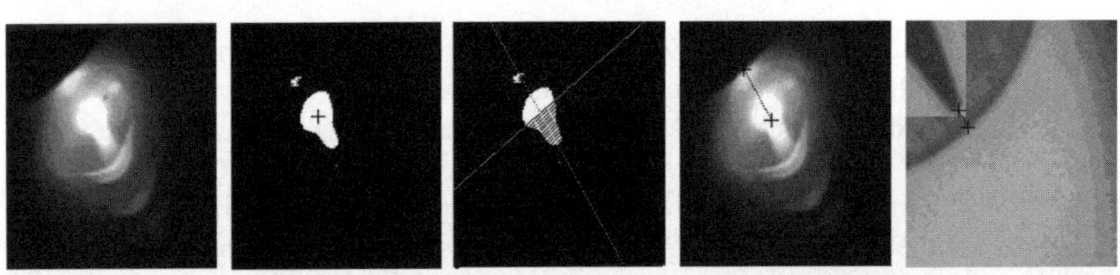

Fig. 3.26 The height result from image processing topside weld pool (**a**) Initial image (**b**) binary image (**c**) width calculation (**d**) height extraction (**e**) distance from the tip to nozzlef

$$H_t = L_a - fcal.z^*(L_0 + L_1)/2.0 \tag{3.35}$$

3.2.3 Extracting Surface Height of the Weld Pool by Shape from Shading

Topside height of the weld pool surface is the direct indicator of penetration. However, it is not visible in 2 dimensional weld pool images. Recently, computer three-dimension vision technique is developed to solve the problem of reconstructing the three-dimension data of object surface from the two-dimensional image. According to the three imaging elements (camera, light source, and object), the technique is divided into three methods, namely, binocular vision with two or more cameras (moving camera); image sequence method (moving object); and shape from shading with a single or more light sources (moving light source). Because of the limited sensing space in practical welding process, the application of two or more cameras is unpractical. Image sequence method is also not applicable to the welding process due to the following two reasons; firstly, interval between two images is 40 ms for each frame or 20 ms for each field (PAL), secondly, the surface shape of the weld pool changes a lot during the interval. Shape-from-Shading (SFS) is a technique of recovering 3D information from variation of shading on the image, first introduced by Horn in the 1970s [10]. Exciting achievements are achieved in the human surface reconstruction from two or more images [11], and the earth surface height reconstruction from remote sensing image. The generalized reflection model and the algorithm of the surface height reconstruction are under wide consideration [12–14].

In order to get a better understand of SFS, it is necessary to introduce how the images are formed. A simple model of image formation is the Lambertian model, in which image gray level (or irradiance) at a pixel is corresponding to the surface orientation in a reflectance map. Thus, for fixed illumination and imaging conditions, and for a surface with known reflectance properties, changes in surface orientation translate into corresponding changes in image gray level. And the SFS method is to inverse the problem of recovering surface shape from changes in image gray level. The relationship is given by

$$\boldsymbol{I} = \boldsymbol{R}(\boldsymbol{p},\boldsymbol{q}) \tag{3.36}$$

Where $R(p,q)$ is the reflectance map of the surface point (x,y), $\boldsymbol{p} = \boldsymbol{p}(\boldsymbol{x},\boldsymbol{y}) = \frac{\partial z}{\partial x}$, $\boldsymbol{q} = \boldsymbol{q}(\boldsymbol{x},\boldsymbol{y}) = \frac{\partial z}{\partial y}$, z is surface depth, $\boldsymbol{I}$ is the normalize image gray level at corresponding point in the equation, we have a nonlinear equation with two unknown variables. Therefore, to find unique solution to SFS requires additional constraints. Extensive study and approaches are reported to solve this solution. Ruo Zhang [15] divided these approaches into four groups: minimization, propagation, local and linear. Comparison experiments showed that minimization approaches are more robust. Therefore, SFS techniques applying in welding are all based on minimization theory. Following section will discuss them respectively.

3.2.3.1 SFS by Ideal Reflection Map Model

Due to the impact of arc plasma, the surface of the weld pool is concave in full penetration and it is convex in partial penetration, as shown in Fig. 3.24.

Figure 3.27 shows the different weld pool images under different welding current. The boundary between the weld pool and the solidified metal in Fig. 3.27(a) is much clearer than that in Fig. 3.27(e). The image of the weld pool with high wire feed speed is difficult for image processing.

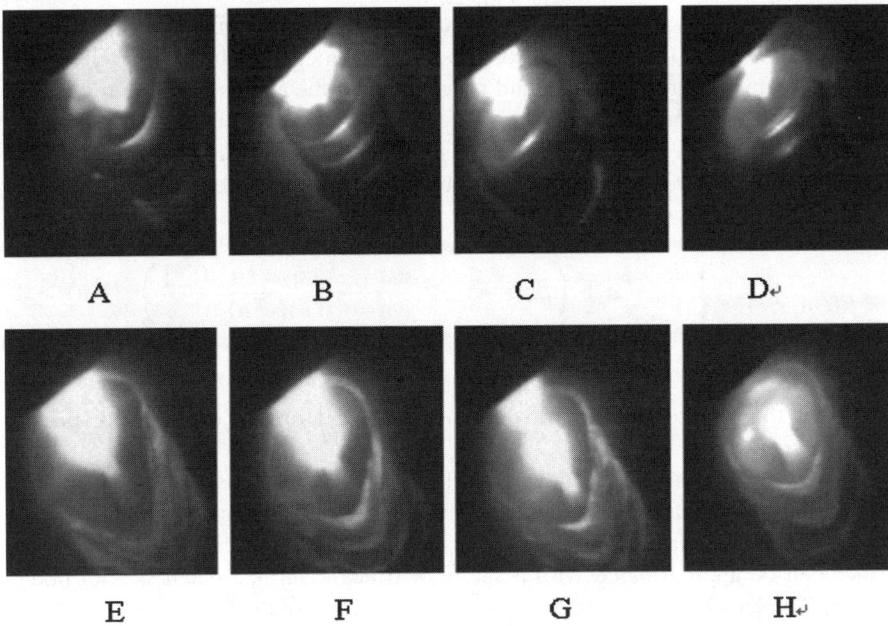

Fig. 3.27 Comparison between the weld pool images with different imaging current **A** – 60A, convex; **B** – 50A, convex; **C** – 40A, convex; **D** – 30A, convex; **E** – 60A, concave; **F** – 50A, concave; **G** – 40A, concave; **H** – 30A, concave

With the imaging current decreasing, the shape of the arc center compressed, and the bright of the weld pool increased. The light source of the electric arc can be thought as a point light source. The weld pool is illuminated by the electric arc to form the image. Therefore, difference in the surface shape of the weld pool makes the images a great difference, such as Fig. 3.27 (d) and (h).

SFS algorithm can be divided into following steps [16],

Step 1. Modeling of generalized reflection

The generalized imaging reflection geometry model is shown in Fig. 3.28. In the figure, ***p*** is a point on object surface, and ***n*** is its normal line. ***S*** is a point light source,

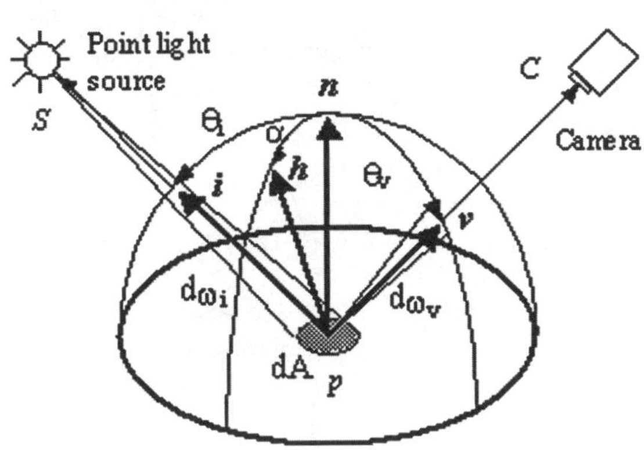

Fig. 3.28 Generalize reflection geometry model

r is the distance from the light source ***S*** to the point ***p***. ***i*** *is* the unit vector toward the light source ***S***, and ***v*** is the unit vector toward the camera ***C***. The specula direction ***h*** is the bisector of ***i*** and ***v***. θi, θv and α represented the angles between ***n*** and ***i***, ***n*** and ***v***, ***n*** and ***h*** respectively.

In consideration of a nearby point light source, perspective projection of camera, and non-lambertain reflection of object surface, the generalized reflection model (image intensity E) is derived as follows

$$E = R(i,n,v,r) = \begin{cases} \frac{I_0}{r^2}(v^T n_z)^4 \left(\beta_d (i^T n) + \beta_s \frac{\exp\{-k[\cos^{-1}(h^T n)]^2\}}{(i^T n)(v^T n)} \right), & (\mathrm{i^T n}) \geq 0 \\ 0, & \text{otherwise} \end{cases} \tag{3.37}$$

Where k is the surface roughness parameter, β_d and β_s are the diffuse and specula reflective coefficients.

Perspective projection of camera on a triangle surface patch is shown in Fig. 3.29 A square image domain can be divided into a set of M_t non-overlapping triangles T_i, $\mathrm{i} = 1,2,\ldots,M_t$, with M_n nodal points P_i, $i = 1,2,\ldots,M_n$. Then the smooth object surface can be approximated with a series of triangle surface patch s_i with nodal point p_i. The lens of the camera is oriented at the origin of the Cartesian coordinate system with the optical axis along $-z$-axis. The image plane is assumed to locate at $z = -f$, then the project relationship between the image point $P(x,y)$ and the surface point $p(x,y)$ is derived.

$$X = -f\frac{x}{z}, Y = -f\frac{y}{z} \tag{3.38}$$

The center point $p_{kc}(x_{kc}, y_{kc}, z_{kc})$ of a triangle s_k is depicted with three vertices $p_i(x_i, y_i, z_i), p_j(x_j, y_j, z_j), p_l(x_l, y_l, z_l)$ of s_k.

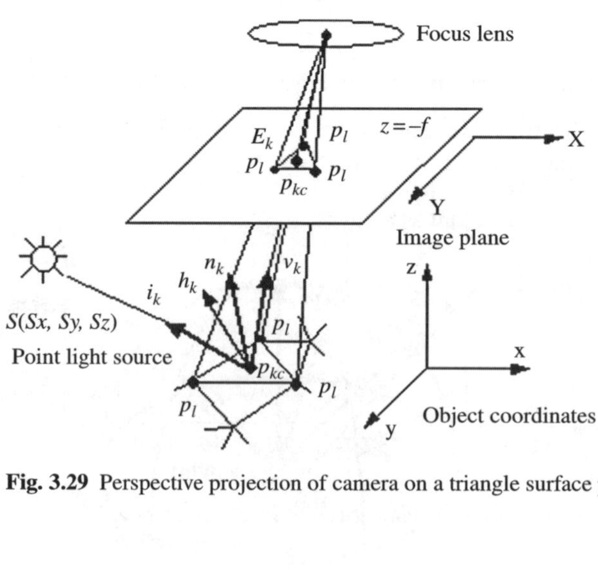

Fig. 3.29 Perspective projection of camera on a triangle surface patch

$$\begin{pmatrix} x_{kc} \\ y_{kc} \\ z_{kc} \end{pmatrix} = \frac{-1}{3f} \begin{pmatrix} X_i & X_j & X_l \\ Y_i & Y_j & Y_l \\ -f & -f & -f \end{pmatrix} \begin{pmatrix} z_i \\ z_j \\ z_l \end{pmatrix} \tag{3.39}$$

Then the vectors $\boldsymbol{n}$, $\boldsymbol{i}$ and $\boldsymbol{v}$ can be described with the surface height of three vertices of each triangle, and the generalized reflection model $R(i_k, n_k, v_k, r_k)$ can be simplified as a function of vertices height $R(z_i, z_j, z_l)$.

Step 2. Solving of the generalized reflection model

The generalized reflection model is nonlinear and hard to solve the height of the nodal points directly, however, it can be linearized with Taylor series expansion.

$$R_k \approx \sum_{m=1}^{Mn} w_{km} z_m + \xi_k \tag{3.40}$$

Where

$$\xi_k = R_k(\mathrm{z}_\mathrm{i}^{\mathrm{n}-1}, \mathrm{z}_\mathrm{j}^{\mathrm{n}-1}, \mathrm{z}_\mathrm{l}^{\mathrm{n}-1}) - \sum_{\mathrm{m}=1}^{\mathrm{Mn}} w_{km} z_m^{n-1},$$

$$w_{km} = \begin{cases} \dfrac{\partial R_k(z_i, z_j, z_l)}{\partial z_m} \Big|_{(z_i^{n-1}, z_j^{n-1}, z_l^{n-1})} & \text{, if m} \in \mathrm{V_k} = \{i, j, l\} \\ 0 & \text{, otherwise} \end{cases},$$

$\mathrm{V_k}$ denoted the vertices of each triangle.
Light error is taken as the cost function.

$$e_b = \sum_{k=1}^{Mt} (E_k - R_k)^2 = \sum_{k=1}^{Mt} \left[E_k - \left(\sum_{m=1}^{Mn} w_{km} z_m + \xi_k \right) \right]^2 \tag{3.41}$$

To minimize the cost quadratic function, a linear formula is derived.

$$\mathbf{Az} = \mathbf{b}, [\mathbf{A}]_{\mathbf{m,n}} = 2 \sum_{k=1}^{Mt} w_{km} w_{kn}, [\mathbf{b}]_{\mathbf{m,n}} = 2 \sum_{k=1}^{Mt} (E_k - \xi_k) w_{km}, 1 \leq m, n \leq M_n \tag{3.42}$$

Where $\mathbf{A}$ is coefficient matrix, sparse and symmetric, $\mathbf{b}$ is load vector.

With the above algorithm, the surface height is reconstructed accurately with stereo two or more images taken by various illumination directions and fixed camera. Extra constraints, such as smoothing constraint, boundary conditions and gray-weighted regulation of light error, are necessary due to the large error of height reconstructed from a single image.

The surface of the weld pool is supposed as smooth, the cost function and its minimization formula are renewed as follows.

$$e = e_b + \lambda e_s, \quad \mathbf{Cz} = \mathbf{b} \tag{3.43}$$

Where e_s is the cost function of smoothing constraint. $\mathbf{C} = \mathbf{A} + \lambda\mathbf{B}$, where λ is a smoothing factor. Smoothing matrix **B** had different stencil forms referring to different position of z_k. The smoothing term can stabilize iteration calculation, and should be reduced gradually as iteration continued. The surface height of the work-piece is known and the weld pool can be considered as a curved surface embedded in the work-piece plane. Grayness-weighted regulation is introduced in the light error.

$$e_b = \sum_{k=1}^{Mt}\sum_{j=1}^{J} w(E_k)\cdot(E_k^j - R_k^j)^2, w(E_k) = \begin{cases} e^{-(E_k-150)^2/5000}, & E_k \geq 150 \\ 1.0 & \text{, otherwise} \end{cases} \quad (3.43)$$

Where $\omega(E_k)$ is the grayness-weighted factor.
The flowchart of the whole algorithm is shown in Fig. 3.30

Step 3. Calculation of surface height of the weld pool

To reconstruct the surface shape of the weld pool, some parameters of the imaging system should be considered:

Point light source: In the analysis of image sensing, electric arc is thought as a point light source, then its position is $s(x_s, y_s, z_s) = (0, 0, l_a)$, where l_a is the arc length. The light intensity is supposed as $I_0 = \gamma \times r_s^2$;

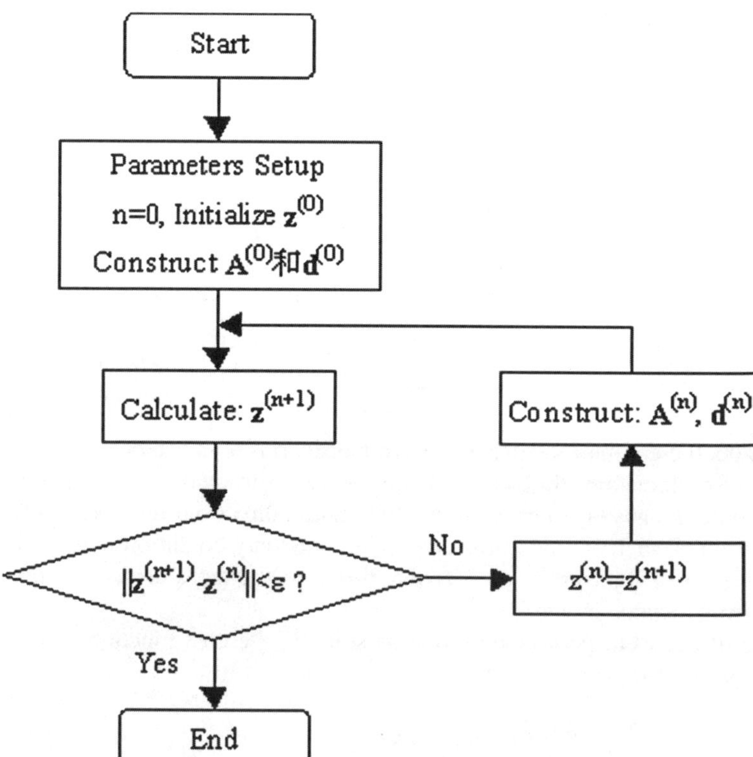

Fig. 3.30 Flowchart of calculating the surface height

Camera: The focal length of the lens f is 50 mm, the target of CCD is 4.8 × 3.6 mm, corresponding to the maximum image size 500 × 592;
Surface reflectivity: The diffuse and specula reflective coefficients and surface roughness parameter of the weld pool are selected as $\beta_s = 0.7$, $\beta_d = 0.3$ and $k = 3.0$ by comparing the synthetic images with various reflectivity to the actual weld pool image. While the surface of the work-piece and solidified are thought as Lambertain surface, so $\beta_s = 0.0$, $\beta_d = 1.0$.

Aiming at the structure of the visual sensing system, coordinate system transformation is introduced. Then with the developed algorithm, the surface shape of the weld pool is calculated. The weld pool images are captured under the same welding conditions except for wire feed speed, shown in Fig. 3.31 The image size is 128 × 128. The reconstructed weld pool surface results are shown in Fig. 3.32.

The height along the x-axis and y-axis of the reconstructed surface are shown in Fig. 3.33. The height parameters of the weld pool, such as the topside height H_t, averaged depression D_w and D_l, are calculated from the reconstructed surface result. It is concluded that the surface height is symmetric along the vertical welding direction, and is irregular along the welding direction. When wire feed speed is larger, the surface depression is smaller and a large convex appeared in the rear part of the weld pool, the corresponding bead is part penetration. With wire feed speed decreasing, the weld pool depressed greatly and the convex in the rear degraded.

In welding process, the molten metal in the weld pool is solidified during the period of pulse base current. Arc impulse decreased and molten metal flowed back to the center of the weld pool, and the surface shape between the fluid and solidified weld pool is different greatly. So the solidified surface is used only as a reference to testify the accuracy of the reconstructed surface. The reconstructed surface height is verified by skilled welder's experience and some effective verifying method is still needed. It cost about ten seconds for the whole iterative calculation, so the algorithm should be optimized in further study for fulfilling the requirement of the real-time extraction and control.

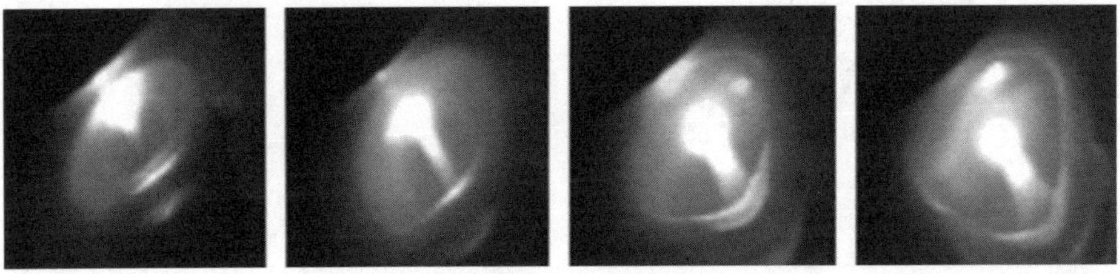

Fig. 3.31 The weld pool images with different wire feed speed during pulsed GTAW with wire filler **(a)** $V_f = 6.0$ mm/s **(b)** $V_f = 4.0$ mm/s **(c)** $V_f = 2.0$ mm/s **(d)** $V_f = 0.0$ mm/s

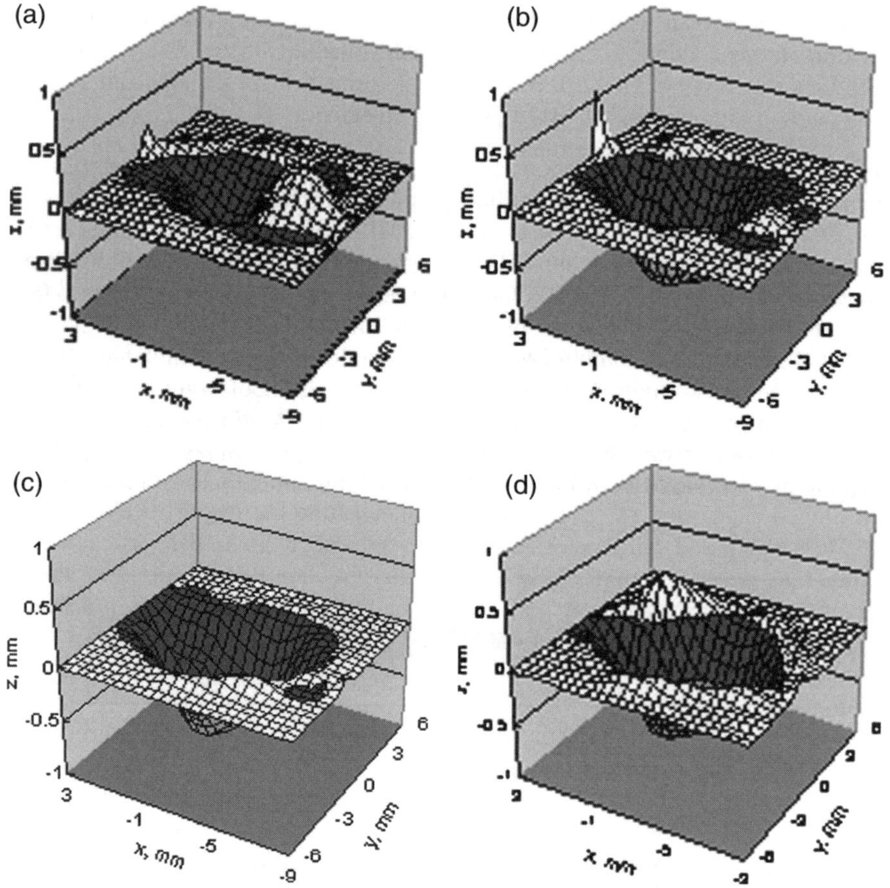

Fig. 3.32 Reconstructed surface height results from single weld pool image (**a**) $V_f = 6.0$ mm/s (**b**) $V_f = 4.0$ mm/s (**c**) $V_f = 2.0$ mm/s (**d**) $V_f = 0.0$ mm/s

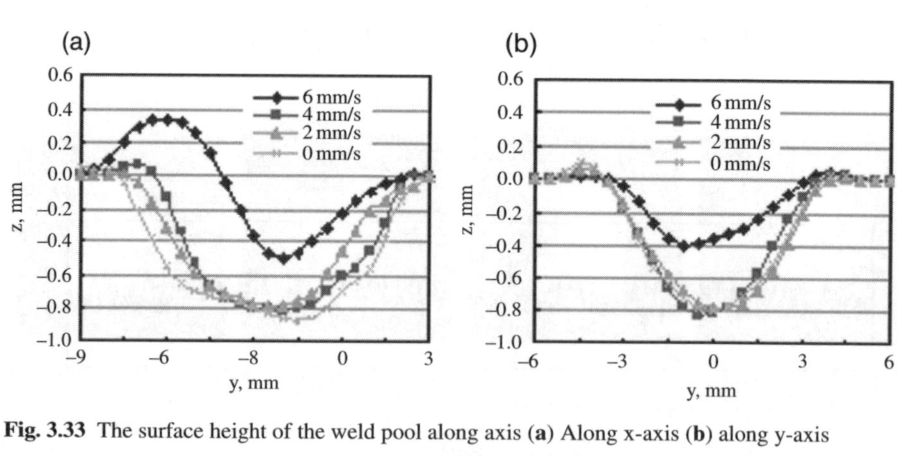

Fig. 3.33 The surface height of the weld pool along axis (**a**) Along x-axis (**b**) along y-axis

3.2.3.2 SFS by Reflection Map Model of Actual Molten Pool Surface

Figure 3.34 are typical images of low carbon steel weld pool, in which Fig. 3.34(a) is a concave type of weld pool and Fig. 3.34(b) is a convex one. The images show that the surface reflection light of weld pool is quite weak, which reflects the weakness of specular reflection.

SFS algorithm is divided into the following steps [17].

Step 1. The reflectanction map model

Lee and Kuo [18] derived a physics-based generalized reflectance map model under three assumptions on surface characteristics and imaging geometry, namely, diffuse and specular reflection effects, a nearby point light source and perspective projection. Figure 3.35 shows reflection geometry used in the generalized

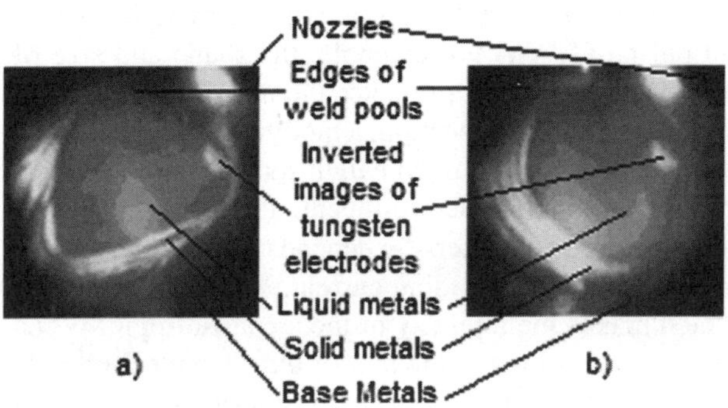

Fig. 3.34 Images of typical weld pools of low carbon steel (a) Concave type (**b**) Convex type

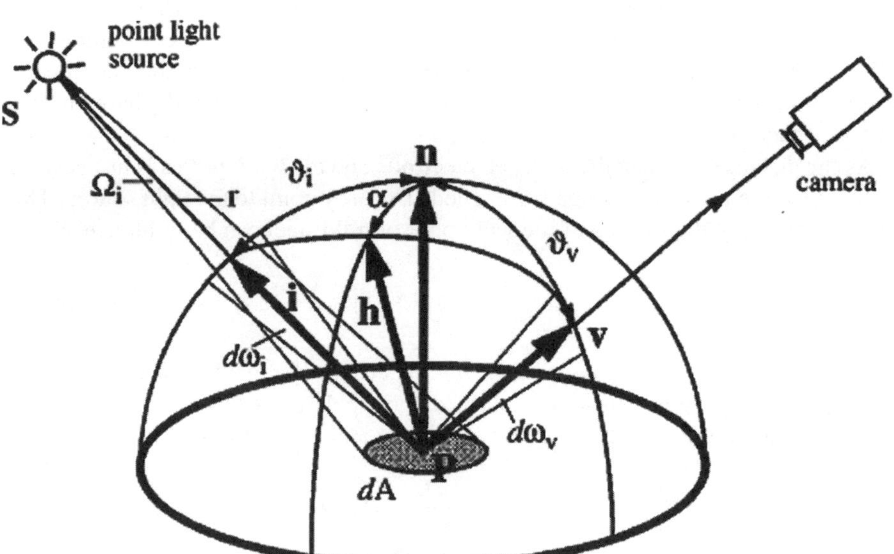

Fig. 3.35 Reflection geometry of generalized reflectance map model

reflectance map model, where r is the distance between a point light source **S** and a surface point **P**, $\vec{n}$ is the unit surface normal at **P**, $\vec{i}$ is the unit vector toward **S**, $\vec{v}$ is the unit vector toward the camera and $\vec{h}$ is the unit vector along the specular direction. ϑ_i is the angle between $\vec{n}$ and $\vec{i}$, ϑ_v is the angle between $\vec{n}$ and $\vec{v}$ and α is the angle between $\vec{n}$ and $\vec{h}$.

The generalized model can be written as:

$$L(\vec{i},\vec{n},\vec{v},\vec{r}) = \beta_d L_d(\vec{i},\vec{n},\vec{v},\vec{r}) + \beta_s L_s(\vec{i},\vec{n},\vec{v},\vec{r}) \tag{3.44}$$

where

$$L_d = \frac{I_0}{r^2\pi}\max[0,\vec{i}^{\,T}\vec{n}] \; L_s = \frac{I_0}{r^2}\frac{\exp\left\{-k[\cos(\vec{h}^{\,T}n)]^2\right\}}{(\vec{i}^{\,T}\vec{n})(\vec{v}^{\,T}\vec{n})}\max[0,\vec{i}^{\,T}\vec{n}]$$

Where k is surface rough factor of the object.

The generalized model is not precise enough to describe the characteristics of a weld pool of low carbon steel pulsed GTAW. For example, the shape and size of light source and the inter-reflection effect, which influence the imaging of the weld pool, are not taken into consideration. So, in this section, a modified reflectance map model of weld pool of mild steel is built by analyzing the light source characteristics, surface characteristics, projection principle and other influential factors.

During pulsed GTAW, the light source characteristic depends on the shape of the arc while the shape of the arc depends on the welding current. When the current is large, the arc is cylindrical, which makes the intensity of the arc anisotropic. When the current is small, the cylindrical arc shrinks to the butt of tungsten electrode. At this time, the intensity is isotropic and the light source can be considered as a point light source with a radius of 0.5 mm.

The large current will not only make the characteristics of the light source complex but also influence the image quality. Figure 3.36 shows images obtained at different periods in a pulse cycle. It can be seen from Fig. 3.36(b–g) that the images obtained at the peak current time are vague while those obtained at the base current time are much better.

As the distance between the tungsten electrode and the work pieces is just several millimeters, the weld pool image is regarded as nearby point to the light source. The intensity reaching the surface of the weld pool of mild steel can be written as E:

$$E = \frac{I_0}{r^2}\max[0,\vec{i}^{\,T}\vec{n}] \tag{3.45}$$

where I_0 – the radiation intensity; and $\vec{i}^T\vec{n}$ – the cosine of angle between $\vec{i}$ and $\vec{n}$.

The reflection characteristics of an object can be described by the bi-directional reflectance distribution function (BRDF), which can be written as f:

$$f = \frac{dL}{dE} \tag{3.46}$$

where L and E denote the reflected radiance in the emitting direction and the irradiance in the direction of incident light, respectively.

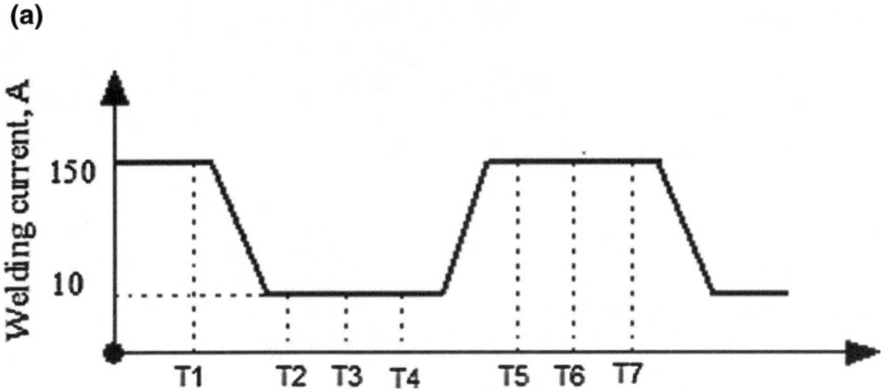

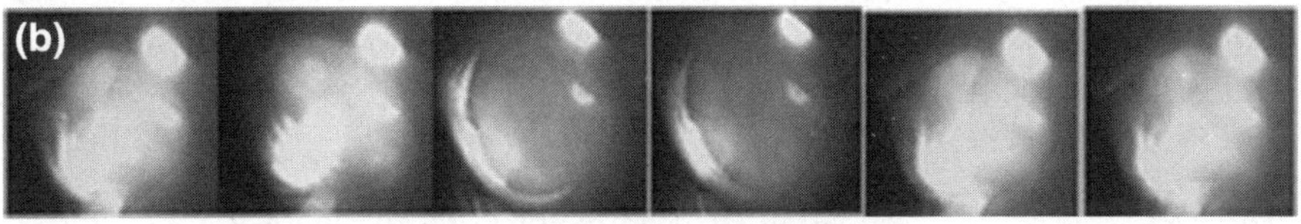

Fig. 3.36 Weld pool images of mild steel obtained at different times in a pulse cycle (convex type) (**a**) curve of welding current in a pulse cycle (**b**) image at T1 (**c**) image at T2 (**d**) image at T3 (**e**) image at T4 (**f**) image at T5 (**g**) image at T6

The BRDF f consists of the diffuse reflection component and the specular reflection component:

$$f = \beta_d f_d + \beta_s f_s \tag{3.47}$$

where β_d and β_s denote the weighting factors of diffuse and specular components, respectively.

Wolff [19] showed that the diffuse reflectance map can be approximated by the Lambertian model when both the angles of incident and emittance are simultaneously less than 50° and the angle between the viewer direction and the illumination direction is less than 60°. They are satisfied in GTAW, so the BRDF of the diffuse reflectance map can be written as f_d:

$$f_d(\vec{i},\vec{n},\vec{v}) = \frac{1}{\pi} \tag{3.48}$$

Torrance and Sparrow developed a model [20] which is widely used for describing the specular reflection. The BRDF for the specular components can be written as f_s:

$$f_s(\vec{i},\vec{n},\vec{v}) = \frac{1}{(\vec{i}^T\vec{n})(\vec{v}^T\vec{n})} FDG \tag{3.49}$$

where

$$F = \frac{1}{2}\frac{(g-c)^2}{(g+c)^2}\left\{1 + \frac{[c(g+c)-1]^2}{[c(g-c)+1]^2}\right\}$$

is the Fresnel function;

$$D = \exp\left\{-k[\cos^{-1}(\vec{h}^T\vec{n})]^2\right\}$$

is the slope distribution function of micro-facets;

$$G = \min\left(1, \frac{2(\vec{h}^T\vec{n})(\vec{v}^T\vec{n})}{(\vec{i}^T\vec{h})}, \frac{2(\vec{h}^T\vec{n})(\vec{i}^T\vec{n})}{(\vec{h}^T\vec{v})}\right)$$

is the geometrical attenuation factor;

$c = \vec{i}^T\vec{n}, g^2 = \eta^2 + c^2 - 1$ and η is the refractive index of object;
k is the surface rough factor of the object.

Figure 3.37 shows the relations between Fresnel function, the incident angle and the refractive index. It indicates that Fresnel function can be simplified to a constant when the incident angle is small. In the imaging of weld pool, the incident angle is usually less than 60°. The refractive index $\eta_{mild} = 1.52$, so F is set as $F_{mild} = 0.05 - 0.1$.

The first term of geometrical attenuation factor G means there is no occlusion when the image is captured, the second term means the reflection light is occluded and the third one means the incident light is occluded. In this imaging system, there is no occlusion. So G is set as $G = 1$.

By substituting (3.48) and (3.49) into (3.47), the BRDF f can be written as:

$$f(\vec{i},\vec{n},\vec{v}) = \beta_d f_d(\vec{i},\vec{n},\vec{v}) + \beta_s f_s(\vec{i},\vec{n},\vec{v}) = \beta_d\frac{1}{\pi} + \beta_s\frac{0.1\times\exp\left\{-k[\arccos(\vec{h}^T n)]^2\right\}}{(\vec{i}^T\vec{n})(\vec{v}^T\vec{n})} \tag{3.50}$$

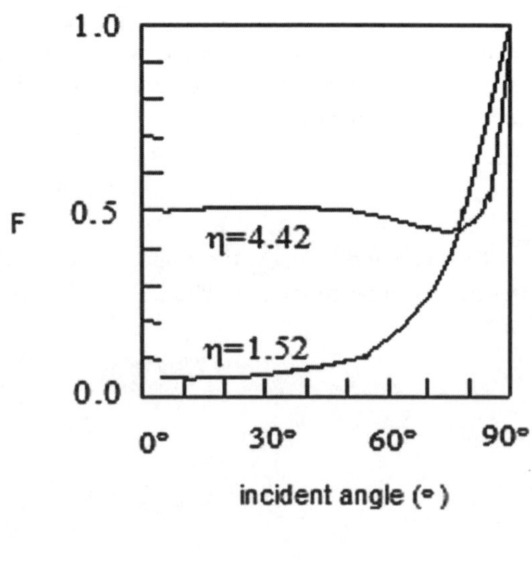

Fig. 3.37 Relations between Fresnel function, incident angle and the refractive index

Thus, reflected radiance in the emitting direction caused by the diffuse and specular reflection can be formulated as L_{ds}:

$$L_{ds}(\vec{i},\vec{n},\vec{v},\vec{r}) = \int_{\omega_x} f(\vec{i},\vec{n},\vec{v})E(\vec{i},\vec{n},\vec{r})d\omega_x = \beta_d L_d(\vec{i},\vec{n},\vec{v},\vec{r}) + \beta_s L_s(\vec{i},\vec{n},\vec{v},\vec{r}) \quad (3.51)$$

where

$$L_d = \frac{I_0}{r^2\pi}\max[0,\vec{i}^T\vec{n}]$$

$$L_s = \frac{I_0}{r^2}\frac{0.1\times\exp\left\{-k[\arccos(\vec{h}^T n)]^2\right\}}{(\vec{i}^T\vec{n})(\vec{v}^T\vec{n})}\max[0,\vec{i}^T\vec{n}]$$

Nayer et al. [21] indicated that most scenes consist of concave surfaces where points reflect light among themselves. Such phenomenon that the light reflected on one point illuminates the other is called inter-reflection.

Weld pool surface sinks by the arc force, which makes the weld pool like an irregular elliptic hemispheroid. So it can be inferred that there exists the inter-reflection in weld pool of mild steel, which is caused by concave liquid melting metal area. Reflected radiance in the emitting direction caused by the inter-reflection can be formulated as L_{inter}:

$$L_{\text{int}\,er} = \int\left[\frac{[\vec{n}_B^T\vec{r}_{AB}][\vec{n}_A^T(-\vec{r}_{AB})]view(A,B)}{[\vec{r}_{AB}^T\vec{r}_{AB}]^2}\right]L_B dB \quad (3.52)$$

Where $\vec{n}_A$ and $\vec{n}_B$ denote the unit surface normal at point ***A*** and point ***B***, respectively;

$\vec{r}_{AB}$ – the unit vector from ***B*** to ***A***;
L_B – the incident light intensity of ***B***;
dB – the element area of ***B***;

The function *view(A,B)* has only two values, namely, 1 and 0. *view(A,B)=1* when *A* and *B* are positioned and oriented so they can illuminate each other. *view(A,B)=0* otherwise. To simplify the calculation, the function of the inter-reflection can be set as $f_{\text{int}\,er(mild)} = f_{d(mild)}$.

Total reflected radiance can be formulated as:

$$L(\vec{i},\vec{n},\vec{v},\vec{r}) = \beta_d L_d(\vec{i},\vec{n},\vec{v},\vec{r}) + \beta_s L_s(\vec{i},\vec{n},\vec{v},\vec{r}) + \beta_{\text{int}\,er} L_{\text{int}\,er}(\vec{i},\vec{n},\vec{v},\vec{r}) \quad (3.53)$$

Some parameters such as the weighting factors and the roughness factor can be set by comparison of the synthetic images. Such parameters are set as follow: $k_{mild} = 0.8$, $\beta_{d(mild)} = 0.8$, $\beta_{s(mild)} = 0.2$, $\beta_{\text{int}\,er(mild)} = 0.8$.

The CCD camera is used to capture the images of weld pool of mild steel. Images formed through a pinhole camera should be modeled by perspective projection. The irradiance *E* at an image point of the film is obtained through the lens system via the radiance *L* of the corresponding surface point . The irradiance *E* can be described as below:

$$E = \frac{\pi}{4}\left(\frac{d}{f}\right)^2 (\vec{v}^T \vec{n}_z)^4 L \tag{3.54}$$

where d – the diameter of the lens;

f – the focal length behind the lens;
$\vec{v}^T \vec{n}_z$ – the cosine of the angle between the light to the lens and the optical axis.

Considering the intrinsic characteristics of camera, the converted intensity of the image, i.e. gray value through a camera can be written as I:

$$I = gE + b \tag{3.55}$$

where g and b denote the camera gain and the bias, respectively.

So far, the surface reflectance map model of the weld pool of mild steel during pulsed GTAW is established. It can be written as:

$$I = g\frac{\pi}{4}\left(\frac{d}{f}\right)^2 (\vec{v}^T \vec{n}_z)^4 L(\vec{i}, \vec{n}, \vec{v}, \vec{r}) + b \tag{3.56}$$

Existence of high intensity of argon arc will not only damage the lens of the camera but also decrease the quality of the captured images, so a primary filter is used as a protecting method, as mentioned in the above section. The frequency response function of the narrow band filter can be written as $W_{mild}(\lambda)$:

$$W_{mild}(\lambda) = 0.288 \cdot \frac{1}{3\sqrt{2\pi}} \cdot e^{\frac{(\lambda-661)^2}{18}}, 651 \le \lambda \le 671 \tag{3.57}$$

Figure 3.38 shows the gray scale histogram of images of weld pool. It can be observed that the intensity of arc light arise the nonlinear increase of the grey value in some regions.

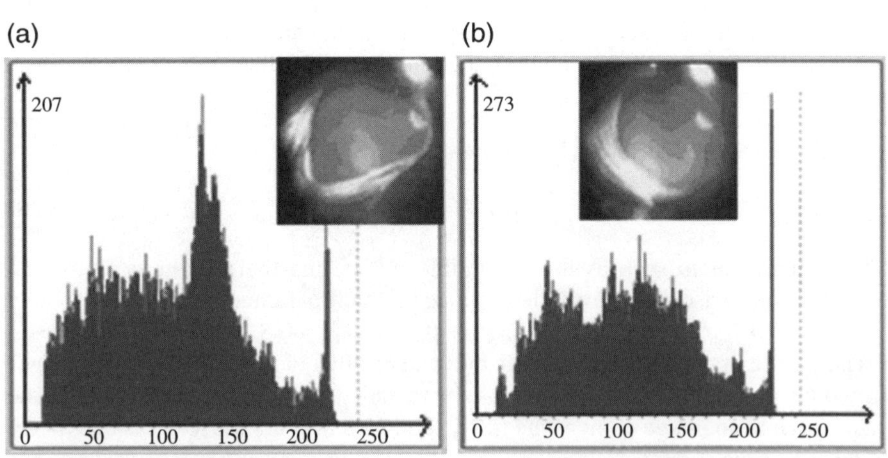

Fig. 3.38 Gray scale histogram of images of weld pool of low carbon steel **(a)** Concave type **(b)** Convex type

To eliminate the light error, light intensity weighting factor is introduced as the error compensation. It can be written as:

$$w_{mild}(E_k) = \begin{cases} e^{-(E_k-210)^2/3000}, & E_k \geq 210 \\ 1.0, & E_k < 210 \end{cases} \tag{3.58}$$

Step 2. Reflection model solving

Shape from shading (SFS) is used as a solving algorithm of the corresponding irradiance equation of the surface reflectance map model of weld pool. By SFS method, the height is calculated by minimizing an error cost functional which consists of the brightness constraint function and the smoothness constraint function. The surface is discretized by a union of triangular surface patches. The vertices of the triangles are called nodal points. Only depths at the nodal points are calculated and others are obtained by interpolation. For each triangular path, the intensity of the triangle is taken as the average intensity of all pixels in the triangle and the gradient of the triangle is approximated by the cross product of any two adjacent edges of the triangle. This establishes a relationship between the intensity and the surface height at its three nodal points:

$$\begin{aligned} E_k = R_k(z_i, z_j, z_l) = b + g\frac{\pi}{4}\left(\frac{d}{f}\right)^2 (\upsilon_k^T n_z)^4 \Bigg[\frac{I_0}{r_k^2} \max(i_k^T n) \\ \times \left(\frac{\beta_d}{\pi} + \beta_s \frac{\exp(-k[\arccos(h_k^T n)]^2)}{(i_k^T n)(v_k^T n)} FG \right) + \beta_{\mathrm{int}\ er} L_{\mathrm{int}\ er} \Bigg] \end{aligned} \tag{3.59}$$

The brightness constraint indicates the total brightness error of the reflectance map model with the measured intensity. It can be written as ε_b:

$$\varepsilon_b = \sum_{k=1}^{M_t} (E_k - R_k)^2 = \frac{1}{2} z^T A_b z - b^T z + c \tag{3.60}$$

where

$$R_k \approx \sum_{m=1}^{M_n} w_{km} z_m + \xi_k$$

$$\xi_k = R_k(z_i^{n-1}, z_j^{n-1}, z_l^{n-1}) - \sum_{m=1}^{M_n} w_{km} z_m^{n-1}$$

$$w_{km} = \begin{cases} \dfrac{\partial R_k(z_i, z_j, z_l)}{\partial z_m} \Big| (z_i^{n-1}, z_j^{n-1}, z_l^{n-1}), & m = i, j, l \\ 0, & \text{otherwise} \end{cases}$$

$a_{i,j}$ – the element of stiffness matrix A_b,

$$a_{i,j} = 2\sum_{k=1}^{M_t} w_{km}w_{kn}, \quad 1 \le m,n \le M_n;$$

$b_{i,j}$ – the element of load vector b,

$$b_{i,j} = 2\sum_{k=1}^{M_t} (E_k - \xi_k)w_{km}, \quad 1 \le m,n \le M_n;$$

M_t – number of triangles in the image;
z – nodal variable

The smoothness constraint ensures a smooth surface to stabilize the convergence to a unique solution. It is written as:

$$\varepsilon_s = \frac{1}{2}\iint_{\Omega} \left[\rho(z_x^2 + z_y^2) + (1-\rho)(z_{xx}^2 + 2z_{xy}^2 + z_{yy}^2)\right] dxdy \tag{3.61}$$

where ρ – the smoothness factor, $0 < \rho < 1$

The first term assures the continuity of the height of the recovered surface, named membrane energy function. The second term assures the second derivative continuity of the height, named thin plate energy function. By combining the membrane energy function and the thin plate energy function, a smoothness characteristic of the image surface can be described. The function can be formulated as a matrix.

The whole error cost function can be written as:

$$\varepsilon = \frac{1}{2}z^T Az - b^T z + c \tag{3.62}$$

where $A = A_b + \lambda B$,

B is the matrix of the smoothness constraint functional.
λ is the smoothness factor.

Figure 3.39 shows the flow chart of the algorithm.

Step 3. Calculation of surface height of the weld pool

The experimental parameters are set as follow, shown is Table 3.1.

The frequency response function of the narrow band filter $W_{mild}(\lambda)$ is written as equation (3.12) and light intensity weighting factor $w_{mild}(E_k)$ is written as equation (3.13).

Figure 3.40 shows entire surface height recovered. Figure 3.41 shows the center cross-section height of weld pool along X axis and that along Y axis.

All these results accords with the facts.

Because the workpiece contracts when it cools down, direct validation of the calculation is impossible. However, an indirect method can be used. As the topside height of the welding pool changes proportionally with the wire feeding velocity, observing the changing tendency caused by changes of weld parameters can be used.

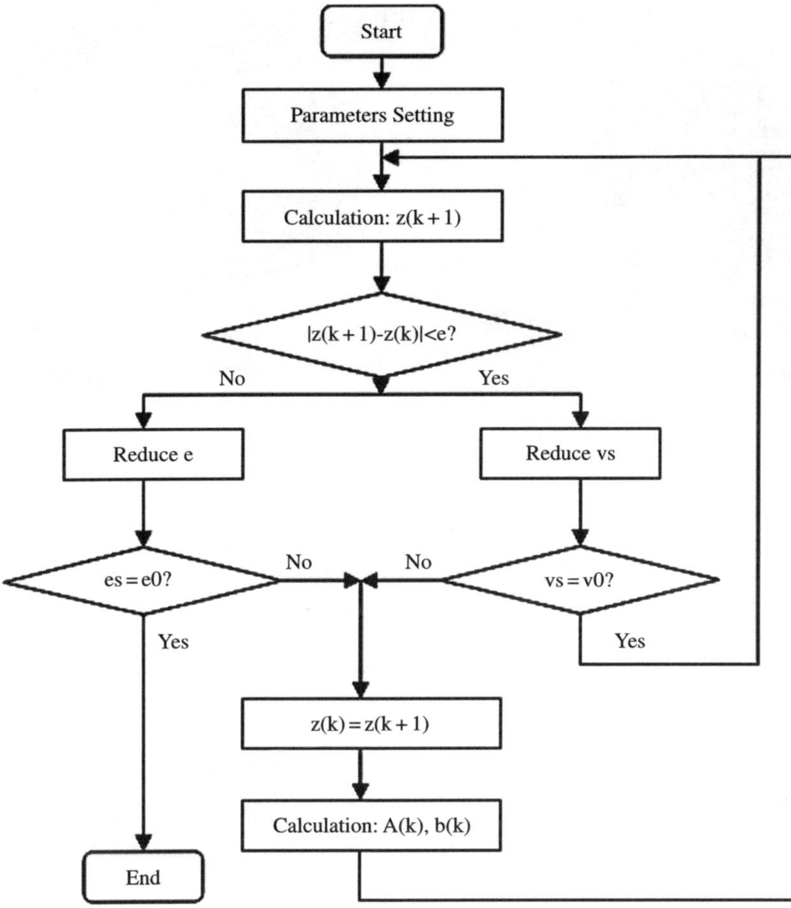

Fig. 3.39 Flow Chart of the SFS algorithm

In this experiment, the heights of weld pool with different wire feeding velocity are compared. Figure 3.42 shows weld pool images with various wire feeding velocity. Figure 3.43 shows the height of the corresponding images and Fig. 3.44 shows a changing tendency of the center cross-section height with different wire feeding velocity along X axis and Y axis, respectively.

It shows that:

Table 3.1 Experimental parameters

Diameter of lens (inch)	Focus lens (mm)	Camera gain	Camera bias
1/3	30	0.45	0.9
Imaging size (pixel*pixel)	base current (A)	Semi-diameter of arc (mm)	Length of arc (mm)
128*128	20	0.5	3

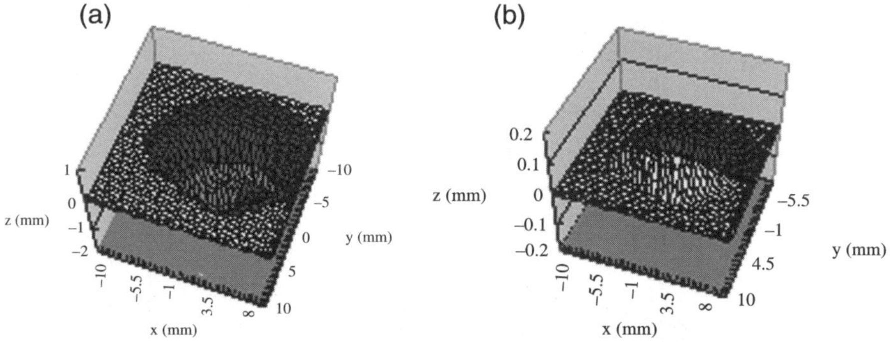

Fig. 3.40 Calculation results of low carbon steel weld pool during pulsed GTAW (**a**) Concave type (**b**) Convex type

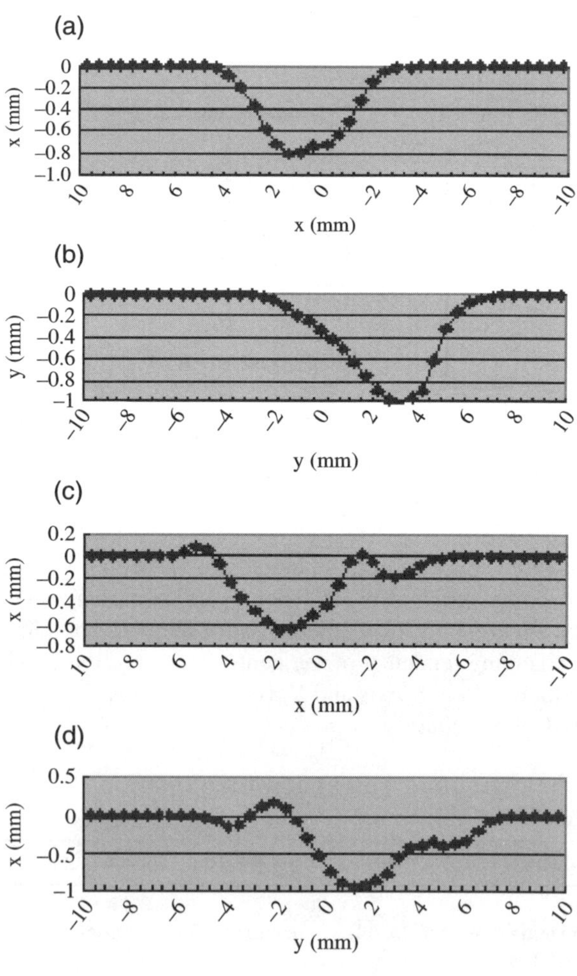

Fig. 3.41 Section height of low carbon steel weld pool during pulsed GTAW (**a**) x axis direction of concave type (**b**) y axis direction of concave type (**c**) x axis direction of convex type (**d**) y axis direction of convex type

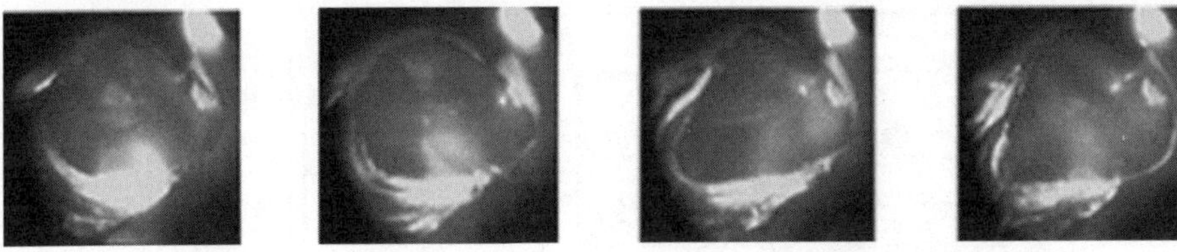

Fig. 3.42 Weld pool Images of low carbon steel with various wire feeding velocity (**a**) $v_f = 7.0\,mm/s$ (**b**) $v_f = 5.00\,mm/s$ (**c**) $v_f = 3.00\,mm/s$ (**d**) $v_f = 0.0\,mm/s$

(1) With the decrease of the wire feeding velocity, height of the rearward of the welding pool decreases and shape of that changes from convex to concave;
(2) The height of the welding pool decreases when the wire feeding velocity is decreasing;
(3) The welding pool heights of center cross-section along X axis and Y axis are symmetry on the whole, expecting a few fluctuations along Y axis.

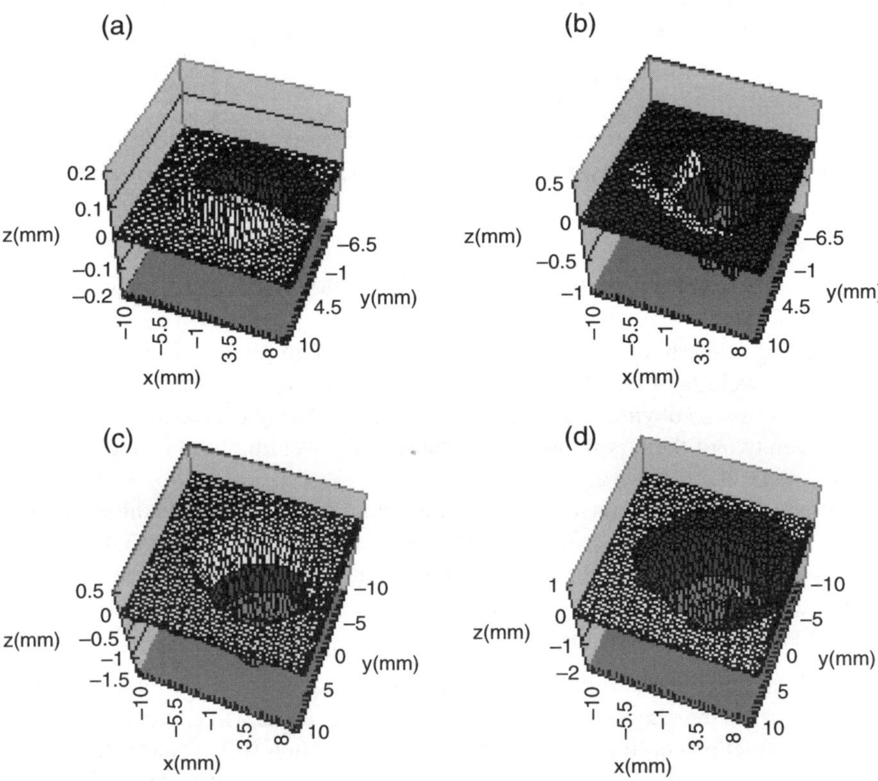

Fig. 3.43 Recovery shape of low carbon steel weld pool during pulsed GTAW (**a**) $v_f = 7.0\,mm/s$ (**b**) $v_f = 5.00\,mm/s$ (**c**) $v_f = 3.00\,mm/s$ (**d**) $v_f = 0.0\,mm/s$

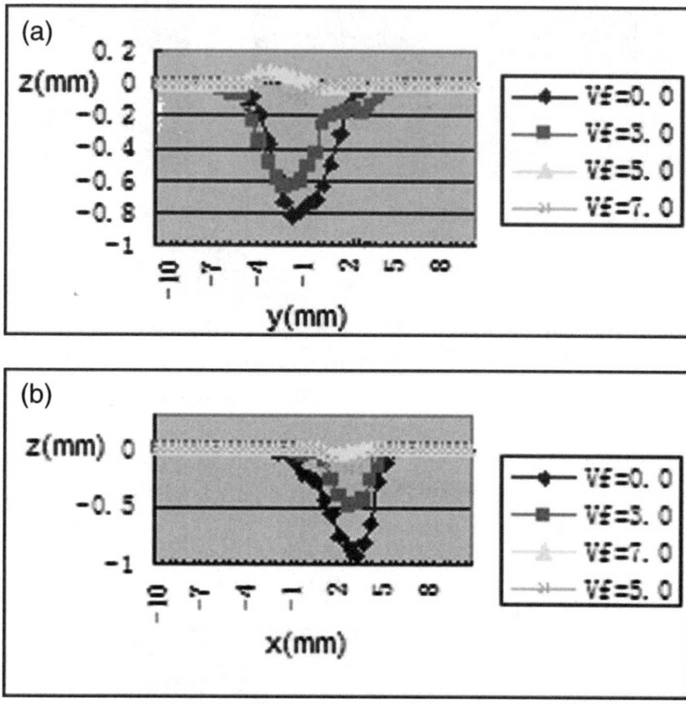

Fig. 3.44 Height of center section of weld pool of mild steel pulsed GTAW (**a**) y axis direction (**b**) x axis direction

All these are according with the real situation and the effectiveness of this method can be verified.

According to the relative theory, there exists a certain relationship between the height of the weld pool and that of its frozen state, showed as a welding bead. This relation becomes so obvious when the wire feeding velocity is zero that a comparison between two of them is showed. The shape of the welding bead is traced out by topology method.

Figure 3.45 (a) is the image of the welding pool with the wire feeding velocity as zero, Fig. 3.45 (b) is the cross-section of the welding bead and Fig. 3.45 (c) is a comparison between the height of the weld pool and that of welding bead.

Table 3.2 shows the errors between the calculated height and measured one.

There are several reasons that cause the error:

(1) The image is obtained during the welding process while the welding bead is drawn in the freezing state. Due to the fact that the metal will expand on heating and contract on cooling, there exists some deformation which causes the error;
(2) The molten metal will flux during the welding process, so two images captured in different time will have some difference;
(3) The measurement method has the system errors.

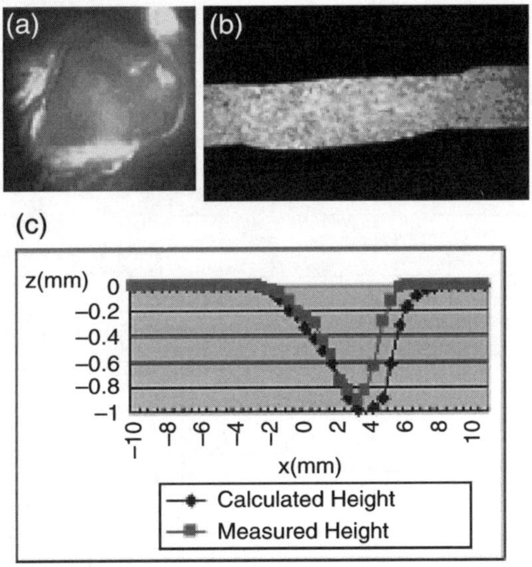

Fig. 3.45 Height by calculation and measurement (**a**) Image of weld pool of mild steel during pulsed GTAW with $v_f = 0.0$ mm/s (**b**) Image of a weld beam (frozen state of the weld pool) (**c**) Comparison of the calculated height of the weld pool and the measured one of the weld beam

Table 3.2 Error between the calculated height and measured one

Times of Calculation (ms)	120.36	Times of iteration	543
Maximum value by measurement (mm)	Maximum value by calculation (mm)	Mean error	Mean square error
−0.91	−1.01	0.12	0.08

3.3 The Software of Image Processing and Characteristic Extracting of Weld Pool During Pulsed GTAW

Based on the above mentioned algorithms for both 2 dimensional and 3 dimensional weld pool image processing, a software, named Obtaining and Image Processing of Weld Pool Vision Characteristics System (OIPWPVCS) [22, 23], is developed for general use to offer accurate information for further modeling and intelligent control. The software is widely adaptive, fast and precise. Because many algorithms are integrated to deal with different situations, the software can process different weld pool images. Meanwhile, the software is also adaptive to different DDC equipment with various precise and experiment parameters.

3.3.1 The Framework and Function of the Software System

For the better management and coordination of different algorithms, a module-based software framework is designed. It includes six modules, namely, files management,

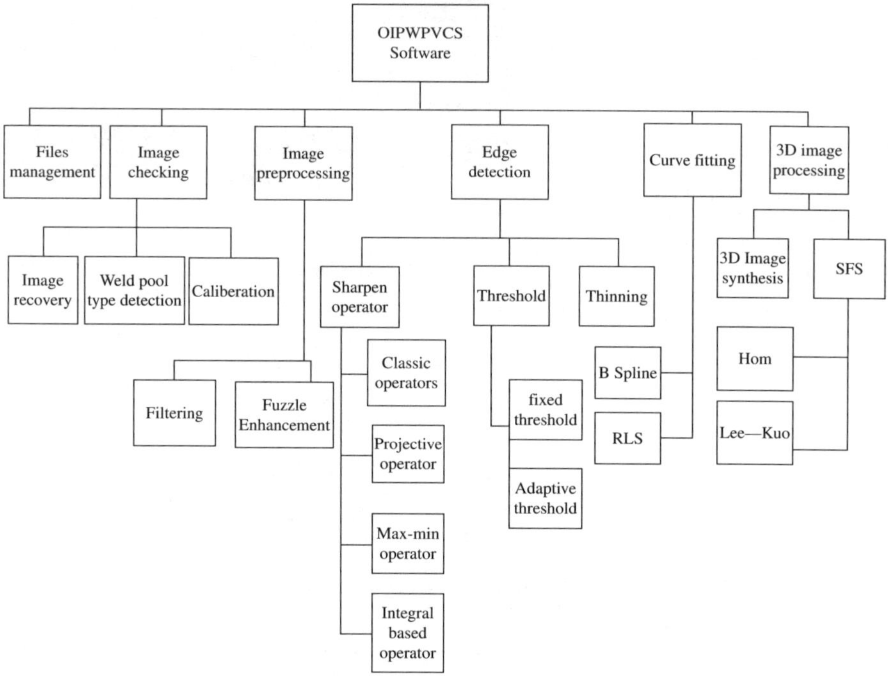

Fig. 3.46 Software architecture

image checking, preprocessing, edge detecting, curve fitting and 3D information acquisition. The architecture of the software is shown in Fig. 3.46.

On the whole, the software is divided into four parts: preprocess of image, edge detection, binary processing and post-process of image.

3.3.2 The Directions for Using the Software System

Figure 3.47 shows the user interface of the software. And in the following part, the use of the software in different modules will be discussed.

3.3.2.1 File Management Module

This module, corresponding to the file menu in the main window, provides functions of file open, save, print and pixel extraction, as shown in Fig. 3.48.

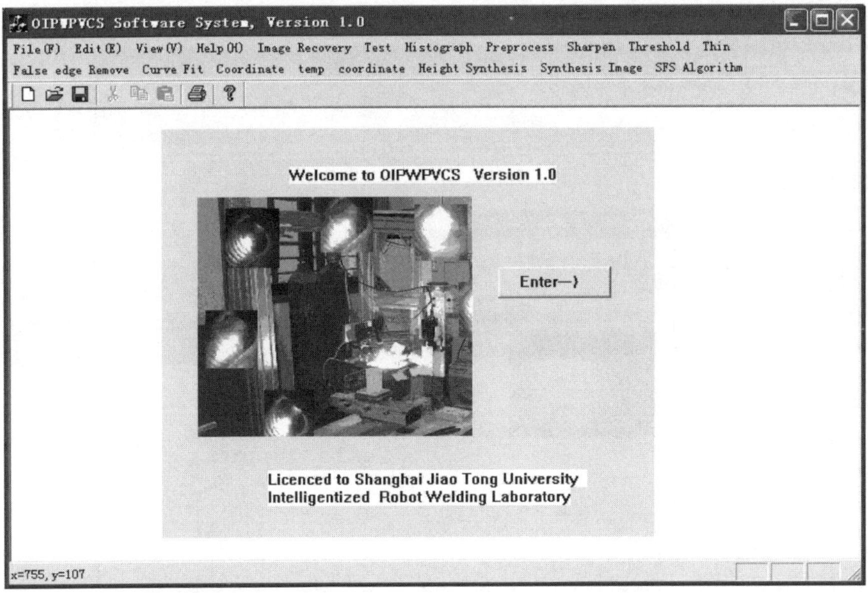

Fig. 3.47 User interface of the software

3.3.2.2 Image Checking Module

This module includes three parts, namely, image recovery, weld pool type detection and calibration.

For image recovery, the software can automatically judge whether the image is needed to be recovered or not, as shown in Fig. 3.49, where (a) is the degraded image and (b) is the result of image recovery.

For weld pool type detection, this part can detect the shape of the weld pool and then use this information to judge penetration of the work piece. As the weld pool can be divided into two types, penetrated and un-penetrated. The penetrated weld pool looks heart-shape (peach-shape) and un-penetrated weld pool looks ellipse-shape for steel work pieces. Therefore, the weld pool type detection can predict the penetration of the work piece. And the result of the image processing is shown in Fig. 3.50.

For calibration, this part is to correspond the real coordinate system with the image coordinate system. Here we define the real coordinate system in Fig. 3.51(a) and the image coordinate system in Fig. 3.51(b).

(1) Real coordinate system-xyz: The origin is the workpiece point under the tungsten tip. In the coordinate system, z axis is in the direction of tungsten tip; x axis is in the direction of welding and y axis is perpendicular to the direction of welding.

(a)

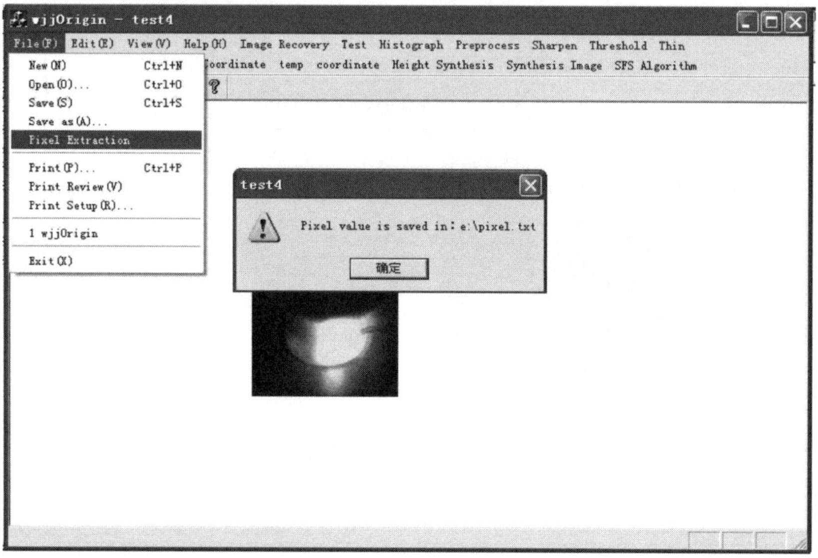

(b)

62	62	61	62	60	60	60	59	60	127	60	59
87	83	80	79	82	82	84	132	157	79	83	81
58	56	56	169	69	59	57	59	58	58	43	60
63	63	74	147	162	62	62	48	47	63	63	70
56	58	58	71	60	58	58	57	57	93	97	60
68	68	67	68	58	71	91	81	81	82	82	83
131	131	132	126	127	126	81	78	80	176	81	97
134	131	133	133	187	130	76	76	78	78	124	124
130	131	130	137	139	152	137	89	81	62	152	62
142	139	133	133	136	135	136	83	87	131	131	194
147	151	151	148	151	151	157	150	80	77	172	193
156	155	155	164	158	163	160	165	87	160	162	156

Fig. 3.48 File management module (**a**) File management menu (**b**) Extraction of pixel of the image (in the pixel files, the red part refers to noises and the blue part refers to the edge of the weld pool)

(2) Image coordinate system O-XY: It is fixed on the image plane, with the origin in the top left corner of the image; x axis in the right direction and y axis in the left direction.

In this software, a circle with known size and direction is used for calibration. After entering the calibration menu, first open the circle and then click the point p1, p2, p3, p4 in sequence, finally the system can calculate the calibration parameters and save them in the memory. The calibration process is shown in Fig. 3.52.

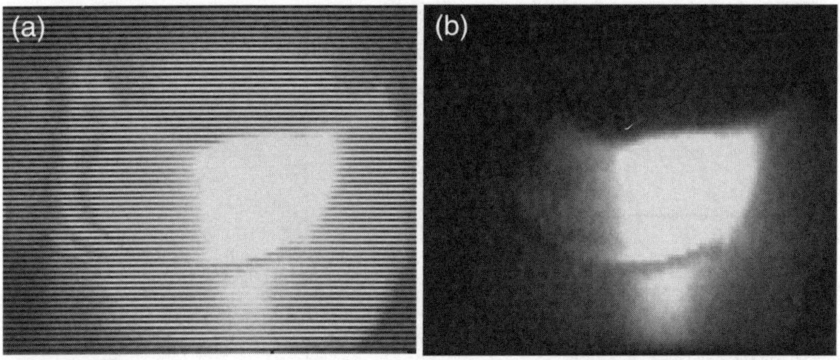

Fig. 3.49 Image recovery (**a**) Image before recovery (**b**) Image after recovery

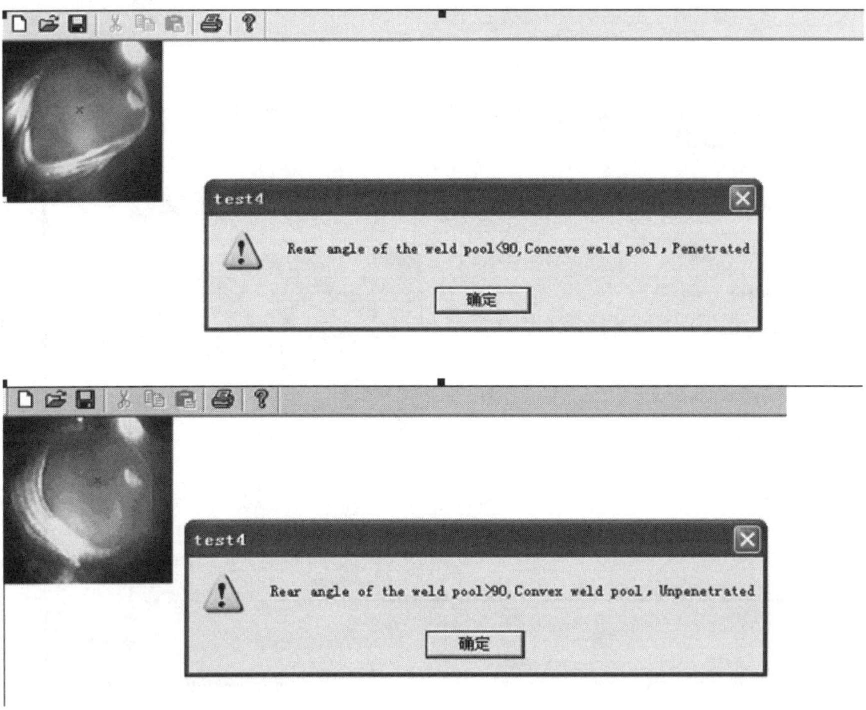

Fig. 3.50 Weld pool type detection

3.3.2.3 Image Preprocessing Module

This part correspond to the submenu, as shown in Fig. 3.53, including gray level changing, filtering, enhancement, color invert and image rotate.

Gray level changing part is to adjust the brightness of the image so as to improve the quality of the image. This part can selectively change the gray level of the image.

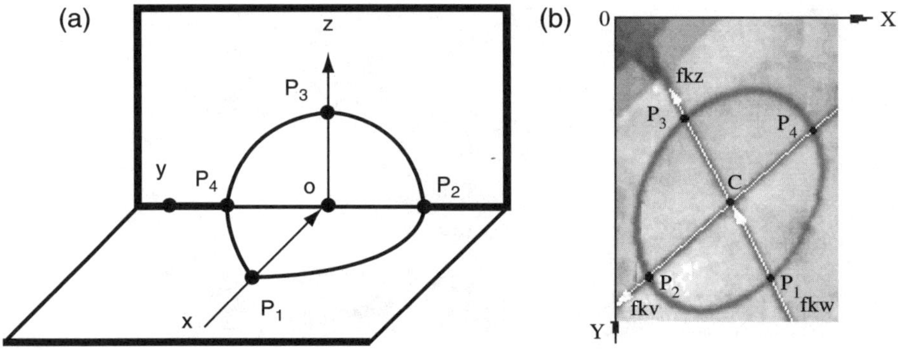

Fig. 3.51 Coordinate system definition (**a**) Workpiece coordinate system (**b**) Image coordinate system

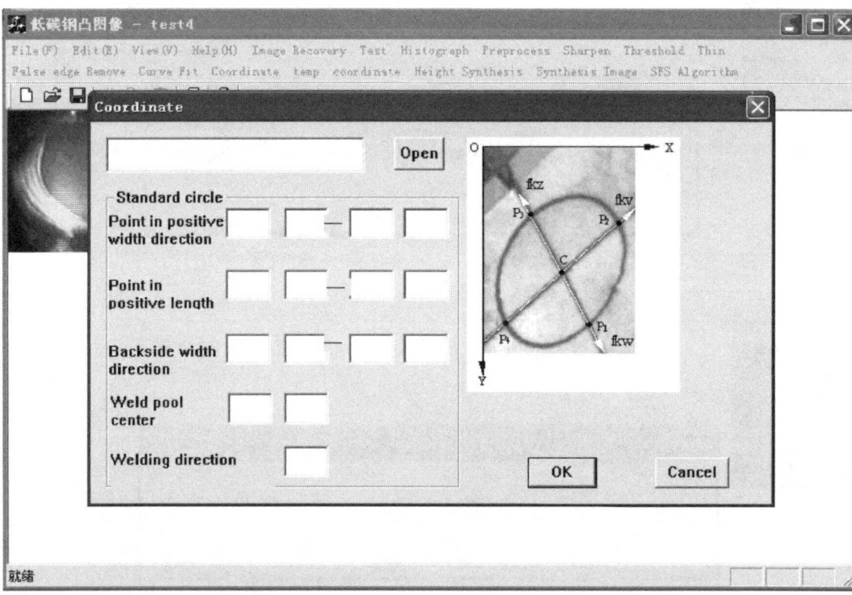

Fig. 3.52 Weld pool calibration

In the dark area the slope of the gray line is greater than 1, thus enlarging the gray level; while in the bright area the gray is less than 1, thus lower the gray level. The gray level changing processing is shown in Fig. 3.54.

Here, 3 kind of filtering method, namely, smooth filtering, Guass filtering, and user define filtering is used in the software, as shown in Fig. 3.55.

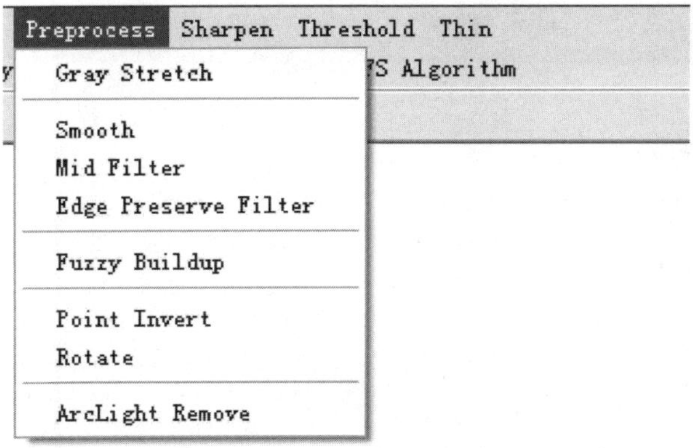

Fig. 3.53 Image preprocessing module

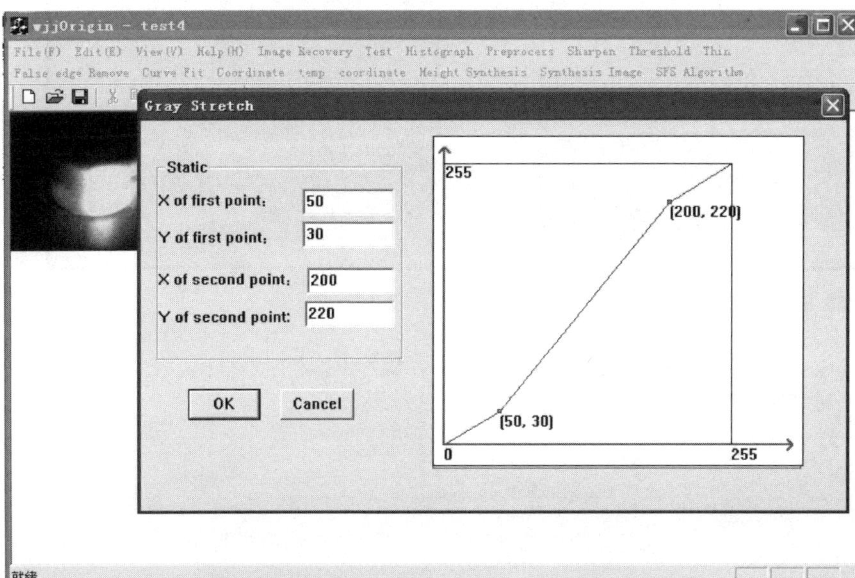

Fig. 3.54 Gray level changing

3.3.2.4 Edge Detection Module

In this software, edge detection module includes sharpening operator, threshold algorithm and thinning algorithm.

Figure 3.56 shows the sharpening menu and Fig. 3.57 shows the thresholding menu.

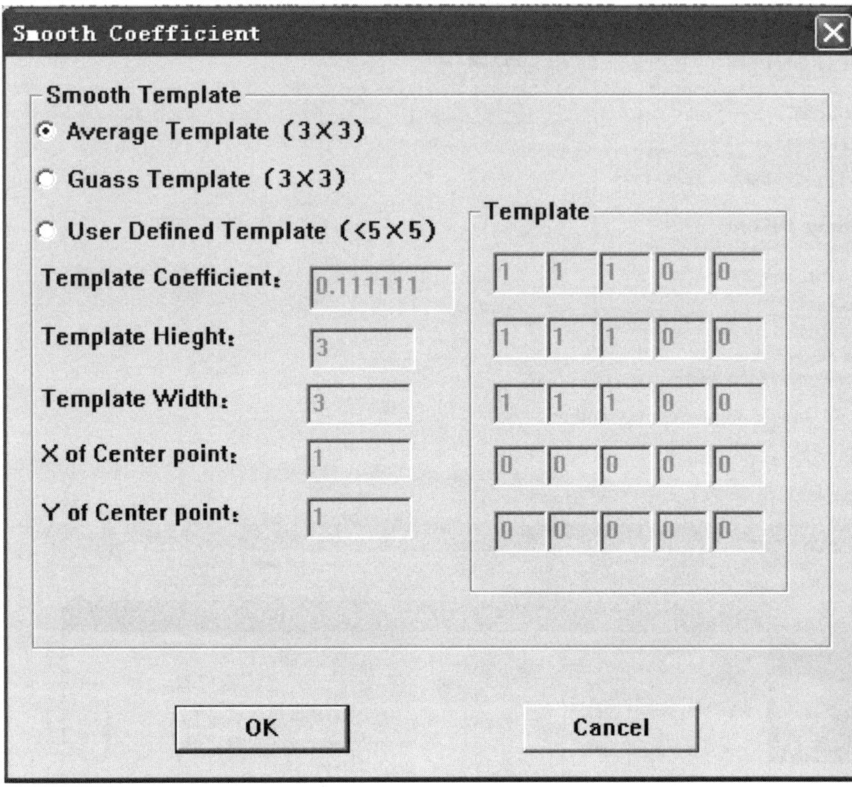

Fig. 3.55 Smooth coefficient setting

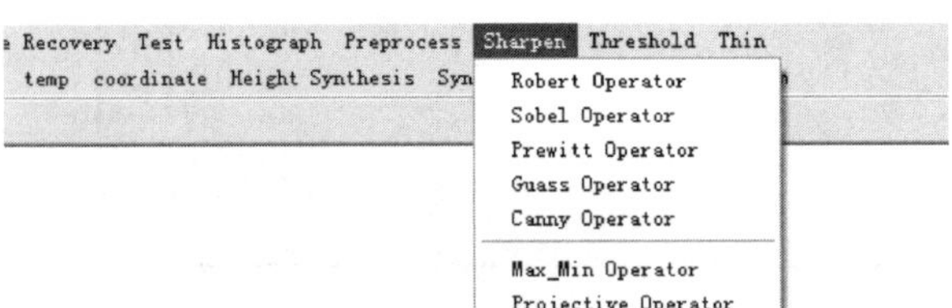

Fig. 3.56 Sharpening menu

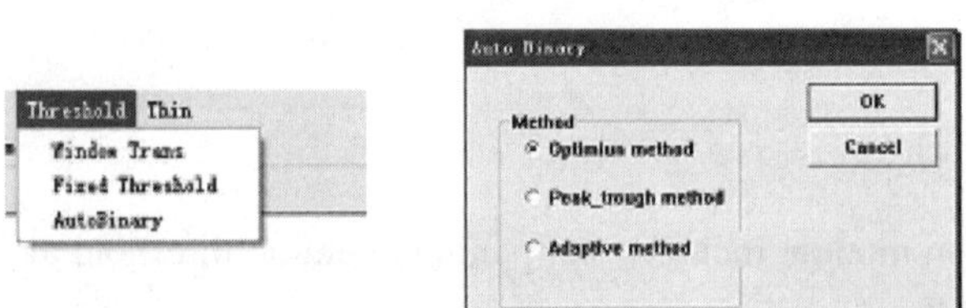

Fig. 3.57 Thresholding menu

3.3.2.5 Curve Fitting Module

To get a complete edge of the weld pool, curve fitting is necessary. In this software, two kind of curve fitting method is provided, i.e. B Spline and nonlinear RLS methods, as shown in Fig. 3.58.

And the result of nonlinear RLS method is shown respectively in Fig. 3.59.

After curve fitting, the measurement of 2D weld pool information is available. And the Fig. 3.60 shows the size of weld pool in image plane.

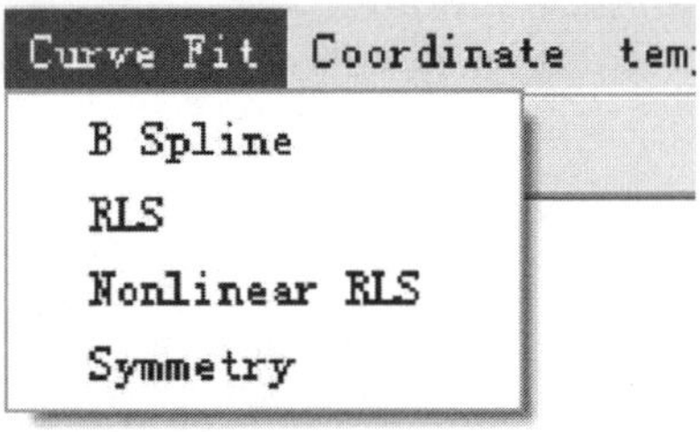

Fig. 3.58 Curve fitting menu

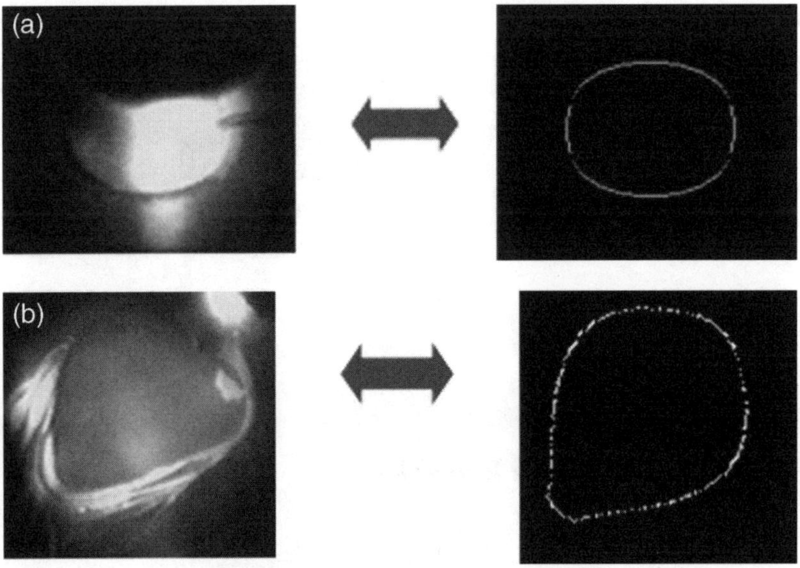

Fig. 3.59 Result of nonlinear curve fitting (a) ellipse-shaped weld pool (b) heart-shaped weld pool

3.3.2.6 3D Image Processing Module

In this module, three different algorithms are provided, namely, Horn algorithm, Lee-Kuo algorithm and llp algorithm [17], as shown in Fig. 3.61. And the result after 3D image processing is saved as pixel gray values in the specified files.

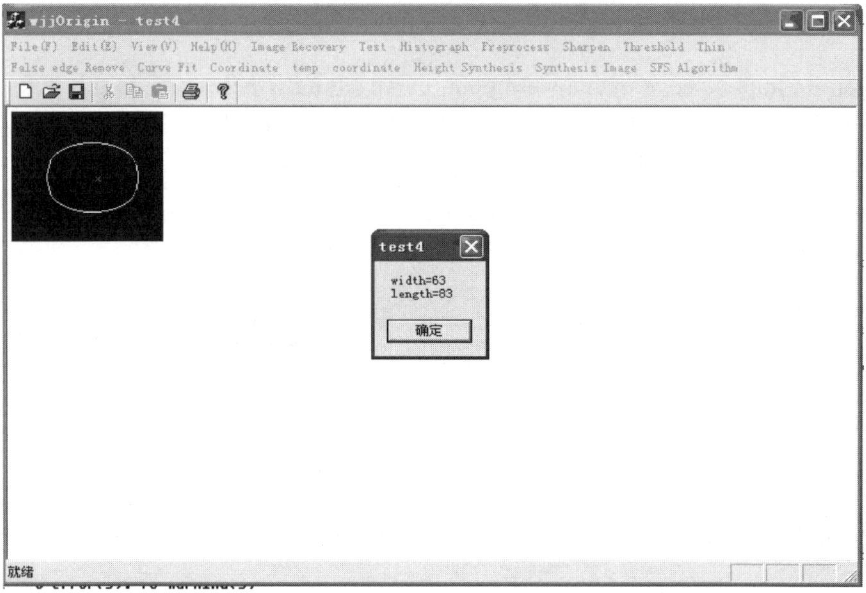

Fig. 3.60 Weld pool measurements

Fig. 3.61 3D image processing menu

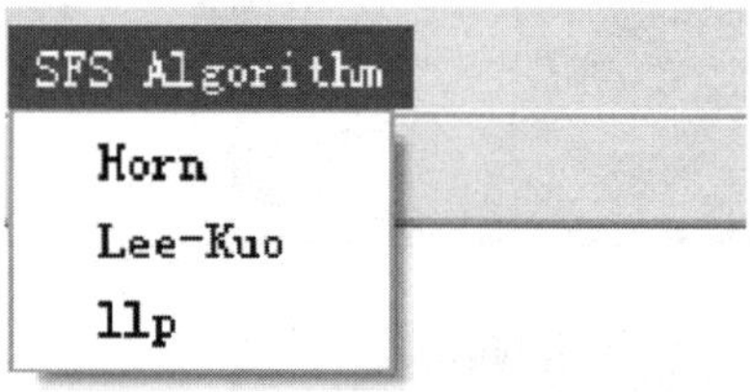

3.4 The Chapter Conclusion Remarks

Real time control of weld pool characteristics during pulsed GTAW process is very important for weld quality. And the control strategies should be based on the precise measurement of the weld pool size. Sensing technology and the intelligent control of dynamic processes have been the focus of much research, however, advance sensing methods are the most important elements. In recent years, the use of sensing technology in the welding process has made great progress. In this chapter, both 2D and 3D image processing methods are described with the experiments in low carbon steel, stainless steel and aluminium alloy. In the last part of the chapter, image processing software is developed and the direction for its usage is described.

References

1. S.B. Chen et al. Intelligentlized technologies for robotic welding. Series Lecture Notes in Control and Information Sciences. 2004, 299:123–143
2. C. Shanben et al. On intelligentized technologies for modern welding manufacturing. Chinese Journal of Mechanical Engineering. 2003, 16(4):367–370
3. R.J. Beatlie, S.K. Cheng, P.S. Logue. The use of vision sensors in multipass welding applications. Welding Journal. 1988, 67(11):28–33
4. Y. Suga, M. Naruse. Application of neural network to visual sensing of weld line and automatic tracking in robot welding. Welding in the World. 1994, 34:275–284
5. Y.M. Zhang, L. Li, R. Kovacevic. Dynamic estimation of full penetration using geometry of adjacent weld pools. ASME Journal of Manufacturing Science and Engineering. 1997, 119:631–643
6. Y.J. Lou. Intelligent control for pulsed GTAW dynamic process based on image sensing of weld pool. PhD dissertation, Harbin Institute of Technology, 1998
7. D. Marr, E.C. Hildreth. Theory of edge detection. Processing of Royal Society London. 1980, B207:187–217
8. R. Gordon, R.M. Rangayan. Feature enhancement of film mammograms using fixed and adaptive neighborhood. Applied Optics. 1984, 23(5):560–564
9. A. Beghdadi, A.L. Negrate. Contrast enhancement technique based on local detection of edges. Computer Vision Graphics Image Process. 1986, 46(9):162–174
10. B.K.P. Horn. Height and gradient form shading. International Journal of Compute Vision. 1990, 5:37–75
11. K.M. Lee, C.J. Kuo. Surface reconstruction from photometric stereo images Journal of Optical Society of America. 1993, 10:855–868
12. M. Oren, S.K. Nayar, Generalization of the Lambertian Model and Implication for machine vision. International Journal of Computer Vision. 1995, 14:227–251
13. Q. Zheng, R. Chellappa, Estimation of illuminant direction, Albedo, and Shape from Shading. IEEE Transactions on Pattern Analysis and Machine Intelligence. 1991, 13:680–702
14. J.K. Hasegawa, C.L. Tozzi. Shape from shading with perspective projection and camera calibration. Computer and Graphics, 1996, 20:351–364
15. R. Zhang, P. Tsai et al. Shape from shading: A survey. IEEE Transactions on Pattern Analysis and Machine Intelligence. 1999, 21(8):690–706
16. Z. Dongbin. Dynamic Intelligent Control for Weld Pool Shape during Pulsed GTAW with Wire Filler Based on Three-Dimension Visual Sensing, [Doctor thesis] Harbin Institute of Technology, 2000
17. Li Laiping Research on modeling of surface reflectance map and height calculation of pulse GTAW molten pool, [3.Doctor thesis] Shanhai Jiao Tong Universit, 2005
18. K.M. Lee, C.J. Kuo. Surface reconstruction from photometric stereo images. Journal of Optical Society of America. 1993, 10(5):855–868
19. L.B. Wolff. Diffuse-reflectance model for smooth dielectric surface. Journal of Optical Society of America. 1994, A11:2956–2968
20. K.E. Torrance, E.M. Sparrow, Theory for off-specular reflection from roughened surfaces. Journal of the Optical Society of America. 1967, 57, 1,105–1,114
21. H. Murase, S.K. Nayer. Visual learning and recognition of 3-D objects from appearance. International Journal of Computer Vision. 1995, 14(1):5–24
22. J. Wu, S.B. Chen, "Software System Designs of Real-time Image Processing of Weld Pool Dynamic Characteristics", Proceedings on '2006 International Conference on Robotic Welding, Intelligence and Automation, RWIA'2006. Lecture Notes in Control and Information Sciences, Springer Verlag, 2007, LNCIS 362:303–310
23. J. Wu. Obtaining and image processing of weld pool vision characteristics system design. [Master thesis] Shanghai Jiaotong University, 2008
24. J.J. Wang. Visual information acquisition and adaptive control of weld pool dynamics of Aluminum alloy during pulsed TIG welding. PhD dissertation, Shanghai Jiao Tong University, 2003

Chapter 4
Modeling Methods of Weld Pool Dynamics During Pulsed GTAW

Abstract GTAW is a thermal process during which the workpiece melts, solidifies and finally forms the welding seam. As is well known, arc welding is influenced by many complex factors, such as material metallurgy, heat conduction, physical chemistry reactions, etc. Due to its multi-variable coupling, nonlinear, time-varying, random and uncertain properties, GTAW dynamics is difficult to be modelled by classical linear system theory. In this chapter, analysis on the welding dynamics is made to understand the process of welding. Based on the analysis, both identification models and intelligent models, e.g. ANN, fuzzy rules model and RS-based model are discussed. ANN model is a "black box" and it is impossible to directly revise the model. For fuzzy rules model, the number of inputs, outputs and their linguistic variables cannot be too large, or it will lead to "rule explosion". RS model is promising for welding process modeling because compared with NN model, RS model is close in predictive ability; and however its complexity is much lower.

GTAW is a thermal process during which the workpiece melts, solidifies and finally forms the welding seam. As is well known, arc welding is influenced by many complex factors, such as material metallurgy, heat conduction, physical chemistry reactions, etc. Due to its multi-variable coupling, nonlinear, time-varying, random and uncertain properties, it is very difficult to model welding dynamics by classical linear system theory. In recent years, some intelligent modeling methods have been introduced to welding. References [1–3] investigated fuzzy reasoning application in modeling, and Refs. [4–8] studied artificial neural networks for modeling. In this chapter, both identification models and intelligent models are discussed for the weld pool dynamics during pulsed GTAW.

4.1 Analysis on Welding Dynamics

Bead-on-plate experiments are conducted on low carbon steel during pulsed GTAW using the double-side imaging system (Fig. 2.5) to analyze the welding process, and the experiment conditions are tabulated in Table 4.1. The peak value of pulsed current is set at 120 Ampere, welding velocity is 2.5 mm/s.

S.-B. Chen, J. Wu, *Intelligentized Methodology for Arc Welding Dynamical Processes*, Lecture Notes in Electrical Engineering 29,

Table 4.1 Experimental conditions of pulsed GTAW

Welding conditions	Pulse frequency	Pulse duty ratio	Base current	Electrode diameter	Tip angle	Arc length	Flow rate	Specimen dimension
Unit	f(Hz)	δ(%)	I_b(A)	ϕ(mm)	θ(°)	l(mm)	L(l/min)	mm × mm × mm
Value	1	45	60	3.2	30	3.5	8.0	280 × 50 × 2

The characteristic parameters of weld pool are defined as described in the Sect. 3.1.1. According to the variation of shape parameters of weld pool, W_b; W_t and H_t are selected as candidate to describe the status of penetration because they are strongly related to the variation of welding parameter in Fig. 4.1, here δ is the pulse duty ratio of welding current.

To design a suitable control system, the dynamic characteristics of welding process must be recognized, and the correlation between weld pool geometry and welding parameters must be established. Step function is used as the input to derive the transfer function. Roughly, welding process is considered as first-order dynamical system with a steady structure, so identification concentrates on determining the model parameters.

In pulsed GTAW, a skilled operator can make the perfect weld by regulating welding parameters such as pulse duty ratio and welding velocity. Therefore, pulse duty ratio (δ)•, welding velocity (V_w), peak current (I_p) and wire feeding velocity (V_f) are adopted as the input signals designed with step function for identifying the transfer functions among welding parameters and the topside and backside geometry parameters. Through experiment data, transient response of weld pool sizes with welding parameters are derived using the algorithm of the so-called area method developed with the Matlab software.

The designed step function and results are shown in Figs. 4.2 and 4.3. All the model parameters of the topside and backside geometry parameters are identified.

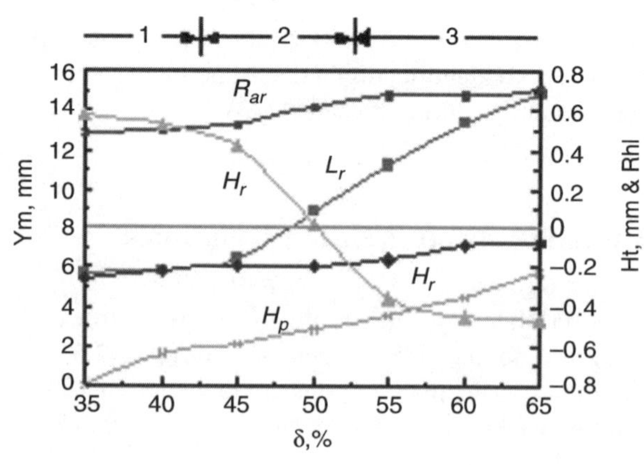

Fig. 4.1 The variation of shape parameters of weld pool

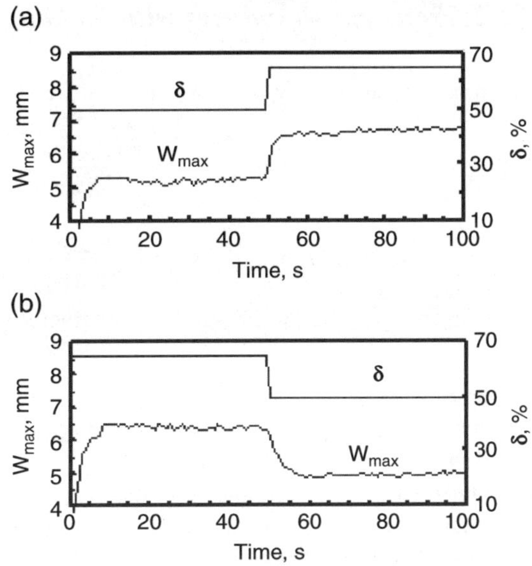

Fig. 4.2 Transient response of backside width with pulse duty ratio (**a**) Positive step (**b**) negative step

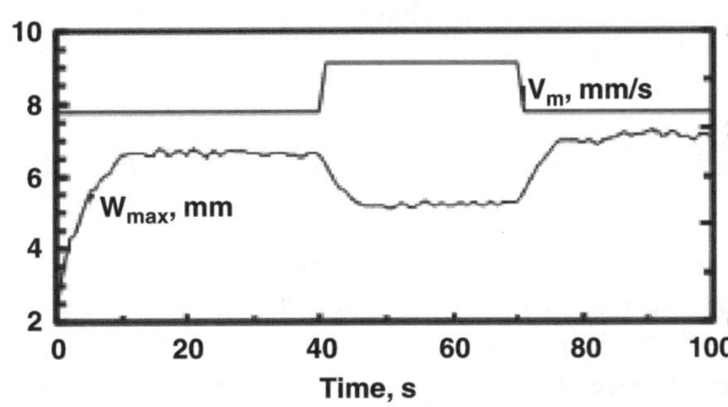

Fig. 4.3 Transient response of backside width with welding velocity

4.1.1 Transient Responses with Pulse Duty Ratio Step Changes

The transfer functions of pulse duty ratio are shown as follows:

$$G_{W_{b\max}}(s) = \frac{W_{b\max}(s)}{\delta(s)} = \frac{0.101}{2.068s+1} \tag{4.1}$$

$$G_{W_{b\max}}(s) = \frac{W_{b\max}(s)}{\delta(s)} = \frac{0.097}{2.785s+1} \tag{4.2}$$

4.1.2 Transient Responses with Welding Velocity Step Changes

The transfer functions of welding velocity are shown as follows:

$$G_{W_{b\max}}(s) = \frac{W_{b\max}(s)}{V_w(s)} = \frac{-2.911}{3.328s+1} \tag{4.3}$$

$$G_{W_{b\max}}(s) = \frac{W_{b\max}(s)}{V_w(s)} = \frac{-3.242}{4.108s+1} \tag{4.4}$$

From the identification results, the characteristic of welding process can be derived as follows:

(1) During the initiation period, the three backside parameters responds with V_w slowly than with δ. This shows that $\bullet\delta$ should be thought high of controller design during the initiation period.
(2) During steady period, the responses of negative and positive step are different, indicating the variable and nonlinear of the process.
(3) From the above explanation, it can be concluded that the affects on geometry with W_b and V_w are not separated, but strong coupled with each other.
(4) Topside parameters has some relationship with backside parameters of the weld pool, thus a predictive model with topside parameters as its input signals can be used to predict the backside parameters.

4.1.3 Transient Responses with Peak Current Step Changes

The transient responses of topside and backside width with peak value current positive and negative step changes are shown as Fig. 4.4.

According to the above step responses, the following transfer function model is adopted for investigating regulated characteristic of weld pool under welding current changes.

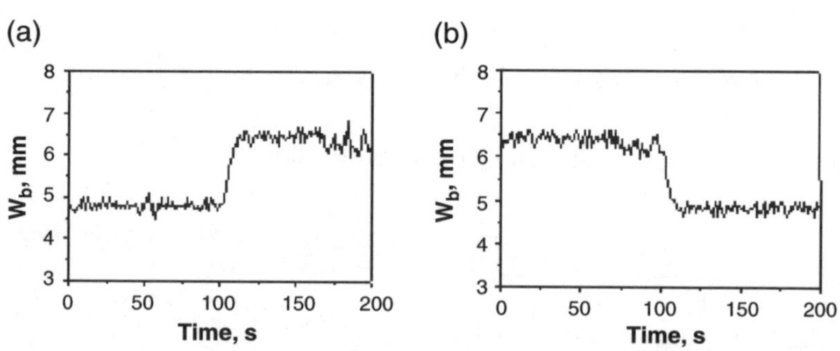

Fig. 4.4 Transient response of backside width with welding current step (**a**) Positive step response (**b**) Negative step response

$$G_{W_{b\max}}(s) = \frac{W_{b\max}(s)}{I_p(s)} = \frac{0.168}{3.972s+1}e^{-0.54s} \tag{4.5}$$

$$G_{W_{b\max}}(s) = \frac{W_{b\max}(s)}{I_p(s)} = \frac{-0.179}{1.239s+1} \tag{4.6}$$

From the above experiment and identification results, we come to the conclusions as follows:

(1) Under the positive step current input, the dynamic response of the backside width W_b has a evident time delay, but it is almost no any time delay under the negative step current input, which says that the weld pool is more sensitive to reducing current heat input due to larger coefficient of heat conduction and faster heat dispersion.
(2) Under the negative step current input, the dynamic response of the backside width W_b has a lesser response time constant than that under the positive step current input due to the above similar cause in *b*.

4.1.4 Transient Responses with Wire Feeding Velocity Step Changes

Wire feeding velocity is a direct and key variable for regulating the shaped features of the weld seam. The experiment conditions are determined as following:

The transient responses of topside and backside width of the weld pool dynamics with wire feeding speed positive and negative step changes are shown as Fig. 4.5.

According to the above step responses, the transfer function model is also adopted for investigating regulated characteristic of weld pool under wire feeding speed changes as follows.

(1) Positive step in steady welding period.

$$G_{W_{b\max}}(s) = \frac{W_{b\max}(s)}{V_f(s)} = \frac{-0.497}{1.653s+1} \tag{4.7}$$

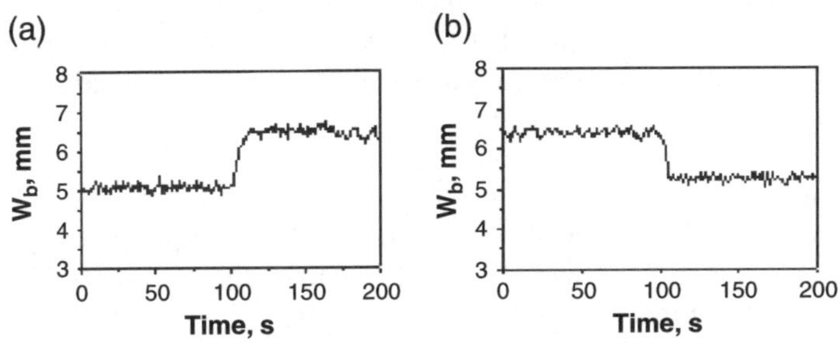

Fig. 4.5 Transient response of the backside width of weld pool with wire feeding speed step (**a**) Positive step response (**b**) Negative step response

(2) Negative step in steady welding period.

$$G_{W_{b\max}}(s) = \frac{W_{b\max}(s)}{V_f(s)} = \frac{0.462}{2.069s+1}e^{-0.54s} \tag{4.8}$$

From the above experiment and identification results, we come to the conclusions as follows,

(1) Under the positive step wire feeding speed input, the dynamic response of the backside width W_b has a lesser time constant than that under the negative step wire feeding speed, increasing wire feeding speed means decreasing corresponding heat input, which also indicates that the weld pool is more sensitive to reducing heat input due to larger coefficient of heat conduction and faster heat dispersion.
(2) Under the negative step wire feeding speed input, the dynamic response of the topside width W_b has a lesser response time constant than that under the positive step wire feeding speed input due to the above similar cause.

Moreover, the differences of the time constants and gain coefficients between the positive step and negative step wire feeding speed input responses also indicate the nonlinearity and time-variety in weld pool dynamics during stable pulse GTAW.

4.2 Identification Models of Weld Pool Dynamics

To design a suitable control system, the dynamic characteristics of welding process must be recognized, and the correlation between weld pool geometry and welding parameters must be established.

4.2.1 Linear Stochastic Models of Aluminium Alloy Weld Pool Dynamics

Linear model is used to describe welding pool dynamics because of its simplicity in form and fast in calculation. In this part a stochastic models of aluminium alloy weld pool during pulsed GTAW will be investigated. The stochastic models of aluminium alloy weld pool during pulsed GTAW are built.

Referring to the above analysis on welding dynamics, W_f and W_b of the weld pool are denoted as outputs of the welding, welding current *I(k)* and wire feeding speed $V_f(k)$ as the inputs, and time-delay of the system is *d*, the orders of the model are *n* and *m*. The model structure is shown in Fig. 4.6 for predicting and control of aluminium alloy welding dynamics.

Where the *u(k)* is input vector, which can be defined by welding current *I(k)* and wire feeding speed $V_f(k)$, *y(k)* is output vector, which can be defined by the topside

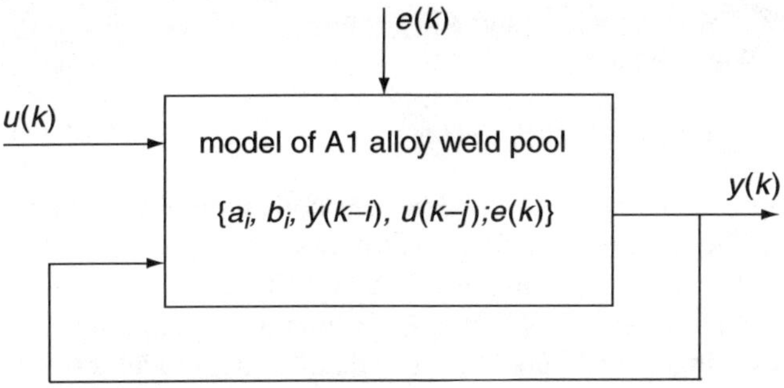

Fig. 4.6 Stochastic model for Al alloy weld pool during pulsed GTAW

width W_f (k) and the backside width W_b (k) of the weld pool, *e(k)* is a noise in the welding process, $\{a_i, b_j\}$ is mapping parameters in the model, which should be identified in real time, *k* is a discrete time variable, and the *u(k-i)* and *y(k-j)* are the *i*th and *j*th time delay for *u(k)* and *y(k)*.

Now we have the following characteristic models for aluminium alloy weld pool during pulsed GTAW.

(a) The model relation between the backside weld pool width $W_b(k)$ and topside weld pool width W_f (k):

$$W_b(k) = \sum_{i=0}^{n} a_i W_b(k-i) + \sum_{j=0}^{m} b_j W_f(k-j-d) + e(k) \tag{4.9}$$

(b) The model relation between the backside weld pool width $W_b(k)$ and welding current *I(k)*:

$$W_b(k) = \sum_{i=0}^{n} a_i W_b(k-i) + \sum_{j=0}^{m} b_j I(k-j-d) + e(k) \tag{4.10}$$

(c) The model relation between the backside weld pool width $W_b(k)$ and wire feeding speed $V_f(k)$:

$$W_b(k) = \sum_{i=0}^{n} a_i W_b(k-i) + \sum_{j=0}^{m} b_j V_f(k-j-d) + e(k) \tag{4.11}$$

The above models includes the welding parameter and state values at the historical time for describing effectiveness of heat inertia and parameter changing, where ***e(k)*** is process noise, for simplify, ***e(k)*** is defined as the white noise. The models (4.9), (4.10), (4.11) can be unified as a discrete formal equation for identification algorithms:

$$\begin{aligned} y(k) = & -a_1 y(k-1) - \cdots - a_n y(k-n) + b_0 u(k-1) \\ & + b_1 u(k-2) + \ldots b_m u(k-m-1) + e(k) \end{aligned} \tag{4.12}$$

Defined $\theta = [a_1, a_2, \ldots a_n, b_0, b_1, \ldots b_m]^T$, and $\phi(k) = [-y(k-1), \cdots - y(k-n), u(k-1), \ldots, u(k-m-1)]$, then we have

$$y(k) = \phi^T(k)\theta + e(k) \tag{4.13}$$

In order to identify the parameters in the models, the corresponding welding experiments are conducted. The welding conditions as follows: The pulse current is 230 A and variable range at 210–250 A, welding welding velocity is 3.3 mm/s, wire feeding speed is 12 mm/s changing range at 4.26 mm/s. In the total 12 experiments, 500 data is obtained for modeling of weld pool and 100 data for the model testing. Testing signal is designed an irrelevant random serial instead of ideal white noise for exciting dynamic process of aluminium alloy weld pool during pulse GTAW.

The Fig. 4.7 is a stochastic input signal of welding current for testing aluminium alloy welding, and Fig. 4.8 are the output signals of the process corresponding to welding current exciting input. The Fig. 4.9 is a stochastic input signal of wire feeding speed for testing aluminium alloy welding, and Fig. 4.10 are the output signals

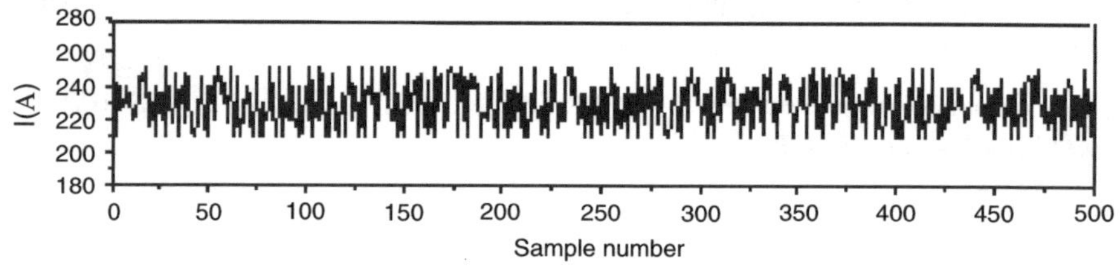

Fig. 4.7 The stochastic input signal of the welding current

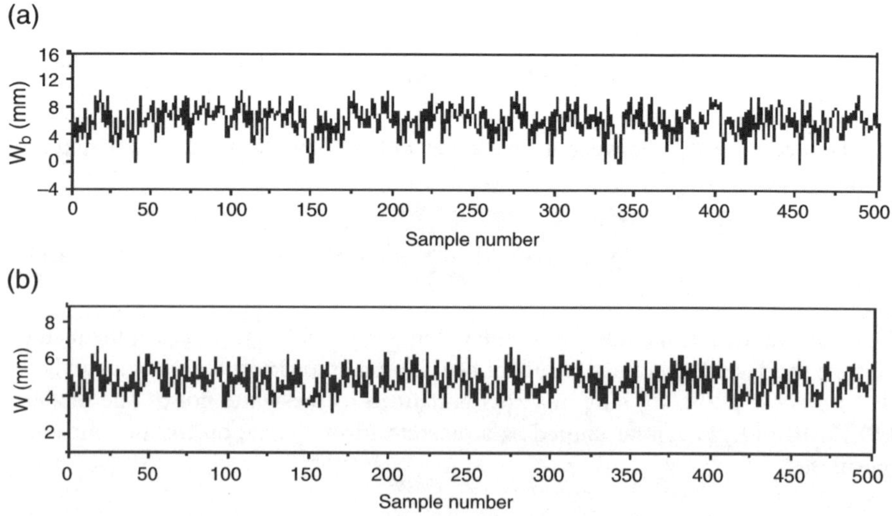

Fig. 4.8 Al alloy weld pool characteristics under the stochastic current input (**a**) The backside width under the stochastic welding current (**b**) The topside width under the stochastic welding current

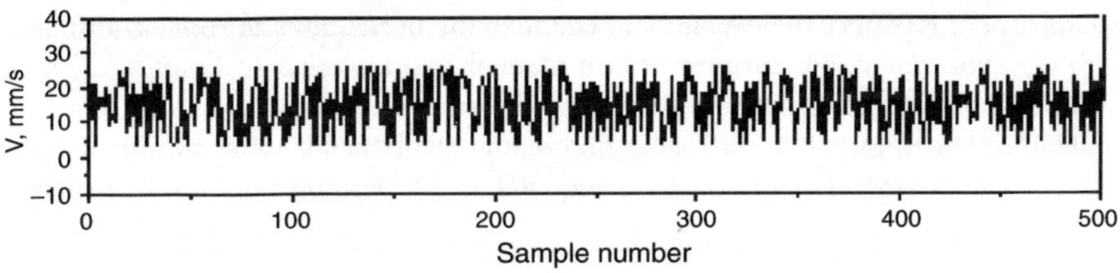

Fig. 4.9 The stochastic input signal of the wire feeding speed

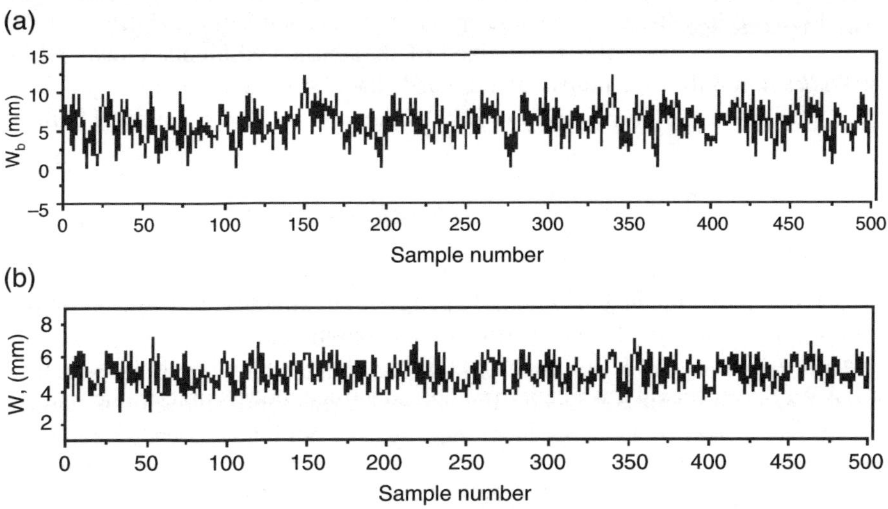

Fig. 4.10 Al alloy weld pool characteristics under the stochastic wire feeding speed input (**a**) The backside width under the stochastic wire feeding speed (**b**) The topside width under the stochastic wire feeding speed

of the process corresponding to stochastic wire feeding speed exciting input. The characteristics of the weld pool are obtained by the visual sensing and image processing algorithms in Ref. [9].

Based on the acquired data from the testing weld pool dynamics, the least square algorithm is utilized to identify the parameters in the model (4.9), the structure parameters d, n, m are determined by the FPE criterion (final predict error), the model identification are completed by the ARX in MATLAB6.1, the algorithm details are omitted here. We have obtained the identified results for dynamic models of aluminium alloy weld pool during pulsed GTAW as following:

(a) The model BWTWC for the relation between the backside weld pool width and topside weld pool width with welding current regulation:

$$\begin{aligned} W_b(k) &= a_1 W_b(k-1) + a_2 W_b(k-2) + a_3 W_b(k-3) + a_4 W_b(k-4) \\ &\quad + b_1 W_f(k-1) + b_2 W_f(k-2) + b_3 W_f(k-3) + e(k) \end{aligned} \tag{4.14}$$

$$\hat{\theta}(0) = [0.09286, 0.06695, -0.07916, 0.002295, 3.387, -0.3545, -0.2812]$$

(b) The model BWPPC for the relation between the backside weld pool width and peak value of welding current

$$\begin{aligned} W_b(k) = {} & a_1 W_b(k-1) + a_2 W_b(k-2) + a_3 W_b(k-3) + a_4 W_b(k-4) \\ & + b_1 I(k-1) + b_2 I(k-2) + b_3 I(k-3) + b_4 I(k-4) + e(k) \end{aligned} \tag{4.15}$$

$$\begin{aligned} \hat{\theta}(0) = {} & [0.2608, 0.0478, -0.0745, 0.03611, 0.1911, -0.02017, \\ & 0.003491, 0.009843] \end{aligned}$$

(c) The model BWWFS for the relation between the backside weld pool width and wire feeding speed

$$\begin{aligned} W_b(k) = {} & a_1 W_b(k-1) + a_2 W_b(k-2) + a_3 W_b(k-3) + a_4 W_b(k-4) \\ & + b_0 V_f(k-2) + b_1 V_f(k-3) + b_2 V_f(k-4) + b_3 V_f(k-5) + e(k) \end{aligned} \tag{4.16}$$

$$\begin{aligned} \hat{\theta}(0) = {} & [0.2238, 0.0588, -0.03576, -0.001274, -0.3739, 0.002593, \\ & 0.00796, 0.00997] \end{aligned}$$

In practical welding, the backside weld pool can't be observed directly, the above models can be used to predict changes of weld pool on line.

Figures 4.11, 4.12 and 4.13 show testing results of the models BWTWC, BWPPC and BWWFS, respectively, which indicate the identified models practicable.

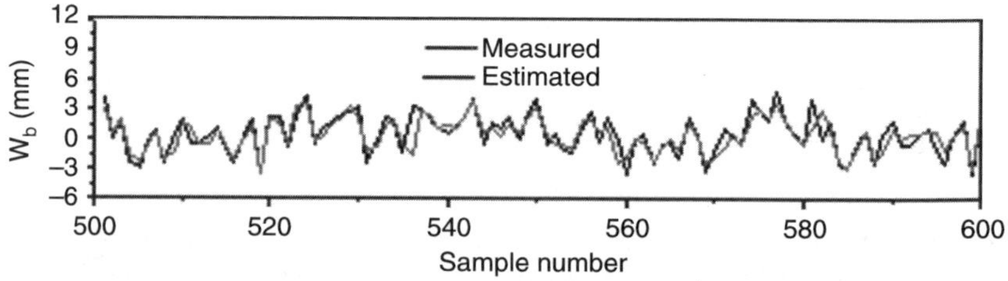

Fig. 4.11 The test result of BWTWC model

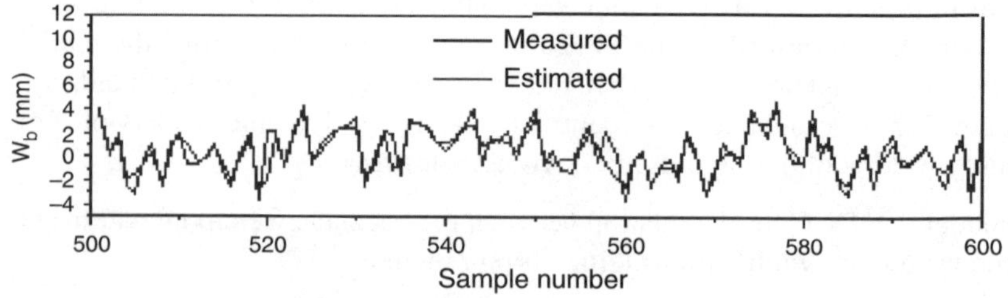

Fig. 4.12 The testing result of BWPPC model

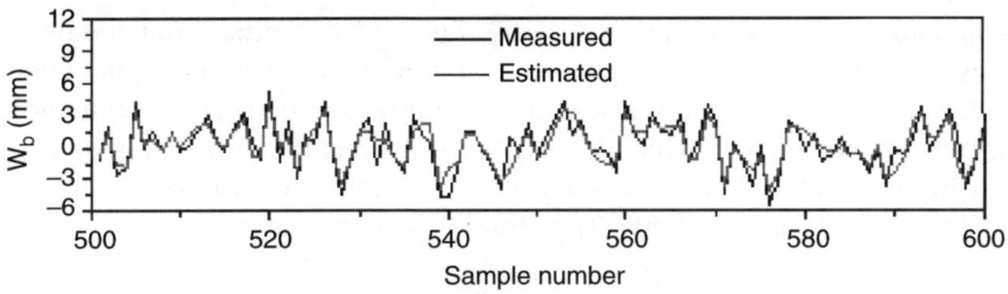

Fig. 4.13 The test result of BWWFS model

4.2.2 Nonlinear Models of Low Carbon Steel Weld Pool Dynamics

GTAW is a complicated thermal history with many uncertainties during the workpiece is heated, melts, solidifies and finally forms the welding seam. This process includes the complex nonlinear sections as discussed in the following part.

4.2.2.1 Analysis on Heat Source Model in Arc Welding

The quantity of arc heat effect during welding is usually described by Gaussian distribution model [10]. Thus the heat transfer rate at Point A away from the heating center is expressed as follows

$$q(r) = q_m \exp\left(-\frac{3r^2}{\overline{R}^2}\right) \tag{4.17}$$

where q_m is the largest heat flux density at the heating center, $\overline{R}$ is effective heating radius, mainly determined by the welding procedure and welding condition, r is the distance from Point A to the heating center.

Equation (4.17) shows that arc emitting heat is a nonlinear process, and independent of time. Thus the heating process can be described as a static and nonlinear function.

4.2.2.2 Analysis on Thermal Conduction Model in Arc Welding

Thermal conduction differential equation during welding process is established based on Fourier Formula and energy conservation law, shown as follows,

$$\frac{\partial T}{\partial t} = \frac{\lambda}{c\rho}\left(\frac{\partial^2 T}{\partial x^2} + \frac{\partial^2 T}{\partial y^2} + \frac{\partial^2 T}{\partial z^2}\right) = a\nabla^2 T \tag{4.18}$$

Here, the boundary condition includes initial temperature and thermal exchange condition on the surface of the workpiece. For the simplicity, the workpiece is assumed as temperature uniformed before welding and the initial temperature is assumed as constant. The thermal exchange condition on the surface is quite complex in actual welding process because a great number factors (such as ventilation in the surrounding and contact with clamping fixture) affect the distribution of temperature field in the workpiece. In this paper, the workpiece is assumed as infinite sheet metal for the sake of simplicity. Therefore, (4.18) becomes

$$\frac{\partial T}{\partial t} = \frac{\lambda}{c\rho}\left(\frac{\partial^2 T}{\partial x^2} + \frac{\partial^2 T}{\partial y^2}\right) - bT \tag{4.19}$$

Equation (4.19) shows that the process of thermal conduction is dynamic and linear with time variable in a fixed-point.

According to the heat source model and thermal conduction model in arc welding, conclusion can be made that welding process approximates the combination of two process, namely arc heating and thermal conduction, in which heating can be described by a static, nonlinear formula, and thermal conduction described by a dynamic, linear formula with time variable. Therefore, the welding process can be approximatively described as a combination of a static nonlinear process and a dynamic linear process, which can be expressed by the so-called Hammerstein Model, as shown in Fig. 4.14.
where $L = q^{-d}$ is pure delay segment, $f(t) = r_1 u + r_2 u^2 + \ldots + r_p u^p$ is time independent nonlinear segment, $\frac{B(z)}{A(z)} = \frac{b_0 + b_1 q^{-1} + b_2 q^{-2} + \ldots + b_m q^{-m}}{1 + a_1 q^{-1} + a_2 q^{-2} + \ldots + a_n q^{-n}}$ is time dependent linear segment, $e(t)$ is white noise with its mean value of zero, $u(t)$ is the input of the model.

The expression of Hammerstein model described in Fig. 4.14 in time domain is described as follows:

$$A(q^{-1})y(k) = B(q^{-1})\sum_{i=1}^{p} r_i u^i(k-d) + C(q^{-1})e(k) \tag{4.20}$$

Where

$$A(q^{-1}) = 1 + a_1 q^{-1} + a_2 q^{-2} + \cdots + a_n q^{-n}$$
$$B(q^{-1}) = 1 + b_1 q^{-1} + b_2 q^{-2} + \cdots + b_m q^{-m}$$
$$C(q^{-1}) = 1 + c_1 q^{-1} + c_2 q^{-2} + \cdots + c_k q^{-k}$$

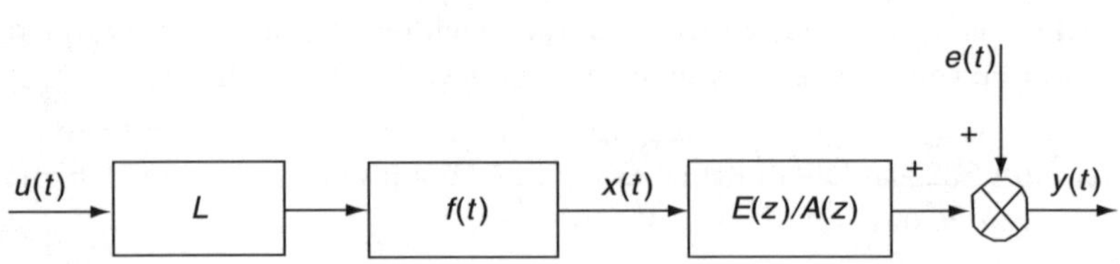

Fig. 4.14 Hammerstein model

Equation (4.20) can be rewritten as

$$y(k) = \varphi(k)\theta + e(k) \tag{4.21}$$

where

$$\begin{aligned}\varphi(k) = [&-y(k-1),...,-y(k-n) \mid u(k-d),...,u(k-m-d) \mid \\ &u^2(k-d),...,u^p(k-m-d) \mid ... \mid u^p(k-d),...,u^p(k-n-d)]\end{aligned}$$

is the vector of measured input-output values

$\theta = [a_1,...,a_n \mid b_0,...,b_n \mid b_0r_2,...,b_mr_2 \mid ... \mid b_0r_p,...,b_mr_p]^T$ is the vector of model parameters to be identified.

$C(q^{-1}) = 1$ because e(k) is white noise.

The structure parameters of the model are the order in the model's linear part: m and n, the order in the model's nonlinear part: p and the pure delay parameter: d, which are trained by the sequence of the input-output samples off-line. The training results are: d = 1, p = 2, m = 3, n = 3.

Let $y = [y(1), y(2), \ldots y(N)]^T$, $\Phi = [\phi^T(1), \phi^T(2), \ldots \phi^T(N)]^T$, $e = [e(1), e(2), \ldots, e(N)]^T$, the index of model error is defined as follows:

$$J = (y - \phi^T(t)\overline{\theta})^T(y - \phi^T(t)\overline{\theta}) \tag{4.22}$$

Where y is the measured value, $\phi^T(t)\overline{\theta}$ is the calculated value and $\overline{\theta}$ is the estimated value of θ. The model is then identified by the recursive least square (RLS) method on line. The recurrence expression shown as follows

$$\hat{\theta}_{N+1} = \hat{\theta}_N + P_N\varphi_{N+1}^T(1 + \varphi_{N+1}P_N\varphi_{N+1}^T)^{-1}\left(y(k_0+N+1) - \varphi_{N+1}\hat{\theta}_N\right) \tag{4.23}$$

$$P_{N+1} = P_N - P_N\varphi_{N+1}^T(1 + \varphi_{N+1}P_N\varphi_{N+1}^T)^{-1}\varphi_{N+1}P_N$$

$$\begin{cases} \hat{\theta}_{N+1} = \hat{\theta}_N + K(N+1)[y(k_0+N+1) - \varphi_{N+1}\hat{\theta}_N] \\ K(N+1) = P_N\varphi_{N+1}^T(1 + \varphi_{N+1}P_N\varphi_{N+1}^T)^{-1} \\ P_{N+1} = [I - K(N+1)\varphi_{N+1}]P_N \end{cases}$$

The initial values and stop threshold are set as follow,

$$\hat{\theta}_0 = 0$$

$$\mathrm{P}_0 = \alpha^2\mathrm{I} = \begin{pmatrix} a^2 & & 0 \\ & \ddots & \\ 0 & & a^2 \end{pmatrix}, 10^5 \le \alpha^2 \le 10^{10}$$

$$\max\left|\frac{\hat{\theta}_{N+1} - \hat{\theta}_N}{\hat{\theta}_N}\right| < \varepsilon \qquad \varepsilon = 0.00001$$

The result of the structure and parameter identification of the I_p-W_b Model (BWHM) is expressed as follows,

The structure parameters are $d = 1, p = 2, m = 3, n = 3$

$$r_1 = 1.0, r_2 = 0.0194, a_1 = -1.1359, a_2 = 0.2700, a_3 = 0.0273$$
$$b_0 = 0.0041, b_1 = 0.0023, b_2 = -0.0009, b_3 = 0.0004$$

$$\begin{aligned} y(k) &= -a_1 y(k-1) - a_2 y(k-2) - a_3 y(k-3) + \\ &b_0 u(k-1) + b_1 u(k-2) + b_2 u(k-3) + b_3 u(k-4) + \\ &r_2\left(b_0 u^2(k-1) + b_1 u^2(k-2) + b_2 u^2(k-3) + b_3 u^2(k-4)\right) \end{aligned} \tag{4.24}$$

where y is W_b, and u is I_p

Figure 4.15 shows the test result of BWHM model

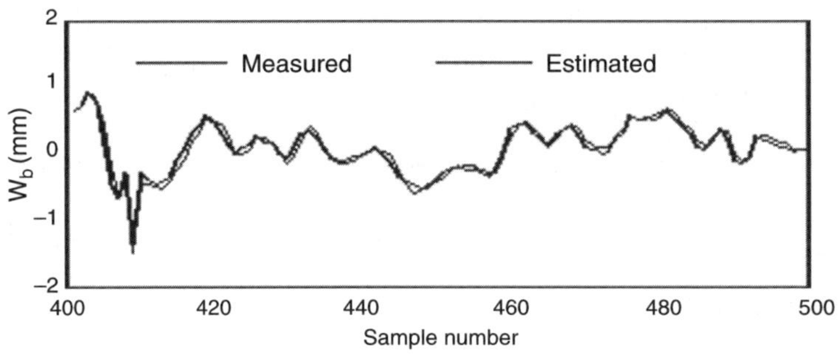

Fig. 4.15 The test result of BWHM model

4.3 Artificial Neural Network Models of Weld Pool Dynamics

Since artificial neural networks (ANN) is able to preferably model complex non-linearity and uncertainty of unknown object, it is suitable to describe weld pool dynamics. Generally, ANN model is to directly learn the input/output data from the system experiment. The index model convergence is called cost function, which need to be minimized.

Identification modeling process by ANN usually includes designing exciting persist signal determine model structure, choosing identifying algorithms, training the model and verifying model precise.

As an important indicator of welding penetration, the backside width is not visible in practical welding. Therefore, it needs to be estimated by topside weld pool information. The correlation among welding parameters, topside geometry and backside geometry should be established.

4.3.1 BWHDNNM Model for Predicting Backside Width and Topside Height During Butt Pulsed GTAW

In this section, Backside Width and Topside Height Dynamic Neural Network Model(BWHDNNM) is established for mapping the correlations between welding parameters and weld pool dynamics. Variations in weld pool and penetration are generated by changing welding parameters. Based on the analysis on the welding process, peak current (I_p), pulse duty ratio(δ), travel speed (V_w) and wire filling velocity (V_f) are selected as the input signal for exciting the characteristics of welding process. According to the theory of system identification, random signals are considered as the input exciting signals for identifying model of the welding process, partly shown in Fig. 4.16.

Other welding parameters are considered in the identification model. The weld pool size parameters except the topside height H_t are measured in real-time during the experiments with the double-side visual sensing system. Topside height H_t is calculated off line by the method in [11], H_t in the closed system is predicted by the neural model. All 30 separate experiments have been carried out to produce 2500 input-output samples. The first 5 data points of each experiment are omitted to avoid the effect of transition process during initiation period.

The back propagation (BP) neural network provides a uniform model frame for almost all types of nonlinear models. In this study, a model BWHDNNM was established for mapping the correlations between welding parameters and weld pool dynamics, shown as Fig. 4.17.

Welding parameters, such as pulse peak current I_p, pulse duty ratio δ, gap g, travel speed V_w and wire filling velocity V_f are the major factors to affect the heat input, which are included in the model inputs. There is a time delay in the response of the size parameters of the welding pool due to heat inertia, the time delay parameters was approximately determined by the step response experiments on the welding system. For this reason, the history information should be included. i.e, $I_p(t)$ means the peak value of current pulse, $I_p\ (t-1)$ means the value of last pulse, $I_p\ (t-2)$ means the value of the last before last pulse. The modeling principle for weld pool by neural networks is indicated as Fig. 4.18. The TSP in Fig. 4.18 is time–delay signal processor. The learning algorithm of modeling weld pool dynamics could be realized in on-line or off-line.

For most applications, one hidden layer in ANN model is sufficient, and the number of nodes in the hidden layer is selected by the principle of minimum mean square error(RMS). In the model BWHDNNM, an appropriate number in hidden layer nodes is determined as 10 by the minimum mean square error principle.

The training of the neural model is performed by the neural network software in the MATLAB tools. The sigmoid function is selected as the nonlinear function of the neuron. The algorithm Levenberg-Marquardt is used to overcome the slow convergence associated with the conventional back propagation algorithm. The learning coefficients and momentum ratios are automatically determined by the algorithm in each training cycles.

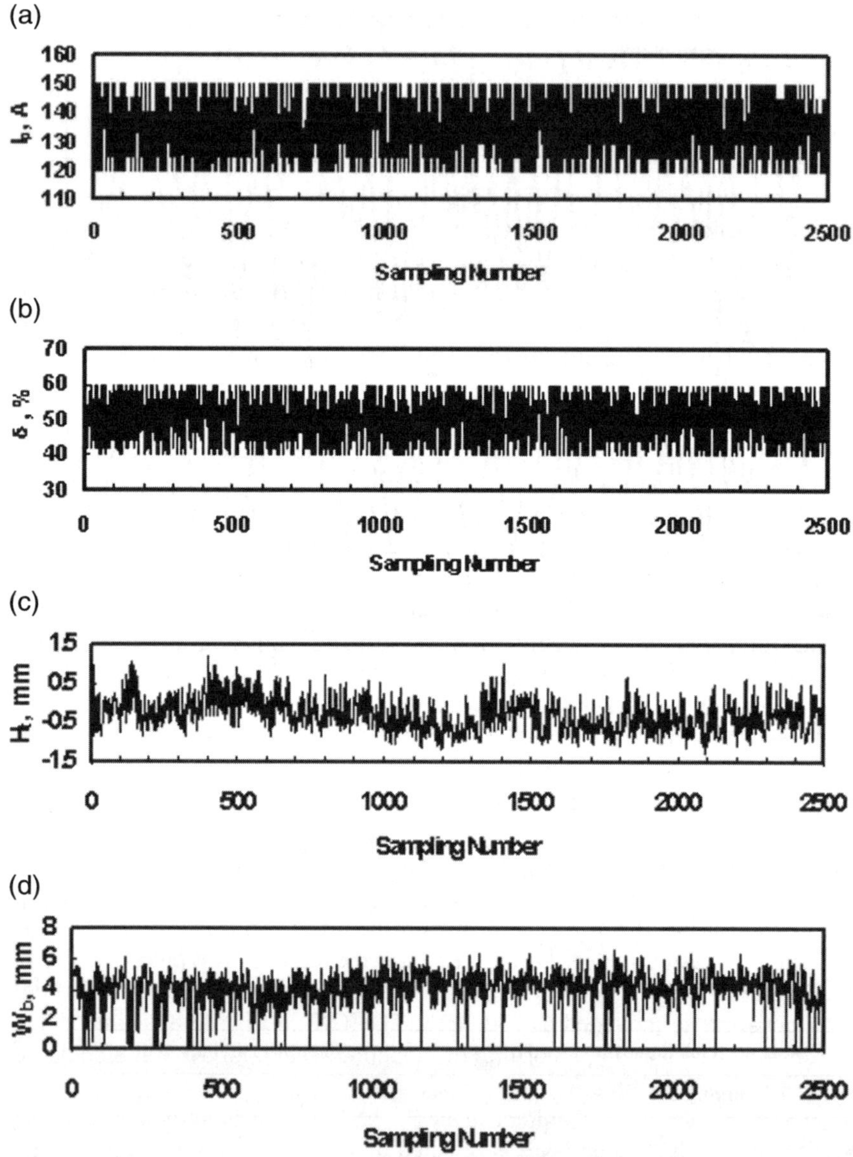

Fig. 4.16 Input random signals and measured shape parameters of weld pool dynamics (**a**) Peak current (**b**) Pulse duty ratio (**c**) Topside height H_t (**d**) Backside width W_b

The RMS of the model BWHDNNM decreased to attain the stable value with the increase of training number. 200 samples without training were used to verify the accuracy results as following: the RMS of W_b is 4.63%, the RMS of H_t is 6.7%, shown as Fig. 4.19.

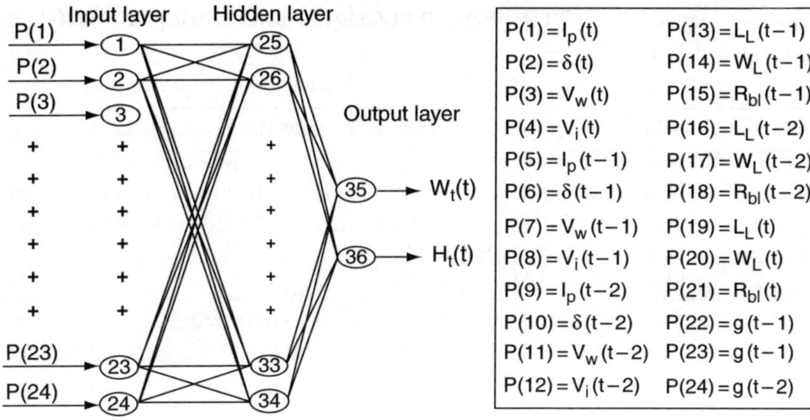

Fig. 4.17 The architecture of neural network dynamic model BWHDNNM

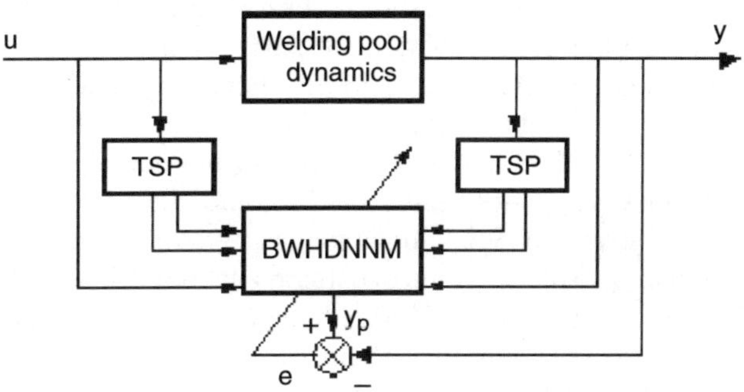

Fig. 4.18 The principle of modeling weld pool with neural network

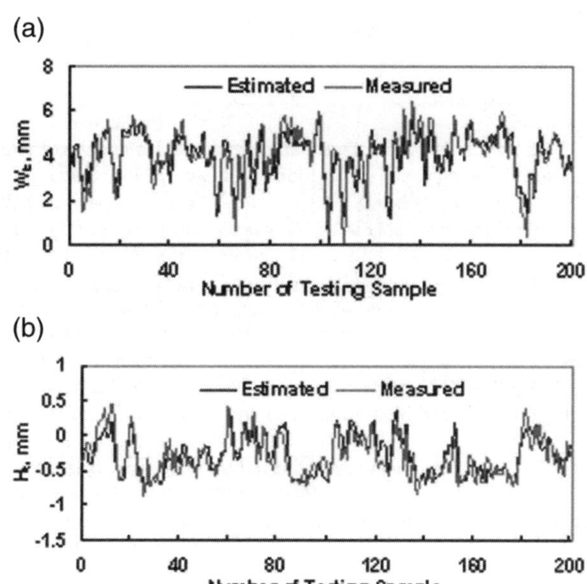

Fig. 4.19 Testing results of BWHDNNM

4.3.2 BNNM Model for Predicting Backside Width During Butt Pulsed GTAW

Backside Neural Network Model (BNNM) is built to map the correlations between welding parameters and weld pool dynamics. δ and V_w are selected as the input signals for exciting the characteristic of welding, designed in Fig. 4.20. According to the theory of system identification, random and step signal are considered as the best input signals as to the welding. Other welding parameters such as peak current, base current, etc. are considered in the identification model.

Experiment conditions are the same in Table 4.1. 24 experiments have been carried out to receive 2350 input samples. The first ten data of each experiment are

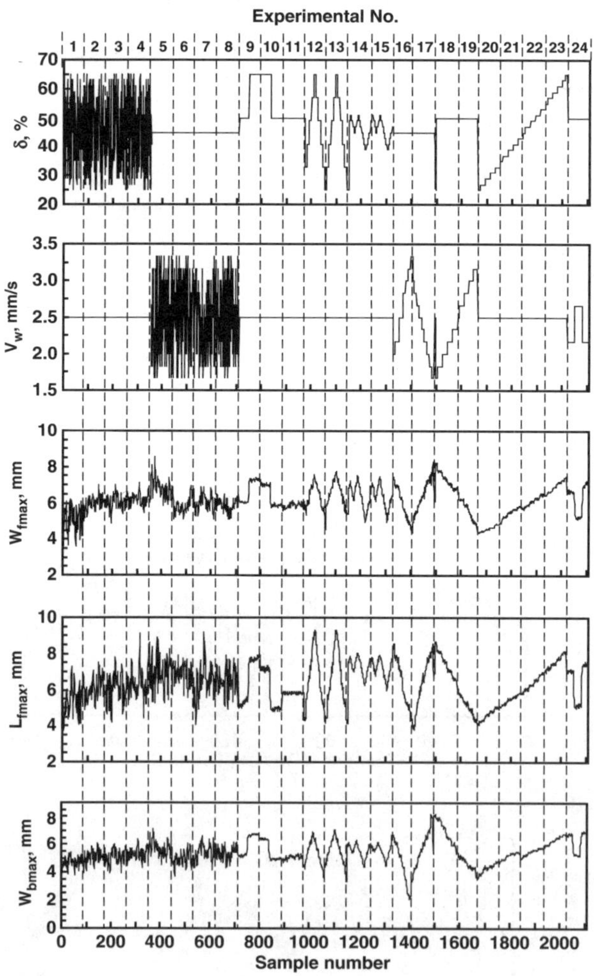

Fig. 4.20 Experimental input signal and resultant considered in experiments

omitted to avoid the effect of transition process during initiation period, so that 2110 samples are left behind. The experiment results are also seen in Fig. 4.20. The data are arranged according to their serial number in the experiments.

It can be seen that the backside maximum width of weld pool varies in a wide range from 2 to 7.5 mm. The topside maximum width varies in a range from 3.5 to 8.5 mm and the topside maximum half-length is from 3.5 to 9.5 mm correspondingly. The variations are caused by the different welding parameters or conditions.

Back propagation (BP) neural network provides a uniform model frame for almost all types of nonlinear model. The actual inputs and outputs are taken as the training samples for the determination of neurons' weight with back propagation algorithm.

Welding parameters, such as pulse duty ratio, peak current, base current, arc voltage, and welding speed are the major factors to affect the heat input, which are included in the model inputs. Because the heat inertia of welding, size parameters respond to welding parameters with time delay, the history information should be included. For example, $V_w\ (t)$ means the value of current pulse, $V_w\ (t-1)$ means the value of last pulse, $V_w(t-2)$ means the value of last before last pulse.

The ANN model principle is shown in Fig. 4.21, where u is actual input variables of the system, y_t is actual output, y_t^m is the output of the model. The error e_t is used for adjusting neuron weight by training off-line.

In BNNM, the model includes 17 inputs, 24 hide layer units, the output of the model, $W_{b\max}$, for the back maximum width of the weld pool. The examined results of the model is shown as Fig. 4.22. The average relative error of the model output $W_{b\max}$ was 3.32%, mean-square error was 4.01%.

4.3.3 BHDNNM Model for Predicting Backside Width and Topside Height During Butt Pulsed GTAW Based on Three-Dimensional Image Processing

Based on extracting three-dimensional characters of the weld pool images, the relationship among the top shape and height parameters and back width of the weld pool with wire filler is modeled by the Backside and Height Dynamic Neural Network Model(BHDNNM), as shown in Fig. 4.23. The model includes the current values of the top pool shape and welding parameters, such as welding current or pulse duty ratio , filling rate, travel speed, and their former moment values, 21 input variables, 5 hide nodes.

With the trained BHDNNM and inputs of the model, backside width and topside height could be predicted, and the predicted and measured values are shown in Fig. 4.24. The results showed that mean errors of backside width and topside height were 0.004 mm and 0.042 mm respectively, and the relative mean square error were 5.54% and 7.83% respectively. The statistic results verified the validity of BHDNNM.

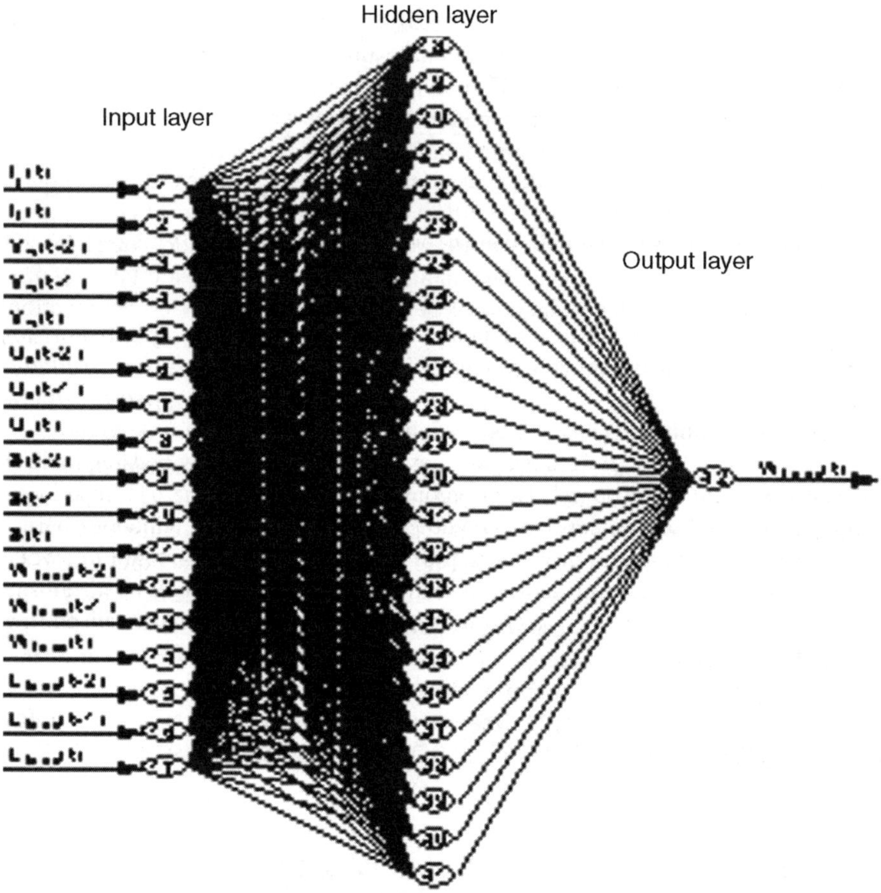

Fig. 4.21 The architecture of neural network dynamic model

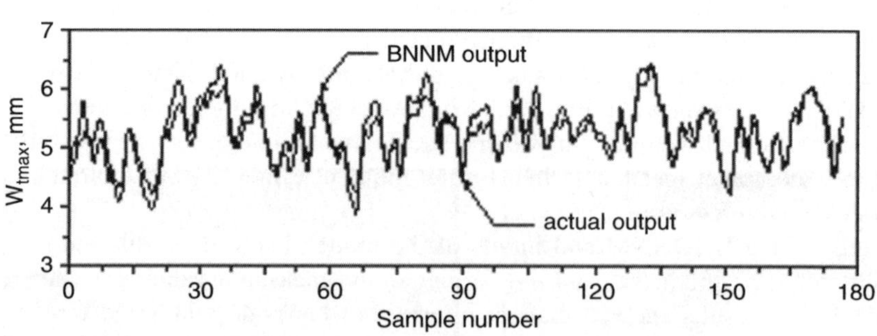

Fig. 4.22 The result of detecting BNNM model

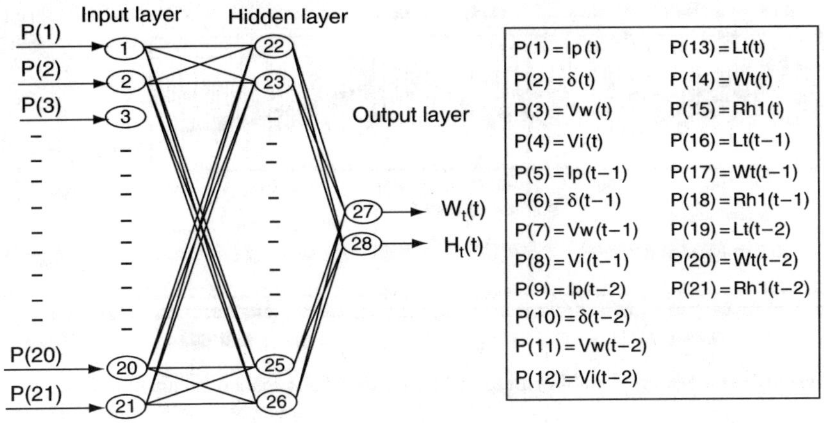

Fig. 4.23 Structure of BHDNNM

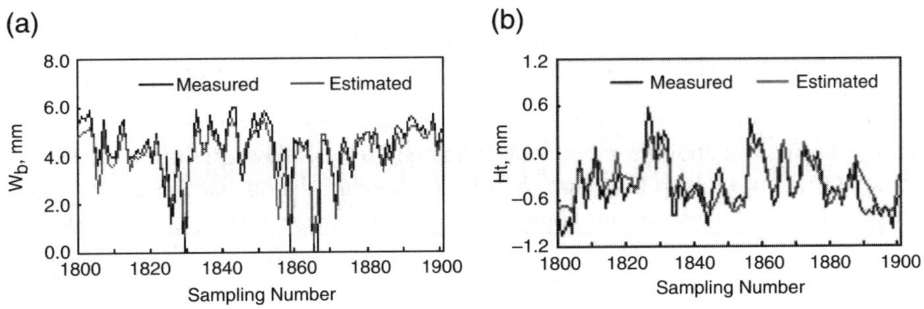

Fig. 4.24 Testing results of BHDNNM

Under the variation of heat-sinking conditions, the backside width and topside height varied irregularly, but the backside width and topside height could be predicted accurately with the developed model BHDNNM, then these predicted values can be input to the controller as feedback for control of bead shape.

4.3.4 SSNNM Model During Butt Pulsed GTAW

A size and shape neural network model (SSNNM) is established to correlate the welding parameters, topside size and shape parameters to the backside size parameters. To excite all characteristics of the dynamic welding process, the inputs are designed with white noise signals, because of the characteristic of widespread spectrum and non-correlation on time. Butt welding experiments are conducted on specimen of 2 mm thickness mild steel without any surface preparation. According to the specialist's experiences, the satisfactory weld bead geometry can be obtained

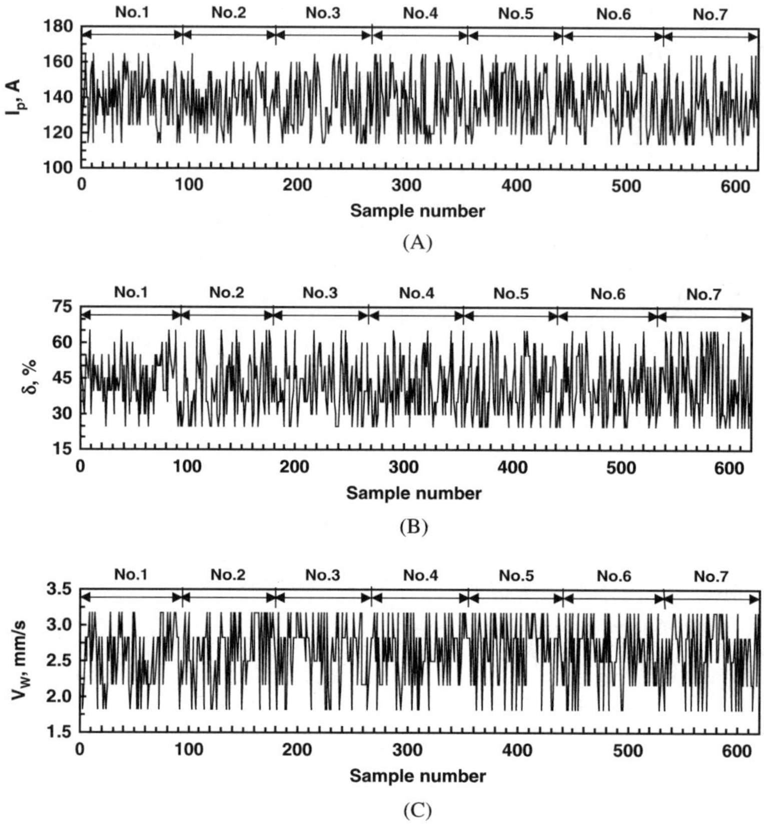

Fig. 4.25 The input signals of white noise (**a**) Pulse peak current (**b**) pulse duty ratio (**c**) welding speed

by the following welding parameters: $I_{p0} = 140\,\text{A}$, $\delta_0 = 45\%$, $V_{w0} = 2.5\,\text{mm/s}$. To receive a complete welding data, the variations of all the inputs are set as follows: $\Delta Ip = \pm 25\,\text{A}$, $\Delta\delta = \pm 20\%$, $\Delta V_w = \pm 0.67\,\text{mm/s}$, the corresponding step variations are 5 A, 5%, and 0.167 mm/s. Then the input signals are generated, totally 28 groups, the first 7 groups are shown in Fig. 4.25.

With the designed input signals welding experiments are conducted, and the size and shape parameters are sensed and extracted on-line. Figure 4.26 shows the curves of double-side size and shape parameters.

From the results, we can see that the size and shape parameters vary in a large range with the different welding conditions. The maximum of $W_{f\max}$ is 8 mm, the maximum of $L_{f\max}$ is 10 mm, and the maximum of S_{fmid} is 45 mm^2. The changes of backside size parameters are as follows: S_b is from 0 to 40 mm^2, $W_{b\max}$ is from 0 to 7.5 mm and $L_{b\max}$ if from 0 to 9 mm. Therefore, all kinds of welding status are excited and the experiment data can be thought as covering completely all the penetration situations.

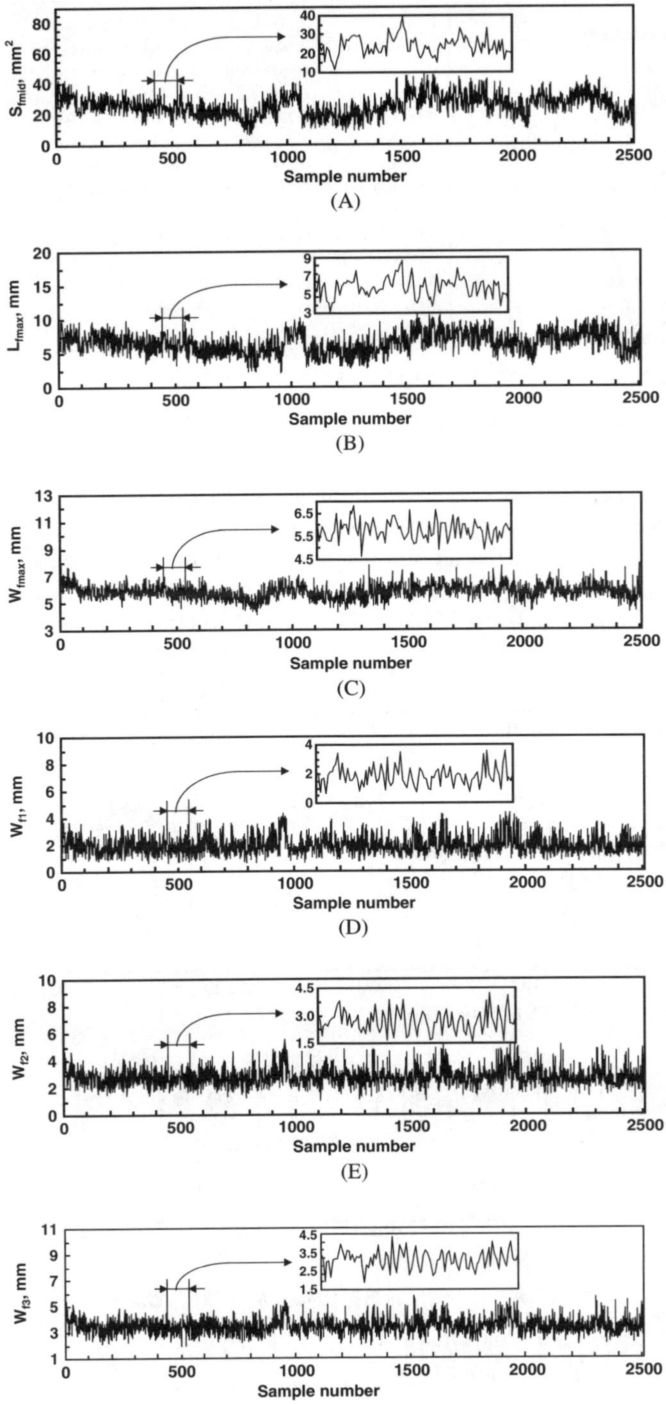

Fig. 4.26 The double-side size and shape parameters of weld pool **(a)** $S_{f\text{mid}}$ **(b)** $L_{f\max}$ **(c)** $W_{f\max}$ **(d)** S_b **(e)** $W_{b\max}$ **(f)** $L_{b\max}$

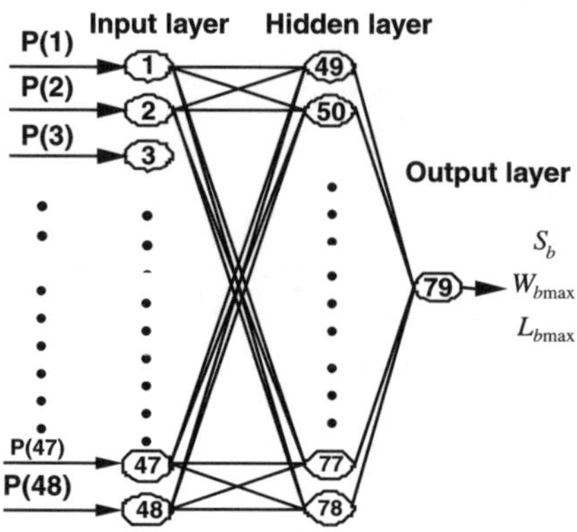

$P(1)=I_p(t-2)$	$P(17)=L_{f\max}(t-1)$	$P(33)=W_{f5}(t)$
$P(2)=I_p(t-1)$	$P(18)=L_{f\max}(t)$	$P(34)=W_{f6}(t-2)$
$P(3)=I_p(t)$	$P(19)=W_{f1}(t-2)$	$P(35)=W_{f6}(t-1)$
$P(4)=\delta(t-2)$	$P(20)=W_{f1}(t-1)$	$P(36)=W_{f6}(t)$
$P(5)=\delta(t-1)$	$P(21)=W_{f1}(t)$	$P(37)=W_{f7}(t-2)$
$P(6)=\delta(t)$	$P(22)=W_{f2}(t-2)$	$P(38)=W_{f7}(t-1)$
$P(7)=V_w(t-2)$	$P(23)=W_{f2}(t-1)$	$P(39)=W_{f7}(t)$
$P(8)=V_w(t-1)$	$P(24)=W_{f2}(t)$	$P(40)=W_{f8}(t-2)$
$P(9)=V_w(t)$	$P(25)=W_{f3}(t-2)$	$P(41)=W_{f8}(t-1)$
$P(10)=U_p(t-2)$	$P(26)=W_{f3}(t-1)$	$P(42)=W_{f8}(t)$
$P(11)=U_p(t-1)$	$P(27)=W_{f3}(t)$	$P(43)=W_{f9}(t-2)$
$P(12)=U_p(t)$	$P(28)=W_{f4}(t-2)$	$P(44)=W_{f9}(t-1)$
$P(13)=S_{fmid}(t-2)$	$P(29)=W_{f4}(t-1)$	$P(45)=W_{f9}(t)$
$P(14)=S_{fmid}(t-1)$	$P(30)=W_{f4}(t)$	$P(46)=W_{f10}(t-2)$
$P(15)=S_{fmid}(t)$	$P(31)=W_{f5}(t-2)$	$P(47)=W_{f10}(t-1)$
$P(16)=L_{f\max}(t-2)$	$P(32)=W_{f5}(t-1)$	$P(48)=W_{f10}(t)$

Fig. 4.27 The structure of SSNNM neural network model

The input variables of SSNNM are composed of the welding parameters, topside size and shape parameters and their last two history values, totally 48. The number of elements in hidden layer is selected as 30. The output variables are backside size parameters S_b, $W_{b\max}$ and $L_{b\max}$, and each of which is relating to each neural network model, totally three models established. The model structure is shown in Fig. 4.27.

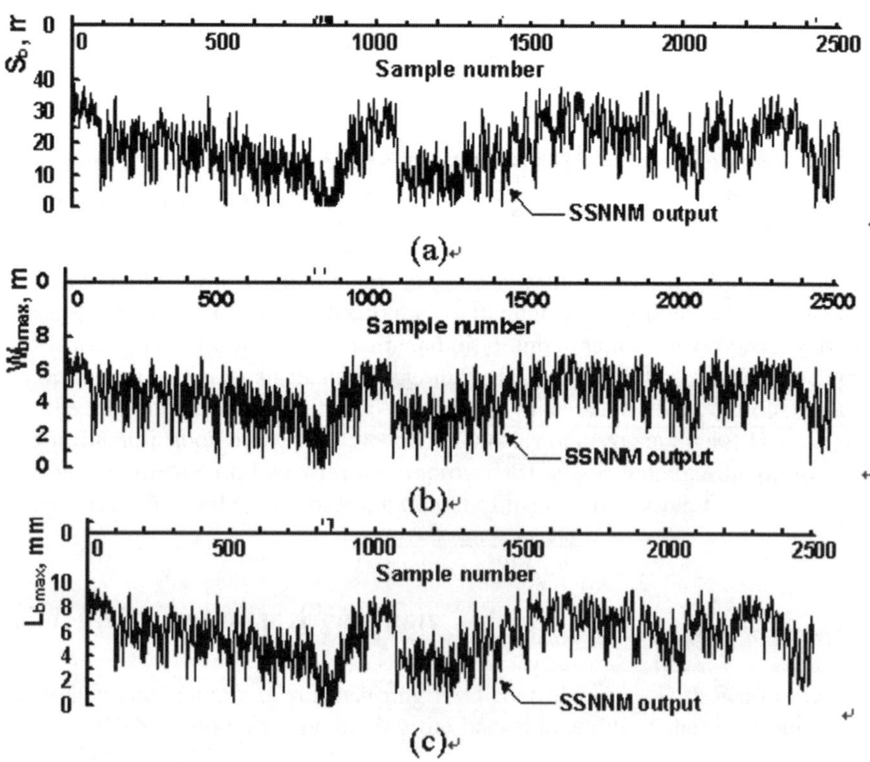

Fig. 4.28 The output of SSNNM model **(a)** S_b **(b)** $W_{b\max}$ **(c)** $L_{b\max}$

Base on the test data, the three models of S_b, $W_{b\max}$ and $L_{b\max}$ are trained. The training is performed using the commercial neural network software, Professional plus II. The sigmoidal function is selected as the nonlinear function of the neuron. The algorithm is the delta-bar-delta (DBD). The learning coefficient and momentum ratio are automatically determined by the algorithm for each 5000 training cycles. The total of training cycle is selected to be 20,000.

The outputs of SSNNM are shown in Fig. 4.28. From the comparison between outputs of the model and the test data, the statistic results can be calculated. The results indicate that SSNNM can predict the backside size parameters accurately.

4.4 Knowledge Models of Weld Pool Dynamical Process

4.4.1 Extraction of Fuzzy Rules Models of Weld Pool Dynamical Process

As is well known, the traditional welding technique is mainly relying on manual experiences, welder's operations in welding could be reduced to experiential rules which can be deduced as so-called fuzzy rules, fuzzy rules express the approximate logical relationship between the inputs and outputs of the system [12–14]. To obtain the fuzzy control rules of pulsed GTAW in butt welding, a C-mean polymerizing algorithm is used for identification of the fuzzy model of the welding. According to actual arc welding, the backside maximum width of weld pool $W_{b\max}$ is defined as controlled variable, and the pulse duty ratio δ of welding current is as regulating variable. To identify the dynamic characteristics of pulsed GTAW in butt welding, the experiments similar to the Table 4.1 conditions without wire feeder are carried out using a pseudo-random sequence of δ as the input of the actual welding system, and acquired 2109 normalizing sample data pairs $\{\delta, W_{b\max}\}$.

The error E and the change in error CE of $W_{b\max}$ are chosen as the input fuzzy variables of the fuzzy control system, and the pulse duty ratio of welding current is chosen as the output fuzzy variable, described as U of the fuzzy control system. The fuzzy subsets of these fuzzy linguistic variables can be described as $\{NB, NM, NS, ZO, PS, PM, PB\}$, where N, ZO and P are meant for negative, zero and positive, and B, M and S for big, medium and small. To extract the fuzzy control rules of pulsed gas tungsten arc butt welding, the C-mean dynamic polymerizing algorithm is described as follows:

Step 1: The initial center of each fuzzy subset of E and CE is chosen as Table 4.3. We define the polymerizing center of ith fuzzy subset of E is $ZE_i(I)$, the polymerizing center of jth fuzzy subset of CE is $ZCE_j(I)$, I means iterating number of the algorithm, where $I = 1$.
Step 2: The Euclidean distance between a sample data pair and a polymerizing center of the fuzzy subset is described as follows:

Table 4.2 The initial polymerizing centers

	NB	NM	NS	ZO	PS	PM	PB
E	−0.95	−0.75	−0.15	0.00	0.25	0.65	0.85
CE	−0.80	−0.60	−0.25	0.00	0.30	0.60	0.90

Table 4.3 The final polymerizing centers

	NB	NM	NS	ZO	PS	PM	PB
E	−0.80	−0.54	−0.29	−0.03	0.21	0.50	0.80
CE	−0.85	−0.58	−0.32	0.04	0.30	0.55	0.82

$$D(e_k, ZE_i(I), ce_k, ZCE_j(I)) = \sqrt{(e_k - ZE_i(I))^2 + (ce_k - ZCE_j(I))^2} \tag{4.25}$$

Where e_k denotes k^{th} sample value of the error, ce^k denotes k^{th} sample value of the change in error. Calculating the Euclidean distance between each sample data pair and each polymerizing center of the fuzzy subset of E and CE, if

$$D(e_k, ZE_l(I), ce_k, ZCE_m(I)) = \min_{\substack{i=1,\dots,7 \\ j=1,\dots,7}} \left\{ D(e_k, ZE_i(I), ce_k, ZCE_j(I)) \right\} \tag{4.26}$$

then e_k belongs to ZE_l, and ce_k belongs to ZCE_m.

Step 3: $I = I + 1$, calculating new polymerizing center of the fuzzy subset of E and CE as follows:

$$ZE_i(I) = \frac{1}{n_i} \sum_{k=1}^{n_i} e_k^i \quad i = 1, 2, \cdots, 7 \tag{4.27}$$

$$ZCE_j(I) = \frac{1}{n_j} \sum_{k=1}^{n_j} ce_k^j \quad j = 1, 2, \cdots, 7 \tag{4.28}$$

where n_i means the sample number of the ith fuzzy subset of E after polymerizing, n_j means the sample number of the jth fuzzy subset of CE after polymerizing, e_k^i means the kth sample of the ith fuzzy subset of E after polymerizing, and ce_k^j means the kth sample of the jth fuzzy subset of CE after polymerizing.

Step 4: Calculating the sum of square-error $J_c(I)$ as follows:

$$J_c(I) = \sum_{i=1}^{7} \sum_{k=1}^{n_i} (e_k^i - ZE_i(I))^2 + \sum_{j=1}^{7} \sum_{k=1}^{n_j} (ce_k^j - ZCE_j(I))^2 \tag{4.29}$$

Step 5: If $|J_c(I) - J_c(I-1)| \leq \varepsilon$, where ε denotes preset polymerizing precision, then polymerizing results has been convergent, and the algorithm is ended. Else, return to Step 2, the algorithm is continued.

Table 4.4 The fuzzy control rules of pulsed gas tungsten arc butt welding

E \ CE	NB	NM	NS	ZO	PS	PM	PP
NB	PB	PB	PB	PM	PS	ZO	ZO
NM	PB	PB	PM	PM	PS	ZO	ZO
NS	PB	PM	PM	PS	ZO	NS	NS
ZO	PM	PM	PS	ZO	NS	NM	NM
PS	PS	PS	ZO	NS	NM	NM	NB
PM	ZO	ZO	NS	NM	NM	NB	NB
PB	ZO	ZO	NS	NM	NB	NB	NB

Using above C-mean dynamic polymerizing algorithm, the final polymerizing centers of the fuzzy subsets of E and CE are shown as Table 4.3. We define the fuzzy subset's centers of the output fuzzy variable U is $\{-0.87, -0.58, -0.29, 0.00, +0.29, +0.58, +0.87\}$. On the basis of the polymerizing results of E and CE, the fuzzy control rules of pulsed gas tungsten arc butt welding are obtained using the principle of nearest distance, tabulated in Table 4.4.

Based on the analysis of butt welding, using fuzzy inference technology and artificial neural network methodology, fuzzy neural network controller can be developed.

4.4.2 Knowledge Models Based-on Rough Sets for Weld Pool Dynamical Process Based on Classic Theory

Rough sets (RS) theory [15] was proposed by Z. Pawlak in 1982. From the viewpoint of RS theory, knowledge has essential relationship with human ability of classifying, with great power to deal with uncertainty. Using RS methodology, we can obtain the rule model of complex processes, moreover, the rule model is understandable for operators and easy to revise directly. RS methodology has been applied in a variety of fields such as data mining, pattern recognition, decision support, fault analysis and so on [16]. In general, main steps of RS methodology include preprocess of raw data; discretization of continuous attributes; condition attribute reduction; condition attribute value reduction and rule reduction.

Among all steps, condition attribute reduction, or attribute reduction for short, is mainly responsible for the complexity of the model. So research on the algorithm of attribute reduction is most important for RS methodology. Now, there are lots of reduction algorithms [17–20]. Most of algorithms are not complete for the definition of attribute reduction given by Z. Pawlak, that is, the output of these algorithms contains redundant condition attributes. As a result, the model will be more complex relatively. On the other hand human experience is not available for these algorithms, which impacts the understanbility of the rule model. RA-order algorithm [21] proposed by J. Wang is complete and can use human experience, but it needs a total order defined on the set of condition attributes, which is too strict for human experience.

Here, a novel complete reduction algorithm is introduced to use human experience conveniently by a partial ordering defined on the set of condition attributes instead of a total order. A Value reduction algorithm and a rule reduction algorithm are generalized from the algorithm. These three algorithms compose the core of RS methodology. Then, we use RS methodology to obtain the model of aluminum alloy pulsed GTAW process. The welding for aluminum alloy is more complex because of oxidation and heat transfer of the material.

4.4.2.1 Basic Notions

RS methodology includes three kinds of reduction: condition attribute reduction (attribute reduction for short), condition attribute value reduction (value reduction for short) and rule reduction (optimization of the set of rules). In this section, some definitions related to these reductions are introduced and more detailed discussion can be found in Ref. [15].

Decision Table

A decision table *DT* consists of five parts, namely $DT = <U,C,D,V,f>$, where U is the universe, disjoint sets of attributes C and D are referred to as the condition and decision attributes respectively, $V = \bigcup_{a \in C \cup D} V_a$ is a set of all attribute values (where V_a is the set of values for attribute a), and f: $U \times C \cup D \to V$ is an information function such that $f\ (x,a) \in V_a$ for each $x \in U$ and each $a \in C \cup D$. It is necessary to point out U, C, D, and V_a for each $a \in C \cup D$ should be non-empty and finite sets. Consider, for example, the following decision table:

In Table 4.5, $U = \{x_1,x_2,x_3,x_4,x_5,x_6\}$, $C = \{a,b,c\}$, $D = \{d\}$, $V = \{1,2,3,4\}$, and $f(x, a)$ equals to the value in row x and column a for each $x \in U$ and each $a \in A$.

Table 4.5 An example of decision table

U	C			D
	a	b	c	d
x_1	1	1	1	1
x_2	1	2	2	2
x_3	2	2	2	3
x_4	3	3	3	3
x_5	3	3	2	4
x_6	3	3	3	4

Indiscemibility Relation

In a decision table $DT = <U,C,D,V,f>$, an indiscernibility relation *IND*(*B*) is a binary relation and

$$IND(B) = \{(x,y) \in U^2 : B \subseteq C \cup D \text{ and for every } a \in B, f(x,a) = f(y,a)\}$$

Obviously, *IND*(*B*) is an equivalence relation and

$$IND(B) = \bigcap_{a \in B} IND(\{a\})$$

For simplicity, if it does not cause confusion, we shall identify $IND(\{a\})$ and *IND*(a). For every $x \in U$, $[x]_{IND(B)}$ is referred to as the equivalence class of *IND*(*B*) containing x and $[x]_{IND(B)} = \bigcap_{a \in B} [x]_{IND(a)}$. *U*/*IND*(*B*) denotes the collection of all equivalence classes of *IND*(*B*). Let $X \subseteq U$, and *IND*(*B*) be an indiscernibility relation. We associate a subset:

$$B_*(X) = \{x \in U : [x]_{IND(B)} \subseteq X\}$$

called the *IND*(*B*)-lower approximation of *X*. Let *P* and *R* be two subsets of $C \cup D$. By *P*-positive region of *R*, denoted by $POS_P(R)$, we understand the set

$$POS_P(R) = \bigcup_{X \in U/IND(R)} P_*(X)$$

It is obvious $[x]_{IND(P)} \subseteq [x]_{IND(D)}$ for every $x \in POS_P(R)$. For example, in Table 4.5, $[x]_{IND(a)} = \{x_1, x_2\}$, $U/IND(C) = \{\{x_1\}, \{x_2\}, \{x_3\}, \{x_4, x_6\}, \{x_5\}\}$, $X = \{x_3, x_4, x_5\}$, $C_*(X) = \{x_3, x_5\}$, and $POS_C(D) = \{x_1, x_2, x_3, x_5\}$.

Attribute Reduction

In a decision table $DT = <U,C,D,V,f>$, let $C' \subseteq C$. C' is the reduction of C, namely an attribute reduction, if $POS_{C'}(D) = POS_C(D)$ and there is no $C'' \subset C'$ such that $POS_{C''}(D) = POS_C(D)$. In other words, the reduction C' is the minimum subset of C such that $POS_{C'}(D) = POS_C(D)$. Obviously, C may have more than one attribute reduction and an attribute reduction maybe C itself. In Table 4.5, $\{a, c\}$ is an attribute reduction because:

$$POS_{\{a,c\}}(D) = \{x_1, x_2, x_3, x_5\} = POS_C(D),$$

$$POS_{\{a\}}(D) = \{x_3\} \neq POS_C(D),$$

$$POS_{\{c\}}(D) = \{x_1\} \neq POS_C(D).$$

Value Reduction

In a decision table $DT = <U,C,D,V,f>$, let $C' \subseteq C$ and $x \in U$. $C'_x = \{(a, f(x,a) : a \in C'\}$. Table 4.5, if let $C' = \{a,b\}$, $C'_{x_2} = \{(a,1),(b,2)\}$. C'_x is an value reduction of x if $\cap[x]_{C'} \subseteq [x]_D$ and there is no $C'' \subset C'$ such that $\cap[x]_{C''} \subseteq [x]_D$. For example, $\{a,b\}_{x_2}$ and $\{a,c\}_{x_2}$ are both value reductions of x_2. Namely, $\{(a,1),(b,2)\}$ and $\{(a,1),(c,2)\}$ are both value reductions of x_2. But $\{(b,2),(c,2)\}$ is not a value reduction of x_2. For x_4 and x_6, they have no value reduction because $[4.x_4]_C \supset [4.x_4]_D$ and $[4.x_5]_C \supset [4.x_5]_D$.

Rule Reduction

In a decision table, let $D_x = \{(a, f(x,a) : a \in D\}$. $\wedge(C'_x) \rightarrow \wedge(D_x)$ will be called a rule of x, where $\wedge(X) = a_1 \wedge a_2 \wedge \ldots \wedge a_i \wedge \ldots, a_i \in X$. We can simply understand the rule of x as an IF-THEN rule. For example, in Table 4.5, $\{a,c\}_{x_2} \rightarrow \{d\}_{x_2}$ is a rule of x_2, namely, $(a,1) \wedge (b,2) \rightarrow (d,2)$. This means if $f(x_2,a) = 1$ and $f(x_2,b) = 2$, then $f(x_2,d) = 2$. It is clear that x may have more than one attribute value reduction. So x may also have more than one rule. Let Rule(x) be the set of rules of x, $x \in U$. For $X \subseteq U$, the collection $\text{Rule}(X) = \{\text{Rule(x)} : x \in X\}$. If x have no value reduction, x have only one rule $C_x \rightarrow D_x$. In a decision table $DT = <U,C,D,V,f>$, let R be a subset of Rule(U), namely, $R \subseteq \text{Rule}(U)$. R is a rule reduction of DT if $R \cap \text{Rule}(x) \neq \emptyset$ for each $x \in U$ and there is no $R' \subset R$ such that $R' \cap \text{Rule}(x) \neq \emptyset$ for each $x \in U$. It is obvious that a decision table can have more than one rule reduction.

4.4.2.2 Rough Set Modeling Method

Modeling can, to some extent, be regarded as decision analyzing as well, that is, to determine the current state of the system by usable parameters of the system. The RS theory is very suitable for decision analysis. Thus, it is also suitable for modeling. In this section we will give main steps of the RS based knowledge modeling method and its key algorithm.

Main Steps

In general, main steps of the RS based knowledge modeling method are as follows:

Step 1: Acquire the raw data;
Step 2: Preprocess the raw data;
Step 3: Discretize continuous attributes;
Step 4: Compute the attribute reduction;
Step 5: Compute attribute value reductions;
Step 6: Obtain the decision algorithm (knowledge model).

The RS based knowledge modeling method begins with acquiring the raw data. The raw data can be results of experiments, history data, and etc.. Usually, the raw data can't be used directly. According to practical situations, corresponding preprocessing methods should be adopted. After preprocessing we get a decision table. Condition attributes will be input variables (parameters of the system) of the model and decision attributes will be output variables (states of the system) of the model. Next, discretization of continuous attributes is needed. Many discretization methods have been proposed such as the equal interval width method, the equal frequency per interval method, the cluster analysis method and so on. Then we will compute the attribute reduction and get a concise decision table. The study on algorithms of computing attribute reduction is one of key problems of the research on the RS theory. In Step 5, we will get an extend decision table by computing attribute value reduction, which is the basis of the next step. At last, in Step 6, we will get the decision algorithm, namely the minimum set of decision rules, which can also be regarded as the knowledge model. Figure 4.29 is the flow chart of the RS based knowledge modeling method.

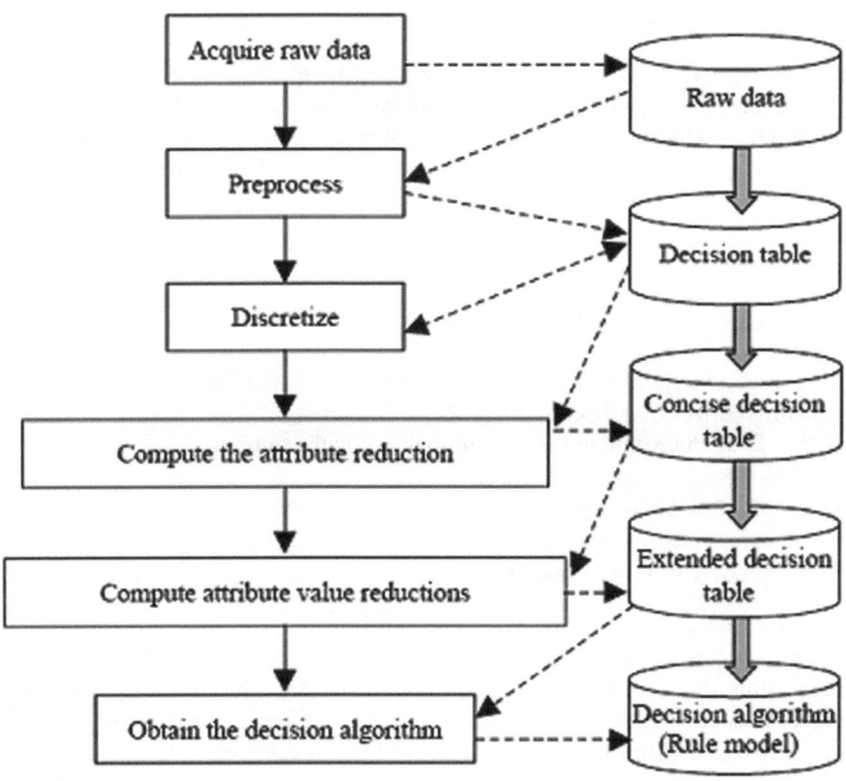

Fig. 4.29 Flow chart of the RS based knowledge modeling method

Key Algorithms of the Modeling Method

We can prove that definitions of the attribute reduction, the attribute value reduction and the decision algorithm are equivalent [22] from the viewpoint of the classical set theory. Therefore the research on their algorithms is uniform, that is, their algorithms can be transformed each other. In this section we give a new algorithm of computing the attribute reduction, which is the key algorithm of our RS based knowledge modeling method. The main idea of the algorithm can be used to construct algorithms of computing the attribute value reduction and obtaining the minimal set of decision rules.

Our attribute reduction algorithm is based on the principle of Skowron's discernibility matrix [23]. Let DT $=$ <U, C, D, V, f> be a decision table. The set of nonempty elements in the discernibility matrix of DT will be called the discernibility set of DT, denoted by ADT. ADT is a subset of P(C), namely $\mathrm{A}_{DT} \subseteq \mathrm{P}(C)$, where P(C) is the power set of C. Let $\mathrm{A} \subseteq \mathrm{P}(C)$ and $B \subseteq C$. Let us define the following two operations:

$$\mathrm{A}^{-}(B) = \{A' : A \in \mathrm{A}, A' = A - B \text{ and } A' \neq \emptyset\},$$
$$\mathrm{A}^{+}(B) = \{A \in \mathrm{A} : A \cap B = \emptyset\}.$$

Let $a \in C$. If $\{a\} \in \mathrm{A}$, a will be called a essential element of A and the set of essential elements of A will be call the essential set of A, denoted by Ess(A). Let $X = \cup\mathrm{A}$ and R be a partial order relation on C. $R' = R \cap (X \times X)$.$<X, R'>$ is a partial order set. The set of the minimal elements of $<X, R'>$ will be called the minimal set of $<X, R'>$, denoted by

Algorithm 1

1. Construct a set of condition attributes P, $P \Leftarrow \emptyset$; (The mark '$\Leftarrow$' means $\emptyset$ be assigned to P.)
2. Construct a temporary set of condition attributes Q, $Q \Leftarrow \emptyset$;
3. Construct a working discernibility set A, $\mathrm{A} \Leftarrow \mathrm{A}_{DT}$;
4. If $\mathrm{A} = \emptyset$, then terminate the algorithm and P is the output;
5. $Q \Leftarrow \mathrm{Ess(A)}$;
6. If $Q \neq \emptyset$, then go to step 8;
7. Randomly select one element α in $\mathrm{Min}_{\mathrm{A}}^{R}$, $\mathrm{A} \Leftarrow \mathrm{A}^{-}(\{a\})$ and go to step 4;
8. $P \Leftarrow P \cup Q$, $\mathrm{A} \Leftarrow \mathrm{A}^{+}(P)$ and go to step 4.

Figure 4.30 is the flow chart of the Algorithm 1.

One character of Algorithm 1 is the completeness [24]. It is sure that the output of the algorithm does not include any redundant condition attributes. But most algorithms presented [17, 18] are incomplete and the outputs of them are redundant. Another character of Algorithm 1 is the using of human experience. Usually, we can use human experience in algorithms by defining an ordering on the set of condition attributes. Take the algorithm in [25] for example. It needs a total order relation but it is too strict to be satisfied. Lost of human experience can't be expressed by a total

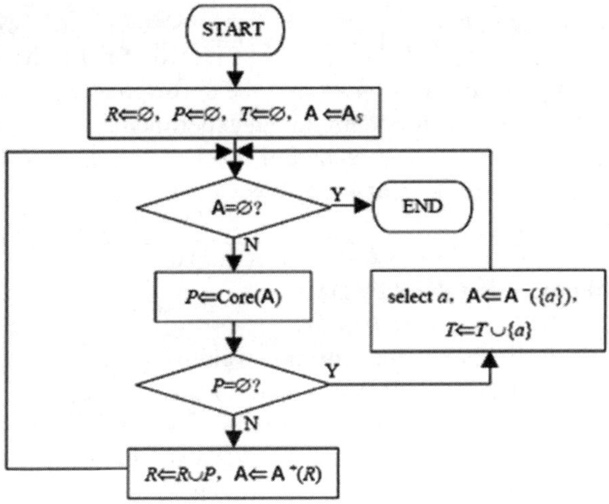

Fig. 4.30 Flow chart of the Algorithm 1

order relation, thus the algorithm can't use these experience. In Algorithm 1, what we need is a partial order relation. It is more suitable to express human experience. It is necessary to point out the human experience is different from the experience containing in the model. The former is used to obtain the model and the latter can be regarded as the model itself.

Algorithm 1 can be changed into Algorithm 2 for computing the attribute value reduction and Algorithm 3 for obtaining the minimal set of decision rules by little modification. In this book, the Algorithm 2 and Algorithm 3 are omitted [24, 26].

4.4.2.3 Modeling for GTAW Process

In this section, we will show how to obtain the model of pulsed GTAW process using RS methodology based on algorithms introduced above. In some real applications, the backside of the weld pool is invisible. However, we have to ensure the appearance of the backside of the weld beam (the width of the backside of the weld beam). It is necessary to obtain the predictive model for the process. With the model we can predict the width of the backside through the information of the topside of the weld beam and other welding parameters. Thus the inputs of the model are the welding current, the maximal width and length of the topside of the weld pool. The output is the maximal width of the backside of the weld pool.

Acquiring the Raw Data

Raw data is obtained by the experiment systems [27] of welding technology for aluminum alloy, which is based on double-side vision sensing and can capture images of the topside and backside of the weld pool at the same time (Fig. 2.16). The welding conditions are also the same as that in Ref. [27]. The maximum width (W_T) and length (L_T) of the topside of the weld pool, the maximum width (W_B) of the backside of the weld pool and the corresponding welding current (I) can be given by the experiment system. In out experiments, we acquire 260 data for 260 continuous weld pools and each datum consists four parts: I, W_T, L_T and W_B. Formally, the ith datum will be denoted by a vector $d^i = \begin{bmatrix} I & L_T^i & W_T^i & W_B^i \end{bmatrix}$.

Preprocessing the Raw Data

It is well known that the arc welding process is a process of heat accumulating. We have to adapt raw data to this character of the welding process by preprocessing. What we need to do is to extend raw data d^i to D^i, where $D_i = \begin{bmatrix} I^{i-k}\ L_T^{i-k}\ W_T^{i-k}\ W_B^{i-k}\ \ldots\ I^i\ L_T^i\ W_T^i\ W_B^i \end{bmatrix}$. In our experiments, when the ith weld pool begins to form, the (i-4)th weld pool has solidified and it nearly has no affection on the ith weld pool. So it is appropriate that k=3. Now we have 257 extended data, since first three data can't be extended. For the purpose of verifying the model, we randomly select a little part of data for verifying and the left data are for modeling. We repeat selecting for 10 times and 10 groups of modeling data and verifying data are got. Every set of the modeling data can be regarded as a decision table containing 15 condition attributes and one decision attribute (the last attribute). Thus, we have now 10 decision tables.

Discretization

In our experiments, the equal interval width method is selected. Intervals of values of I, W_T, L_T and W_B are divided into some subintervals and every subinterval will be assigned a symbol.

I (Welding current)

1-[4.120, 160)
2-[4.160, 200)
3-[4.200, 240)
4.[4.240, 280)
5-[4.280,320)

L_T (Maximum length of the topside of the weld pool)

1-[4.0, 6)
2-[4.6, 8)
3-[4.8, 10)
4.[4.10, 12)
5-[4.12, + ∞)

W_T (Maximum width of the topside of the weld pool)

1-[4.0, 6)
2-[4.6, 9)
3-[4.9, 12)

4.[4.12, 15)
5-[4.15, + ∞)

W_B (Maximum width of the backside of the weld pool)
1-[4.0, 3)
2-[4.3, 6)
3-[4.6, 9)
4.[4.9, 12)
5-[4.12, + ∞)

Then we replace values of condition attributes in all decision tables with these symbols. For the decision attribute, special treatment is needed. We replace the values of the decision attributes with the median of the interval instead of the symbol assigned to the interval. In this way, the output of the final model can be real value of the maximum width of the backside of the weld pool, not the symbol to which the real value belongs. Table 4.6 is an example and the last row in it is the discrete datum.

Table 4.6 An example

Conditions															Decision
250	8	7.61	4	250	10.295	15.22	8	130	9.135	10.05	9.1	290	8	14	5.2
4	3	2	2	4	4	5	3	4	3	3	4	5	3	4	4.5

Knowledge Model of the Aluminum alloy Pulsed GTAW Process

We deal with these 10 discrete decision tables using Algorithm 1, Algorithm 2 and Algorithm 3 [24, 26] in turn. During processing, the partial order relation R can be defined based on welders' experience. Follow experience is obvious:

Experience 1: The smaller the distance between two weld pools is, the greater effect of heat on the latter the former has.
Experience 2: For one weld pool, the current welding current has closer relation with the maximum width of the backside of the weld pool than the maximum width and length of the topside of the weld pool.

By Experience 1, if $i \leq j$, we have $(I^i, I^j) \in R$, $(L_T^i, L_T^j) \in R$, $(W_T^i, W_T^j) \in R$ and $(W_B^i, W_B^j) \in R$. By Experience 2, we have $(I^i, W_T^i)R$ and $(I^i, L_T^i) \in R$. It is easy to validate R is a partial order relation on the set of condition attributes. Using R, the final model obtained will be more understandable.

At last, we obtain 10 minimum sets of decision rules, leading to 10 decision algorithms (knowledge models). One of 10 models consists of 146 decision rules and part of them is listed in Table 4.7. Take the first decision rule in Table 4.8 for example. It means if the welding current of the last one weld pool before the current weld pool is 1, the maximum length of the topside of the current weld pool is 1 and the maximum width of the topside of the current weld pool is 3, then the maximum

Table 4.7 Decision algorithm (Knowledge model)

No.	Conditions (IF)							Decision (THEN)
	I^t	I^{t-1}	I^{t-2}	$L_t{}^t$	$L_t{}^{t-1}$	$W_t{}^t$	$W_t{}^{t-2}$	$W_b{}^t$ (mm)
1	–	1	–	1	–	3	–	4.5
2	–	–	1	–	1	3	2	4.5
3	–	–	–	2	1	–	3	7.5
4	–	–	5	4	–	1	–	10.5
5	4	–	2	–	–	–	–	7.5
					...			
65	4	2	–	2	–	–	–	10.5
66	–	2	1	–	–	–	4	1.5
67	4	1	–	–	1	–	–	4.5
					...			
146	5	–	5	–	2	1	–	10.5

Table 4.8 The executive process

	A	P	Attribute selected
0	$\{\{a, b, d\}, \{a, c, d\}, \{b, c, d\}, \{a, b, c\}, \{e, f\}\}$	Ø	
1	$\{\{b, d\}, \{c, d\}, \{b, c, d\}, \{b, c\}, \{e, f\}\}$	Ø	*a*
2	$\{\{d\}, \{c, d\}, \{c\}, \{e, f\}\}$	Ø	*b*
3	$\{\{e, f\}\}$	$\{c, d\}$	
4	$\{\{f\}\}$	$\{c, d\}$	*e*
5	Ø	$\{c, d, f\}$	

width of the backside of the current weld pool equals 4.5 mm. If we replace symbols in the decision rule with corresponding intervals, the meaning of the decision rule will be more distinct, namely, if the welding current of the last one weld pool before the current weld pool is between 120 A and 160 A, the maximum length of the topside of the current weld pool is between 0 mm and 6 mm, and the maximum width of the topside of the current weld pool is between 9 mm and 12 mm, then the maximum width of the backside of the current weld pool equals 4.5 mm. Obviously, the model is easy to be understood and revised.

Reasoning About the Knowledge Model of the Aluminium Alloy Pulsed GTAW

Theoretically, if the data for modeling can cover the input space, for each input there must be one decision rule such that the input can fully match the rule, and the values of the decision attributes of the rule will be the output of the model. Unfortunately, in real applications the data for modeling could not cover the input space. That means there may be some inputs, which can't fully match any rules in the model. For these inputs the model can't give output directly. So reasoning is necessary to ensure the model can give an output for any input.

Let r be a rule in the model M, where M is the set of rules, and r have i antecedents (conditions). If a new input d satisfies j antecedents of r, the *r-matching* degree of d can be defined as j/i, denoted by $\alpha_d^r (0 \le \alpha_d^r \le 1)$. Let $\mathrm{Max}(d,M) = \{r : r \in M \text{ and } \alpha_d^s \le \alpha_d^r \text{ for any } s \in M\}$ and $\alpha_d^{Max} = \alpha_d^r$, where $r \in \mathrm{Max}(d,M)$. There are different cases and different actions should be taken.

<1> $|\mathrm{Max}(d,M)| = 1$ and $\alpha_d^{Max} \neq 0$
The model outputs the consequents (decisions) of r, where $r \in \mathrm{Max}(d,M)$;
<2> $|\mathrm{Max}(d,M)| > 1$ and $\alpha_d^{Max} \neq 0$
The model outputs the averages of consequents of rules in Max(d, M);
<3> $\alpha_d^{Max} = 0$
The output of the model is the same with the previous output.
We can regard the reasoning as part of the modeling.

Verifying of the Knowledge Model of the Aluminum Alloy Pulsed GTAW

We define the complexity of the model as

$$\eta = \frac{\sum_{i=1}^{m} a_i}{|U| \times |C|} \tag{4.30}$$

where $|U|$ is the number of data for modeling, $|C|$ is the number of condition attributes, m is the number of rules in the model and a_i is the number of antecedents (conditions) of the ith rule. It is obvious that $0 \le \eta \le 1$.

We define the precision of the model as

$$J = \frac{\sum_{i=1}^{n} (\bar{y}_i - y_i)^2}{n} \tag{4.31}$$

where n is the number of data for verifying, y_i is the real output for the ith datum and $\bar{y}_i$ is the desired output of model for the ith datum.

In our experiments, complexities and precisions of 10 models are listed in Table 4.9. Considering the complexity of the aluminum alloy pulsed GTAW process the models are acceptable. Figure 4.31 is the error curve of the first model.

In this section, we propose RS methodology for the modeling of the aluminum alloy pulsed GTAW process. Using human experience during modeling makes the model easy to be understood for engineers and possible to revise the model directly. Our work is the basis of intelligent control of the arc welding. In addition, our work can also propel applications of rough set theory.

Table 4.9 Complexities and precisions

Model	Number of data for modeling	Number of data for verifying	Complexity(η)	Precision (J)
1	217	40	0.1576	2.3378
2	214	43	0.1439	1.8090
3	216	41	0.1627	2.0163
4	206	51	0.1476	1.3239
5	211	46	0.1523	2.5875
6	209	48	0.1579	2.1274
7	203	54	0.1570	2.5142
8	205	52	0.1440	1.7209
9	213	44	0.1383	2.2907
10	212	45	0.1557	1.4288
Average	0.1517	2.0157	–	–

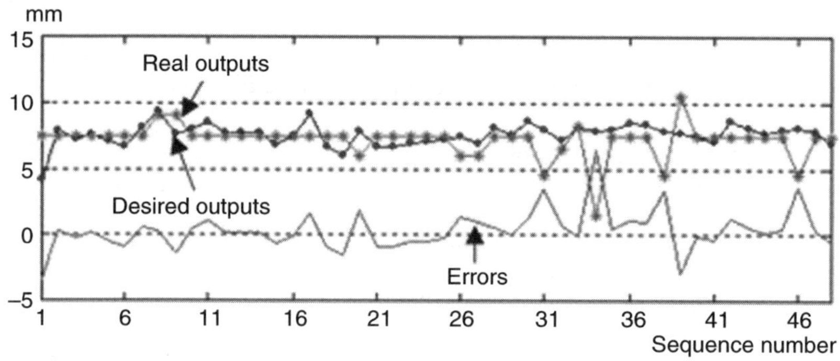

Fig. 4.31 Error curve of the experiment

4.4.3 A Variable Precision Rough Set Based Modeling Method for Pulsed GTAW

To improve the general rough set model, a generalized rough set–variable precision rough set (VPRS) is introduced to welding because of its better noise suppression merits.

4.4.3.1 Brief Description of VPRS

Both VPRS and classic RS operate on what is described as a *decision table* or *information system*. As illustrated in Table 4.10 [28], a set of *objects U* $(o_1,\ldots,o_7)$ are contained in the rows of the table. The columns denote *condition attributes* $C(c_1,\ldots,c_6)$ of these objects and a related *decision attribute D* (*d*). A value denoting the nature of an attribute to an object is called a *descriptor*. A VPRS data requirement

Table 4.10 An example of decision table

Object	Condition attributes(C)						Decision attributes(D)
	c_1	c_2	c_3	c_4	c_5	c_6	d
o_1	1	0	1	1	0	1	0
o_2	1	0	0	0	0	0	0
o_3	0	0	1	0	0	0	0
o_4	1	0	1	1	0	1	1
o_5	0	0	0	0	1	1	1
o_6	1	0	1	1	0	1	1
o_7	0	0	0	0	1	0	1

is that it must be in discrete or categorical form. The table shows that the objects have been classified into one of these decision values.

For the condition attributes in this example, all of the objects (U) can be placed in five groups: $\mathbf{X_1} = \{o_1, o_4, o_6\}$, $\mathbf{X_2} = \{o_2\}$, $\mathbf{X_3} = \{o_3\}$, $\mathbf{X_4} = \{o_5\}$ and $\mathbf{X_5} = \{o_7\}$. The objects within a group are indiscernible from each other. For example objects o_1, o_4 and o_6 in X_1 have the same descriptor values for each of the condition attributes. These groups of objects are referred to as *equivalence classes* or *conditional classes* for their specific attributes. The equivalence classes for the decision attribute are: $\mathbf{Y_1} = \{o_1, o_2, o_3\}$ and $\mathbf{Y_2} = \{o_4, o_5, o_6, o_7\}$. The abbreviation of the set of equivalence classes for the conditional attributes C is denoted by $\mathbf{E(C)} = \{\mathbf{X_1}, \mathbf{X_2}, \mathbf{X_3}, \mathbf{X_4}, \mathbf{X_5}\}$ and for the decision attribute it is defined $\mathbf{E(D)} = \{\mathbf{Y_1}, \mathbf{Y_2}\}$.

VPRS measurement is based on ratios of elements contained in various sets. A case in point is the conditional probability of a concept given a particular set of objects (a condition class). For example: $\Pr(\mathbf{Y_1}|\mathbf{X_1}) = \Pr(\{o_1, o_2, o_3\}|\{o_1, o_4, o_6\}) = |\Pr(\{o_1, o_2, o_3\} \bullet \{o_1, o_4, o_6\} / |\{o_1, o_4, o_6\})| = 0.333$.

It follows that this measures the accuracy of the allocation of the conditional class $\mathbf{X_1}$ to the decision class $\mathbf{Y_1}$. Hence for a given probability value, the β-positive region corresponding to a concept delineated as the set of objects with conditional probabilities of allocation at least equal to or larger than β. More formally, β-positive region of the set $\mathbf{Z} \subseteq \mathbf{U}$

$$POS_P^{\beta}(Z) = \bigcup_{p_r(Z|Xi) \geq \beta} \{X_i \in E(P)\}, \text{ with } P \subseteq C \tag{4.32}$$

Following [23], β is defined to be within range of 0.5 to 1. Hence for the current example, the condition equivalence class $\mathbf{X_1} = \{o_1, o_4, o_6\}$ have a majority inclusion (with at least 60% majority needed, i.e. $\beta = 0.6$) in $\mathbf{Y_2}$, in that most objects (2 out of 3) in $\mathbf{X_1}$ belong in $\mathbf{Y_2}$. Hence $\mathbf{X_1}$ is in $\mathrm{POS_C}^{0.6}$ $(\mathbf{Y_2})$. It follows $\mathrm{POS_C}^{0.6}(\mathbf{Y_2}) = \{o_1, o_4, o_5, o_6, o_7\}$.

Corresponding expressions for the β-boundary and β-negative regions are given by Ziarko [29], as follows: β-boundary region of the set $\mathbf{Z} \subseteq \mathbf{U}$

$$BND_P^{\beta}(Z) = \bigcup_{1-\beta<p_r(Z|Xi)<\beta} \{X_i \in E(P)\},\ \text{with } \mathbf{P} \subseteq \mathbf{C}, \tag{4.33}$$

β-negative region of the set $\mathbf{Z} \subseteq \mathbf{U}$

$$NEG_P^{\beta}(Z) = \bigcup_{p_r(Z|Xi)\leq\beta} \{X_i \in E(P)\},\ \text{with } \mathbf{P} \subseteq \mathbf{C}. \tag{4.34}$$

Using **P** and **Z** from the previous example, if $\beta = 0.6$, then $BND_C{}^{0.6}(\mathbf{Y_2}) = \emptyset$ (empty set) and $NEG_C{}^{0.6}(\mathbf{Y_2}) = \{o_2, o_3\}$ can be obtained. Similarly, for the decision class $\mathbf{Y_1}$, it follows that $\mathrm{POS}_C{}^{0.6}$ $(\mathbf{Y_1}) = \{o_2, o_3\}$, $BND_C{}^{0.6}(\mathbf{Y_1}) = \emptyset$ and $NEG_C{}^{0.6}(\mathbf{Y_1}) = \{o_1, o_4, o_5, o_6, o_7\}$.

VPRS applies these concepts by firstly seeking subsets of the attributes, which are capable (via construction of decision rules) of explaining allocations given by the whole set of condition attributes. These subsets of attributes are termed *β-reducts* or *approximate reducts*, and the process is called *attribute reduction*. Ziarko [29] states that a β-reduct, a subset **P** of the set of conditional attributes **C** with respect to a set of decision attributes **D**, must satisfy the following conditions:

(i) the subset **P** offers the same quality of classification (subject to the same β value) as the whole set of condition attributes **C**; and
(ii) no attribute can be eliminated from the subset **P** without affecting the quality of the classification (subject to the same β value).

The quality of the classification is defined as the proportion of the objects consisted of the union of the β-positive regions of all the decision equivalence classes based on the condition equivalence classes for a subset **P** of the condition attributes **C**. Associated with each conditional class is an upper bound on the β value above which there is no opportunity for majority inclusion and hence not in a β-positive region: in the previous example $\Pr(\mathbf{Y_1}|\mathbf{X_1}) = 0.333$ and $\Pr(\mathbf{Y_2}|\mathbf{X_1}) = 0.667$, hence if $\beta = 0.7$ then $\mathbf{X_1}$ is not in the associated β-positive region, since the upper bound on β (for majority inclusion) is 0.667. The lowest of these upper bounds (amongst the condition classes) on the β values is defined $\beta_{\min}$ and relates to the overall level of confidence in classification by a particular β-reduct. The quality of classification associated with all the condition attributes is dependent on the $\beta_{\min}$ value. Hence for $\beta_{\min} \in (0.500, 0.667]$ the quality of classification is one (with $\mathbf{X_1}$ in $POS_C{}^{\beta}(\mathbf{Y_1})$), for $\beta_{\min} \in (0.667, 1]$ the quality of classification is 0.571 (i.e., classifying four of the seven objects, with $\mathbf{X_1}$ in $BND_C{}^{\beta}$ $(\mathbf{Y_2})$). Table 4.11 provides examples of four such β-reducts based on the condition attributes in Table 4.10.

Table 4.11 Examples of β-reducts from data shown in Table 4.10

$\beta_{\min}$	β-reduct	Classification quality
0.571	$\{c_2\}$	1
0.667	$\{c_1, c_3\}$	1
0.667	$\{c_1, c_4\}$	1
1.000	$\{c_4, c_5\}$	0.571

After the attribute reduction, *attribute value reduction* and *rule reduction* are then to be computed. Attribute value reduction is to remove the superfluous condition attribute in each object for a given β-reduct while preserving the consistency of classifications. After the attribute value reduction, rule reduction is implemented to construct a minimal rule set, which also preserves the consistency of classifications of original decision table.

4.4.3.2 VPRS Modeling Method

When VPRS is applied into welding, the specific characteristics of the welding and VPRS theory should be considered. To begin this section, the procedure of VPRS modeling method is first described. Then, the reduction algorithm in the modeling method is addressed.

Procedure of the VPRS Modeling Method

Figure 4.32 shows the procedure of VPRS modeling method, which begins with acquiring raw data including the results of experiments, history data, etc. Usually, the raw data cannot be directly treated by VPRS, and corresponding preprocessing is needed to improve the quality of data. Furthermore, if data in decision table is continuous, discretization is necessary. After the pre-processing, a decision table can be obtained, where condition attributes are input variables of system and decision attributes are output variables of system. It is followed by attribute reduction, during which computation of attribute value reduction and rule reduction are conducted, the most important part in the modeling method. After that a VPRS model, made up of

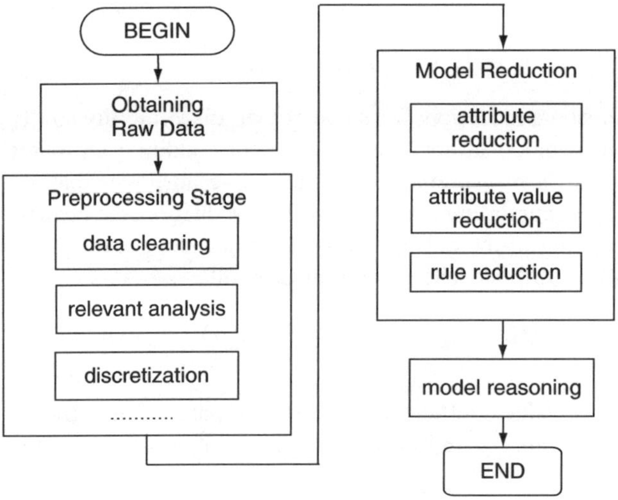

Fig. 4.32 Procedure of the VPRS modeling method

"IF THEN" rules, is built. At last, the VPRS model can be used to predict the unseen objects (samples).

Key Algorithm in VPRS Modeling

As mentioned above, the reduction algorithm which includes attribute reduction, attribute value reduction and rule reduction is the most important part in the modeling method. There might be more than one reduction result. The attribute reduction with the minimal number of attributes for a given decision table is called the *minimal reduction*, and it is usually the optimal solution. However, in RS theory, the optimal solution is NP-hard [30]. Heuristic information has to be used to improve the efficiency of algorithm and only suboptimal solution can be obtained.

<1> Attribute reduction algorithm

In RS theory, there are some condition attributes which appear in all the attribute reduction and they are called *reduction core*. It is obvious that the reduction core is determined first to avoid redundancy in the obtained reduction, and the computing of reduction core is completed according to its definition in VPRS. Then, other attributes should be added into reduction core (or removed out condition attributes) to get the β–attribute reduction that maintained the same discernibility as origin decision table.

Here hieuristic information is used by computing attributes' significance, which decided the attribute reduction strategy, and usually is defined according to the roughness of attributes, the information entropy or attribute frequency function in discernibility matrix (DM) [23]. The entropy based attribute importance is adopted in the work because it is computationally cheap and simple. Here is the definition:

Let H(a/R) be the condition entropy of the attribute a for the attribute set R. Let D be the decision attribute. The attribute significance of decision table can be defined as.

$$SGF(\alpha,R,D) = H(D/R) - H(D/R \cup \{\alpha\}) \tag{4.35}$$

In VPRS theory, all the attribute reductions are same of their discernibility ability for decision table. However in practical application, different condition attributes have different accuracy or cost in measurement. To make more attributes that are easy to measure and have high accuracy appear in the attribute reduction, the hieuristic information is used according to the entropy based significance, accuracy and cost of the measurement of the attribute. The new formula is followed as:

$$SGF(\alpha,R,D) = \kappa_1 \times \Big(H(D/R) - H(D/R \cup \{\alpha\})\Big) + \kappa_2 \times \mathrm{cos}t(\alpha) + \kappa_3 \times precise(\alpha) \tag{4.36}$$

where k_1, k_2, k_3 are the weight value coefficients for different parts. If an attribute is easier to measure, its *cost*(a) value is high. If its value have higher accuracy, the *precise*(a) value is higher. For an attribute with higher *SGF*(a,R,D), it is easier to be selected in the attribute reduction.

<2> Algorithms of Attribute value reduction and rule reduction

It has been shown that there are conditional equivalence relation between attribute reduction, attribute value reduction, rule reduction and minimum set cover [22]. Therefore, the attribute value reduction and rule reduction could be constructed based on minimal set cover, and DM could be used here. In this work, only attribute value reduction algorithm is given, and rule reduction algorithm is similar to that.

Before introducing the algorithm, the following definition is introduced.

(i) In a decision table *DT*, let x be an object. $\mathrm{FS}_{DT}(x)$ is the set of 'attribute set' which discerns object x with other objects in *DT*. For example, in Table 4.16 $\mathrm{FS}_{DT}(o_1) = \{\{c_3,c_4,c_6\}\{c_1,c_4,c_6\}\{c_1,c_3,c_4,c_5\}\{c_1,c_3,c_4,c_5,c_6\}\}$. The first element can discern o_1 with o_2, the second can discern o_1 with o_3, the third can discern o_1 with o_5 and the last can discern o_1 with o_7.

(ii) Let A be set of attributes, eg A $= \{\{a,b,c\}\ \{b,d\}\ \{e\}\{f\}\}$, ES(A) is the union of sets, which have only one element in A. In this case, $\mathrm{ES(A)} = \{e\} \cup \{f\} = \{e,f\}$.

(iii) Let A, B be the attribute set, REM_SET(B, A)=$\{$A':A'=A-B, and A' $\neq \emptyset\}$.For example, if A=$\{\{a,b\}\{a,b,c\}\{a,b,f\}\{h\}\{i\}\}$ and B=$\{a,b\}$ then REM_SET (A,B)=$\{\{c\}\{f\}\{h\}\{i\}\}$.

(iv) Let A, B be the attribute set, EXC_SET(B, A)=$\{$B:B$\cap$A$\neq\emptyset\}$. Take the example in (iii), EXC_SET(A,B)=$\{\{h\}\{i\}\}$. Now the attribute value reduction algorithm is showed below.

Attribute value reduction algorithm:

<1> Assign objects universal U to X;
<2> If X=$\emptyset$, the program ends;
<3> Take a object x from X, $X \Leftarrow X - \{x\}$, computer $\mathrm{FS}_{DT}(x)$;
<4> $i \Leftarrow 0$;
<5> If $i > Kind$, go to (2);
<6> $P \Leftarrow \emptyset$, $Q \Leftarrow \emptyset$, $T \Leftarrow \emptyset$;
<7> $A \Leftarrow \mathrm{FS}_{DT}(x)$;
<8> If A=$\emptyset$, then $i \Leftarrow i+1$ and go to (13), else go to (9);
<9> $Q \Leftarrow \mathrm{ES(A)}$;
<10> If Q=$\emptyset$, then go to (11), else go to step (12);
<11> Randomly select an attribute a from C–$(P \cup T)$, A $\Leftarrow$ REM_SET$(A,\{a\})$, $T \Leftarrow T \cup \{a\}$, go to (8);
<12> $A \Leftarrow$ EXC_SET(A,P), $P \Leftarrow P * Q$, go to (8);
<13> if P have been obtained go to step (2), else note down P and go to step (5).

After the program, P noted down each time is the attribute value reduction. For any object$x$, its attribute value reduction is less than the value of *Kind* in the algorithm.

4.4.3.3 VPRS Model of Pulsed GTAW

Welding experiments are carried out for both low carbon steel and aluminium alloy. The welding conditions for low carbon steel is shown in Table 4.1, and for aluminium alloy is shown in Ref. [27].

Obtaining Raw Date

Before raw date are obtained, the input variables (condition attributes) and output variables(decision attributes) of the predictive model should be selected. According to the character of GTAW process, welding current (I_p), wire feeding velocity (*Vf*), welding velocity (V_w) and topside shape of welding pool are used as input variables and backside width (W_b) of the welding pool as the output variable.

The area of topside pool's area $A(= W_T \times L_T)$ is also added into the decision table. To make the model better reflect the real process, I_p, $\mathrm{V_w}$ and V_f are randomly changed within specified range. Considering the pure delay of welding, the *t*th, $(t-1)$th, $(t-2)$th, $(t-3)$th time's I_p, $\mathrm{V_w}$, V_f, A, W_T, L_T and W_b are used to predict the *t*th time's $W_{b.}$ At last, a decision table which included 1013 samples (objects), with 27 condition attributes and 1 decision attribute are obtained.

Preprocessing

It is known that discretization for attributes (including decision attribute sometimes) in decision table, is usually necessary if their values are continuous. In this case, the equal width method is used to discrete the decision attribute, and each interval is 1 mm in width. Then, MDLP [31], an entropy based method, is used to discrete the condition attributes since the MDLP method has been demonstrated as better than other common algorithms in welding experiments [32].

Model Reduction

The algorithms introduced in Sect. 4.4.3.2 are used to obtain the model. In attribute reduction, when computing the importance according equation (4.35), It is set that $k_1 = 0.5$, $k_1 = 0.25$, $k_1 = 0.25$ and the cost or precision of attributes are showed in Table 4.12. The larger the value *SGF* is, more important the attribute is. In the experiment, the attribute reduction is "${S_F}^{t-3}, {V_f}^{t-1}, {V_W}^{t-2}, {V_W}^{t-3}, {I_p}^{t-1}, {I_p}^{t-2}, {I_p}^{t-3}, {W_T}^t, {W_T}^{t-3}, {L_T}^t, {L_T}^{t-1}, {L_T}^{t-2}, {L_T}^{t-3}, A^t, A^{t-1}, A^{t-2}, A^{t-3}, {W_b}^{t-1}, {W_b}^{t-2}, {W_b}^{t-3}$".

In attribute value reduction, the most reduction for an object is *kind* $= 5$. After attribute reduction and rule reduction, the VPRS model has 262 rules and the average length of rules is 5.25. It is obvious that the VPRS model has good reduction ability.

Table 4.12 The cost and precise value of attributes

	Cost	Precise
I	0.8	1
S_F	0.8	0.5
S_W	0.8	0.6
A	0.3	0.3
W_T	0.3	0.3
L_T	0.3	0.3
W_B	0.1	0.4

Model Reasoning

Approximate reasoning is implemented when the VPRS model is use to predict unseen data. Because welding is very complex, there will be few rules that match the unseen data. In the reasoning, the most matched five rules are selected, and the minimum *matching rate* of the model is 70%. Furthermore in rough set method, the longer a rule is, the better its prediction ability can be predicted. Therefore, a longer rule is prior to other rules which have the same matching rate. One of the obtained VPRS model contained 267 rules and the average length of a rule (including conclusion part) is 6. It indicates that the VPRS model can greatly reduce the redundant information of the origin decision table.

The VPRS model of low carbon steel is shown in Table 4.13 and of aluminium alloy is shown in Table 4.15, and the result of model is shown respectively in Table 4.14 and Fig. 4.33 for low carbon steel and in Table 4.16 and Fig. 4.34 for aluminium alloy.

4.4.3.4 Comparison Between VPRS and Classic RS

Generally, the procedure of classic RS modeling method is similar to that of VPRS [24, 26]. To compare their predictive ability, the two modeling method should use the same raw data, take the same preprocessing procedure, adopt the same reasoning method, and test obtained models on same testing data. Because of the complexity of welding, there are usually many noisy data in raw data. Moreover, RS model usually is obtained based on small part of the sample space and is used to predict more unseen samples in practice. Therefore in the comparison, a small part of raw data is used for modeling and the larger for testing. The raw data is randomly split into 3 parts (2:1:1). The first 2/4 part is used as testing data, and the other 1/4 data is used to obtained classic RS or VPRS model. To better show the comparative result, following parameters are introduced.

(1) Accurate rate R_{right}: the rate of right predict samples in testing data
(2) Mean error:

Table 4.13 Part rules in VPRS model of low carbon steel's random current dataset

C													*D*
I^t	I^{t-1}	I^{t-2}	I^{t-3}	$W_t{}^t$	$W_t{}^{t-1}$	$W_t{}^{t-3}$	$L_t{}^t$	$L_t{}^{t-1}$	A^t	A^{t-1}	A^{t-2}	$W_b{}^{t-1}$	$W_b{}^t$
–	–	–	–	–	–	1	0	1	–	–	–	0	4
0	1	–	–	0	–	–	–	–	–	–	–	0	4
							……						
–	–	–	–	–	–	0	–	0	–	–	–	–	5
0	–	–	1	–	–	–	0	–	–	1	–	1	5
							……						
–	–	–	–	–	–	–	–	–	1	–	1	1	6
–	–	2	0	–	–	–	–	–	–	–	–	1	6
0	–	1	–	1	2	1	–	–	2	2	–	–	6
							……						
1	1	2	–	1	2	–	–	–	1	–	–	2	7
–	–	–	2	0	–	–	2	–	–	–	2	2	7
–	–	–	1	–	1	1	1	2	–	1	–	2	7

Table 4.14 Predicting result of low carbon steel random welding current VPRS model

Expected value/mm	Predicted value/mm				Not matched
	4	5	6	7	
4	3	1	3	1	1
5	1	10	8	2	5
6	6	15	146	39	20
7	0	1	28	85	3

Table 4.15 Part rules in VPRS model of 3 mm aluminium alloy's random current dataset

C														*D*
I^t	I^{t-1}	$W_t{}^t$	$W_t{}^{t-1}$	$L_t{}^t$	$L_t{}^{t-1}$	$L_t{}^{t-2}$	$L_t{}^{t-3}$	A^t	A^{t-1}	A^{t-2}	$W_b{}^{t-1}$	$W_b{}^{t-2}$	$W_b{}^{t-3}$	$W_b{}^t$
–	–	–	–	–	–	–	–	0	–	–	–	–	–	4
–	0	0	–	–	–	–	–	–	–	–	–	0	–	4
–	–	–	1	0	–	–	0	–		–	0	–	–	4
							……							
0	–	–	–	–	–	1	–	1	1	–	1	–	1	5
1	1	–	–	–	–	–	–	–	1	1	1	–	–	5
0	–	1	–	1	1	–	–	–	–	–	–	–	1	5
							……							
–	–	0	–	1	–	–	–	2	1	2	–	3	–	6
0	–	–	–	–	1	–	–	1	1	–	2	–	2	6
1	0	–	–	–	–	–	–	–	–	2	2	2	–	6
1	–	–	–	–	2	–	1	–	2	1		3	1	6
							……							
–	–	1	–	–	–	–	–	–	–		4	2	–	7
0	1	1	–	–	–	–	–	–	2	1	–	3	2	7
–	–	–	–	1	–	–	–	1	1	1	–	–	3	7
							……							
2	–	–	–	–	–	–	–	5	–	–	3	–	–	11

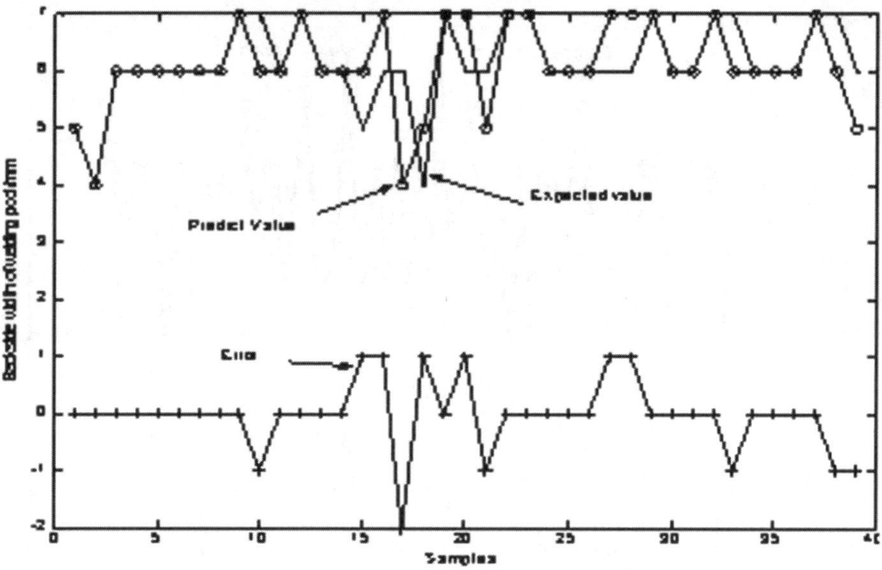

Fig. 4.33 Part validation result of random welding current VPRS model of low carbon steel

Table 4.16 Validation result of 3 mm thickness aluminium alloy VPRS model

Expected value /mm	Predicted value/mm						Not matched
	3.75	5.25	6.75	8.25	9.75	11.25	
3.75	29	24	9	2	0	0	3
5.25	15	61	42	7	0	0	9
6.75	10	45	99	73	11	1	15
8.25	2	3	52	115	37	7	12
9.75	0	0	1	24	37	9	4
11.25	0	0	0	15	7	0	0

$$E_{mean} = \frac{\sum_{i=1}^{N} \left| x_{i,\mathrm{exp}ect} - x_{i,pre} \right|}{N} \tag{4.37}$$

(3) Standard error

$$E_{sd} = \sqrt{\frac{\sum_{i=1}^{N} (x_{i,\mathrm{exp}ect} - x_{i,pre})^2}{N}} \tag{4.38}$$

In Eqs. (4.37) and (4.38), N is the number of testing samples, $x_{i,expect}$ is the expected value of sample x, and $x_{i,pre}$ is the predicted value of sample x.

The results of low carbon steel and aluminium alloy are shown in Tables 4.17 and 4.18 respectively. When $\beta = 1$, VPRS became classic RS.

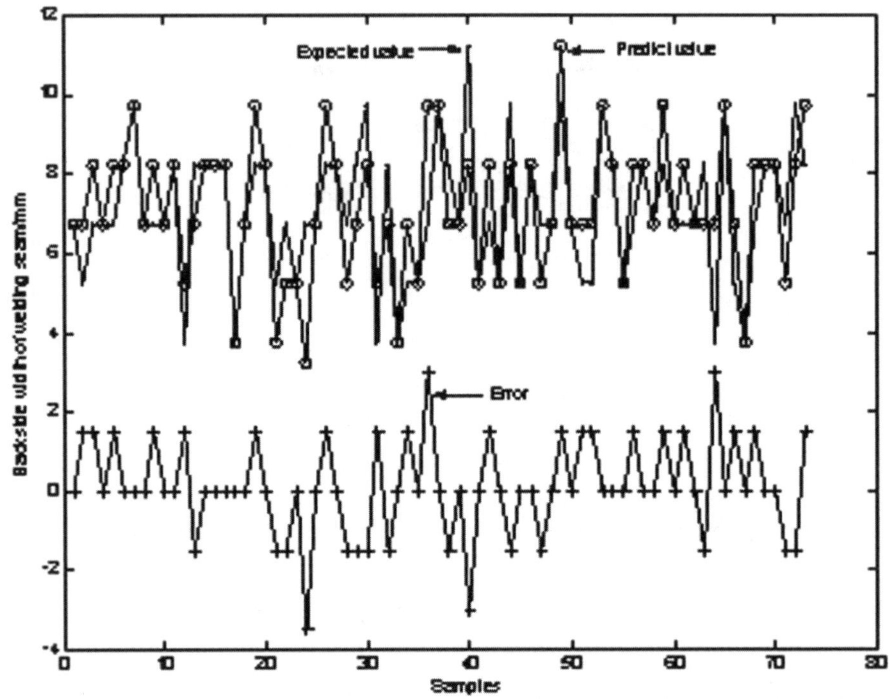

Fig. 4.34 Part validation result of random welding current VPRS model of aluminium alloy

It shows that when the classic RS model ($\beta = 1$) is used, the results accuracy varies within a relative large range, while the results of VPRS are stable. Furthermore, when proper β value is selected, the VPRS model can perform better than classic model.

Table 4.17 Comparative result between VPRS and classic RS on soft steel's multi-variable model

β	The first small decision table			The second small decision table		
	$R_{right}/\%$	E_{mean}/mm	E_{sd}/mm	$R_{right}/\%$	E_{mean}/mm	E_{sd}/mm
1	55.6	0.48	0.79	60.6	0.42	0.72
0.9	62.1	0.41	0.72	60.0	0.42	0.72
0.8	57.7	0.45	0.75	58.5	0.46	0.77
0.7	58.5	0.45	0.75	59.2	0.45	0.75
0.6	61.1	0.42	0.72	61.5	0.41	0.71

Table 4.18 Comparative result between VPRS and classic RS on aluminium alloy random current model

β	The first small decision table			The second small decision table		
	$R_{right}/\%$	E_{mean}/mm	E_{sd}/mm	$R_{right}/\%$	E_{mean}/mm	E_{sd}/mm
1	42.7	0.98	1.39	46.4	1.01	1.28
0.9	45.0	0.94	1.27	47.8	0.99	1.25
0.8	42.7	0.95	1.29	48.0	0.95	1.24
0.7	45.5	0.93	1.25	41.6	1.04	1.32
0.6	44.0	0.94	1.24	39.1	1.10	1.38

4.5 The Chapter Conclusion Remarks

In this chapter, analysis on the welding dynamics is made to understand the process of welding. Based on this analysis, both identification models and intelligent models, such as ANN, fuzzy rules model and RS-based model are discussed. The identification models are easy to be obtained, but it cannot effectively describe the complex process of welding due to its simple structure. ANN model is a "black box" and it is not understandable. It is impossible to directly revise the model obtained by ANN methodology. For FL methodology, the number of inputs, outputs and their linguistic variables cannot be too large, or it will lead to "rule explosion". It is clear that the RS model's predictive ability is close to that of NN while the complexity of RS is much lower than that of NN with good understanbility. RS model is promising for welding process model building. However, it is not fully developed. Further research is necessary, for instance, the method of discretization and the way of expansion of the rule model. The former is concern with the loss of information and the latter relates to knowledge reasoning.

References

1. L. Dimiter. Adaptive robot under fuzzy control. Fuzzy Sets and Systems. 1985, 17(1): 23–28
2. S. Murakami. Weld-line tracking control of arc welding robot using fuzzy logic controller. Fuzzy Sets and Systems. 1989, 32(2): 31–36
3. J.W. Kim, S.J. Na. A self-organizing fuzzy control approach to arc sensor for weld joint tracking in gas metal arc welding of butt joints. Welding Journal. 1993, 72(1):60s–66
4. K. Andersen, G.E. Cook. Gas tungsten arc welding control using artificial neural networks. Proceedings of the 3rd International Conference on Trends in Welding Research, Gatlinburg, Tennessee, USA, 1-5, June, 1992, 135–142
5. T.G. Lim, H.S. Cho. Estimation of weld pool sizes in GMA welding using neural networks. Journal of Systems and Control Engineering. 1993, 207(1):15–26
6. Suga, M. Naruse. Application of neural network to visual sensing of weld line and automatic tracking in robot welding. Welding in the World. 1994, 34:275–284
7. R. Kovacevic, Y.M. Zhang. Neuro-fuzzy model-based weld fusion state estimation. IEEE Transactions on Control Systems Technology. 1997, 5(4):30–42
8. Y. Kaneko, T. Iisaka, K. Oshima. Neuro-fuzzy control of the weld pool in pulsed MIG welding. Quarterly Journal of the Japan Welding Society. 1994, 12(3):374, 378

9. J.J. Wang. "Visual information acquisition and adaptive control of weld pool dynamics of Aluminum alloy during pulsed TIG welding," PhD dissertation, Shanghai Jiao Tong University, 2003
10. W.Y. Zhang. Heat conduction theory in welding process. China Machine Press. 1987, 18–33
11. D.B. Zhao, J.Q. Yi, S.B. Chen, et. al. Extraction of three-dimensional parameters for weld pool surface in pulsed GTAW with wire filler. ASME, Journal of Manufacturing Science and Engineering. 2003, 125(3):493–503
12. S.B. Chen, L. Wu, Q.L. Wang, Y.C. Liu. Self-learning fuzzy neural network and computer vision for control of pulsed GTAW. Welding Journal. 1997, 76(5):201s–209s
13. S.B. Chen. Intelligent methodology for sensing, modeling and control of pulsed GTAW: Part1 – Band-on-plate welding. Welding Journal. 2000, 79(6):151s–163s
14. S.B. Chen. Intelligent methodology for sensing, modeling and control of pulsed GTAW: Part2 – Butt joint welding. Welding Journal. 2000, 79(6):164s–174s
15. Z. Pawlak. Rough sets. International Journal of Computer and Information Science. 1982, 11(5):341–356
16. Z. Pawlak. Rough set approach to knowledge-based decision support. European Journal of Operational Research. 1997, 99:48–57
17. X.H. Hu, N. Cercone. Learning in relational databases: A rough set approach. International Journal of Computational Intelligence. 1995, 11:323–338
18. J. Jelonek et al. Rough set reduction of attributes and their domains for neural networks. International Journal of Computational Intelligence. 1995, 11:339–347
19. J. Wang et al. Data enriching based on rough set theory. Chinese Journal of Computers. 1998, 21(5):393–400
20. D.Q. Miao, G.R. Hu. A heuristic algorithm for reduction of knowledge. Journal of Computer Research and Development. 1999, 36(6):681–684
21. W. Jue, W. Ju. Reduction algorithms based on discernibility matrix: the ordered attributes method. Journal of Computer Science and Technology. 2001, 16(6):489–504
22. B. Wang, S. Chen. Reduction and minimal set cover. Journal of Shanghai Jiaotong University. 2002, 36:106–108
23. A. Skowron, C. Rauszer. The discernibility matrices and functions in information systems. Intelligent Decision Support-Handbook of Application and Advances of the Rough Sets Theory, Kluwer Academic Publishers: Netherlands, 1992
24. B. Wang, S.B. Chen. A complete algorithm for attribute reduction based on discernibility matrix. Shanghai Jiaotong Daxue Xuebao/Journal of Shanghai Jiaotong University. January, 2004, 38(1):43–46
25. A. Wakulicz-Deja, M. Boryczka, P. Paszek, Discretization of continuous attributes on decision systems in mitochondrial encephalomyopathies, Proceedings of RSCTC'98, Warsaw, Poland, June 22–26, 1998:483–490
26. B. Wang, S.B. Chen, W.H. Li, J.J. Wang. Modeling method of the welding process based on rough set theory. Control Theory and Applications. June, 2004, 21(3):411–414
27. J.J. Wang, T. Lin, S.B. Chen. Obtaining of weld pool vision information during aluminum alloy TIG welding. International Journal of Advanced Manufacturing Technology (2005, 26:219–227)
28. M.J. Beynon, M.J. Peel. Variable precision rough set theory and data discretisation: an application to corporate failure prediction. Omega-International Journal of Management Science, 2001, 29(6):561–576
29. W. Ziarko, Variable precision rough set model. Journal of Computer and System Sciences. 1993, 46(1):39–59
30. S.K.M. Wong, W. Ziarko. On optimal rules in decision tables. Bulletin of the Polish Academy of Sciences, Mathmatics. 1985, 33:693–696
31. U.M. Fayyad, K.B. Irani. Multi-interval discretization of continuous-valued attributes for classification learning. Proceedings of the Thirteenth International Joint Conference on Artificial Intelligence. 1993, 2
32. W.H. Li, S.B. Chen, T. Lin. The comparison of discretization method in rough set based modeling method for welding. Journal of Shanghai Jiaotong University, 2006, 40(7):1094, 1097

Chapter 5
Intelligent Control Strategies for Arc Welding Process

Abstract As a complicated process with non-linearity, time varying and uncertainties, GTAW is difficult of modeling and control by classical linear system theory. Intelligent techniques are required to model and control such systems. Among many different "intelligent" approaches, neural network and fuzzy methodologies have emerged as powerful tools owing to their capabilities of emulating human learning processes. And model-free controller is another promising control strategy, which only needs the observed input output data, but can improve the performance of the controller. Based on the visual sensing technology and modeling methods mentioned in the above sections, several different intelligent controllers are described in this chapter, including PSD, Neural Network, model free controller, composite intelligent controller and they are also compared with the open-loop experiment results.

The non-linear behaviour and uncertainties of welding process makes it difficult for conventional methods to model analytically and control this system. Many real industrial processes fall into this category, and hence intelligent techniques are required for such systems. Among many different "intelligent" approaches, neural network and fuzzy methodologies have emerged as very powerful tools for designing intelligent control systems, owing to their capabilities of emulating human learning processes. Model-free controller is another promising control strategy, which only needs the observed input output data. Based on the visual sensing technology and modeling methods mentioned in the above sections, several intelligent control strategies for arc welding will be discussed in this section, and they will be compared with the conventional control strategies.

5.1 Open-Loop Experiments

To investigate the feasibility of the developed control algorithm for practical use, experiments of open-loop during pulsed GTAW are conducted for comparison.

The specimens are aluminium alloy, the thickness is 3 mm, butt joint, pulsed wire feeding, and other conditions shown as Table 2.2. The trapezoid-shaped specimen

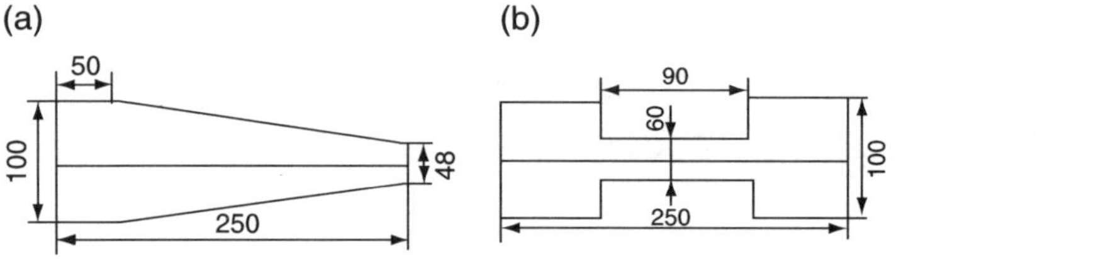

Fig. 5.1 Geometry of specimens (**a**) Trapezoid specimen (**b**) Dumbbell-shaped specimen

for imitating gradient change of heat conduction in the welding process and the dumbbell specimen is used for imitating sudden changes of heat conduction in the welding process, as shown in Fig. 5.1.

From the test results shown in Figs. 5.2 and 5.3, the transition of weld pool size is distinguished when the specimen size change especially for the dumbbell-shaped specimen. The sizes of weld pool become much larger when the heat transfer is varied suddenly, and the weld bead geometry and quality are hardly accepted, as shown in Fig. 5.4. Therefore, control for the welding process is necessary so as to ensure the welding quality. In the following parts, different controlling strategies will be discussed for different welding conditions.

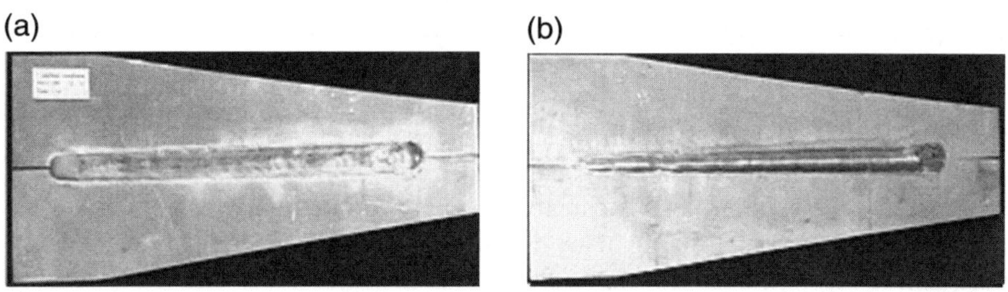

Fig. 5.2 The photographs of trapezoid specimen in constant welding parameters (**a**) Topside (**b**) backside

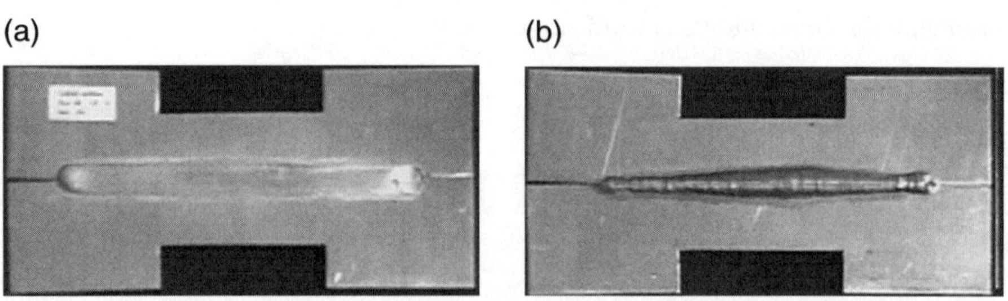

Fig. 5.3 The photographs of dumbbell-shaped specimen in constant welding parameters (**a**) Topside (**b**) Backside

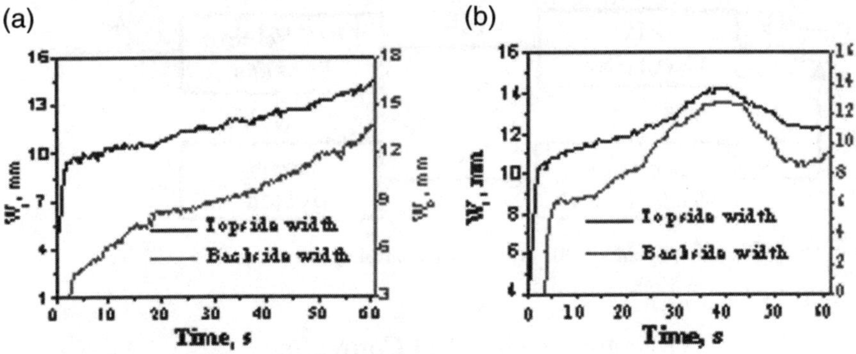

Fig. 5.4 Width curves under varied heat sink in constant welding parameters (**a**) Trapezoid specimen (**b**) Dumbbell-shaped specimen

5.2 PID Controller for Weld Pool Dynamics During Pulsed GTAW

The PID controller is widely used in practice due to its simplicity and robustness. By regulating the proportional, integral and differential item of the error, PID control strategy can regulate the steady-state deviation and transient deviation of the system so as to make the system stable.

5.2.1 PID Control Algorithm

For computer control, increment algorithm is adopted because of the following reasons: taking increment variable as output, calculation without sum, little impulse with hindrance. The increment algorithm with four point difference equations is denoted as follows:

$$\Delta u(k) = q_0 e(k) + q_1 e(k-1) + q_2 e(k-2) + q_3 e(k-3) + q_4 e(k-4) \tag{5.1}$$

$$q_0 = K_p(1 + T / 2T_i + T_d / 6T)$$
$$q_1 = K_p(-1 + T / 2T_i + T_d / 3T)$$
$$q_2 = -K_p \cdot T_d / T$$
$$q_3 = K_p \cdot T_d / 3T$$
$$q_4 = K_p \cdot T_d / 6T$$

where K_p is the coefficient of proportional, T_i is the constant of integral, T_d is the constant of differential, and T is the sampling cycle. The schematic diagram of PID controller is shown is Fig. 5.5.

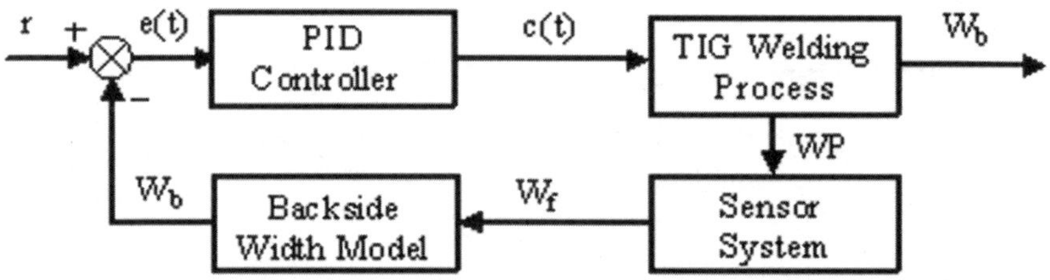

Fig. 5.5 The schematic diagram of PID closed-loop control system for pulsed GTAW

5.2.2 Welding Experiments with PID Controller

The controlled experiment is carried for aluminium alloy pulsed GTAW. The experiment condition is shown in Table 2.2. PID controller is fast in regulating, but large with overshooting. Welding Experiments are carried out for both low carbon steel during butt welding and for aluminium alloy during butt welding to show the effectiveness of PID controllers.

In the controlling diagram, the BWHDNNM model is used for aluminium alloy butter welding. The PID controller parameters K_p, T_i, T_d are determined by the usual optimized method. For current regulating PID controller, $K_p = 0.8941$, $T_i = 1.7833$, $T_d = 1.3575$; for wire feeding velocity regulating PID controller, $K_p = 1.7662$, $T_i = 2.5047$, $T_d = 0.4226$.

The controlled curves are as Figs. 5.8 and 5.11, and the topside and backside photographs are shown in Figs. 5.6, 5.7, 5.9 and 5.10.

Experiments show that PID controller features fast regulating speed, but large overshooting and instable system. From controlling curve, backside width of the trapezoid-shaped specimen fluctuates around 6 mm, but with the variation of heat conduction conditions it fluctuates more greatly with larger steady-state error. Wire feeding velocity regulating PID controller has fast regulating speed, but the variation of wire feeding volume makes disuniform appearance of weld seam. For dumbbell-shaped specimen, both the output of controller and backside width vary greatly when the heating conduction varies violently with large overshooting and

(a)

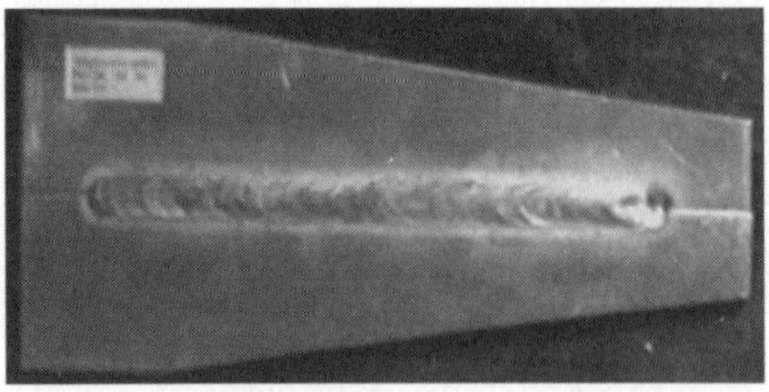

(b)

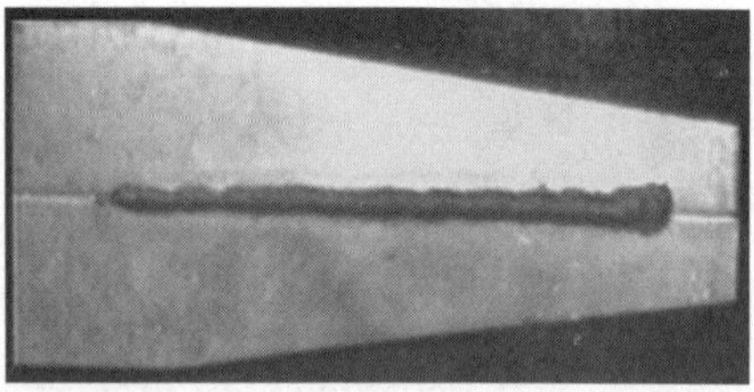

Fig. 5.6 The photographs of trapezoid specimen with PID current controller (**a**) Topside (**b**) Backside

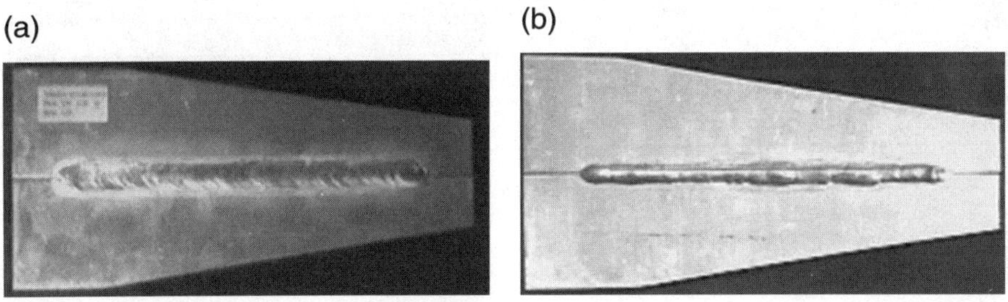

Fig. 5.7 The photographs of trapezoid specimen with PID wire feeding velocity controller (**a**) Top-side (**b**) Backside

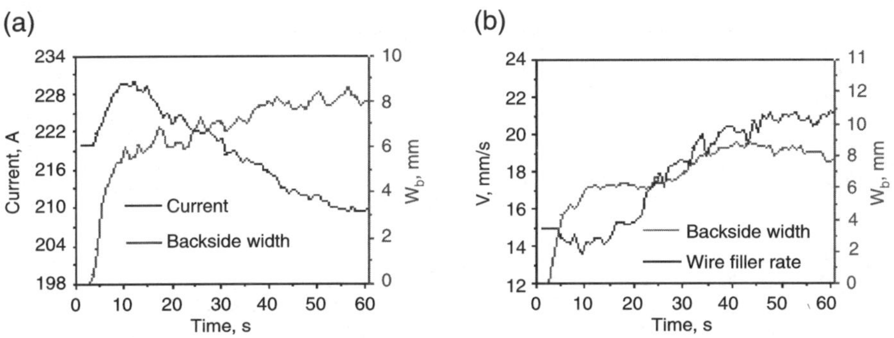

Fig. 5.8 The control curves of trapezoid specimen using PID controller
(**a**) PID current control (**b**) PID wire feeding velocity control

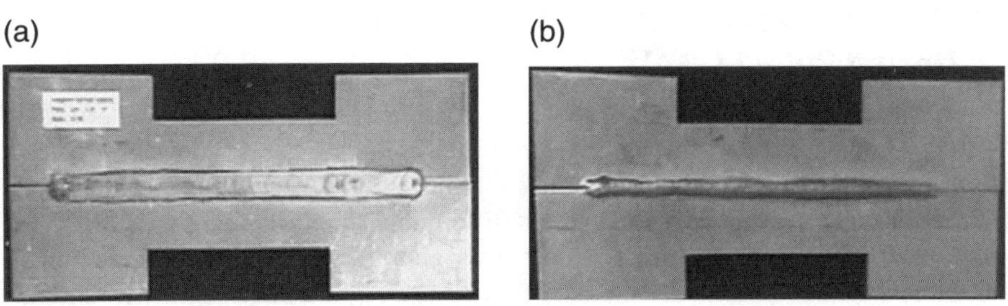

Fig. 5.9 The photographs of dumbbell-shaped specimen with PID current controller (**a**) Topside (**b**) Backside

large steady-state error. Current regulating PID controller is better than wire feeding velocity regulating PID controller when heat conduction is stably pool. However, regulating only current or only wire feeding velocity has different controlling result. Wire feeding velocity regulating is more effective in weld seam cutting because the wire feeding velocity can regulate the volume of melt metal in the weld pool so as to deal with the depression of weld pool.

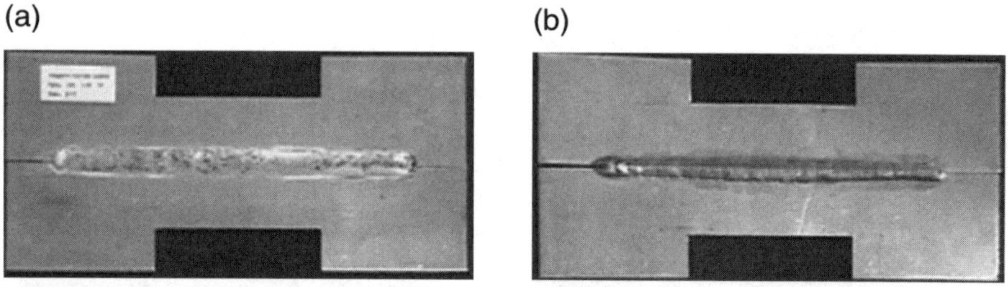

Fig. 5.10 The photographs of dumbbell-shaped specimen with PID wire rate controller (**a**) Topside (**b**) Backside

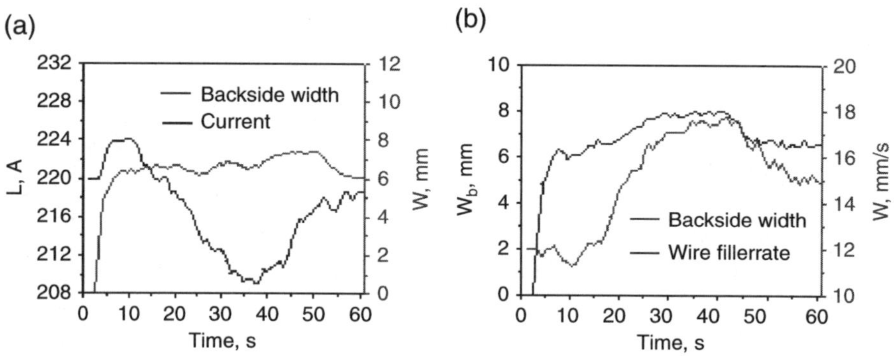

Fig. 5.11 The control curves of dumbbell-shaped specimen using PID controller (**a**) PID current control (**b**) PID wire feeding speed control

5.3 PSD Controller for Weld Pool Dynamics During Pulsed GTAW

As to the multi-variable, time-delay, nonlinear welding process, conventional PID closed-loop control is limited due to the coefficients are hardly regulated on-line. A self-learning PSD (proportional, sum and differential) control is proposed based on single neuron so that it can change structure and coefficients. Only the desired output and the real output on-line detected are needed to design a PSD controller, and the on-line identification of the process coefficients is not necessary here. The weight of neuron is corrected on-line with the improved BP algorithm, and the performance of error is minimized to optimizc the output of the control system.

5.3.1 PSD Controller Algorithms

The schematic diagram of self-learning PSD control system is shown in Fig. 5.12. And the BNNM model for predicting the backside width of the weld pool is discussed in Sect. 4.3.2.

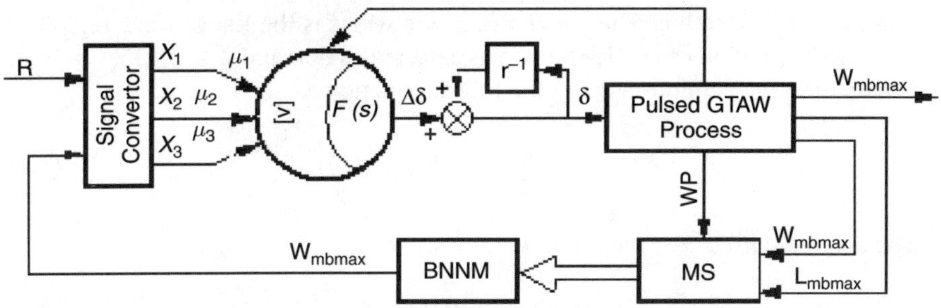

Fig. 5.12 Schematic diagram of single neuron self-learning PSD control system

Where the effect of signal convert is to generate the variables $X = \{X_1, X_2, X_3\}$ as the input of self-learning PSD controller with the desired and predicted backside width. In PSD controller, X is summed and transferred nonlinearly to derive the variation of pulse duty ratio (Δ). The actual outputs are measured by MS and input in BNNM to attain the predicted $W_{mb\max}$. At the same time, the input weights of neuron is adjusted on-line with BP algorithm according to the error of backside width, to keep the control with optimized state.

The input variables $X = \{X_1, X_2, X_3\}$ of controller are the error, error with first order, error with second order between desired backside width and BNNM output, shown as follows.

$$x_1 = \alpha_1 e(t) = \alpha_1 (R - W_{mb\max}(t)) \tag{5.2}$$

$$x_2 = \alpha_2 \Delta e(t) = \alpha_2 (e(t) - e(t-1)) \tag{5.3}$$

$$x_3 = \alpha_3 \Delta^2 e(t) = \alpha_3 (e(t) - 2e(t-1) + e(t-2)) \tag{5.4}$$

Where α_1, α_2 and α_3 are constant indicating the emphasis degree in the controller, selected as α_1 is 1.0, α_2 is 0.3, α_3 is 0.1.

Weight normalization is to avoid the saturation of weight during learning process, shown as follows:

$$w_i' = w_i \Big/ \sqrt{\sum_{j=1}^{3} w_i^2} \tag{5.5}$$

The sum of weighted neuron is:

$$s = \sum_{i=1}^{3} w_i' \cdot x_i \tag{5.6}$$

$F(s)$ is nonlinear transfer function, selected as hyperbolic tangent function.

$$\Delta\delta = F(s) = \gamma(1 - e^{-2\xi S}) / (1 + e^{-2\xi S}) \tag{5.7}$$

Where γ and ξ are two constants, γ sets the saturated value of control variable, and ξ sets the linear degree of control variable. The larger is γ, the more possible is to attain the desired value. The smaller is ξ, the wider is the linear work region to restrain the fluctuation of stable status. γ is selected as 300 and ξ is 0.135.

Object function is minimized by the follow equation.

$$E(w) = \frac{1}{2}\sum_{k}(R - W_{mb\max}(k))^2 \tag{5.8}$$

The derived corrected formula of weight is.

$$\Delta w_i = \eta \varphi x_i \tag{5.9}$$

$$w_i(k+1) = w_i(k) + \Delta w_i, \quad i = 1,2,3 \tag{5.10}$$

Where η is learning rate, equals to 1.0. φ is the equalized output error of neuron, resembled with the actual output error.

$$\varphi = (R - W_{mb\max}(k)) \cdot (1 - \Delta\delta(k)/\gamma) \cdot (1 + \Delta\delta(k)/\gamma) \tag{5.11}$$

5.3.2 Welding Experiments with PSD Controller

Welding experiments are carried out on low carbon steel respectively during bead-on-plate welding and during butt welding so as to validate the effectiveness of the PSD controller. The experiment condition is shown in Table 2.1

Figure 5.13(A) shows the simulation results based on BNNM given backside width is 5.0 mm. The maximum override is 2.81%, the regulating time is 2s, steady error is 0.04 mm, and pulse duty ratio is stabilized at 42%. Figure 5.13(B) shows the variation curve of weight. The different simulations are performed with the output value set on 4.5 mm and 4.0 mm. The results show that the maximum override is similar to the PID, but the regulating time and steady error are smaller, furthermore, the coefficients can be adjusted on-line to make it capable of controlling the non-linear process.

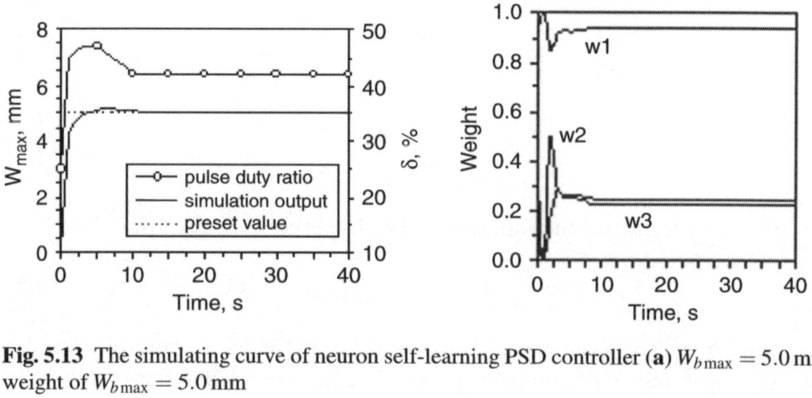

Fig. 5.13 The simulating curve of neuron self-learning PSD controller (**a**) $W_{b\max} = 5.0$ mm (**b**) the weight of $W_{b\max} = 5.0$ mm

5.3.2.1 PSD Controller for Low Carbon Steel During Butt Welding

Butt welding experiment for low carbon steel is conducted. The specimen is dumbbell-shaped mild steel plate with 2 mm thickness. The size and the shape of the work-piece are shown in Fig. 5.14. The varying curves of peak current and maximum width of the weld pool are shown as Fig. 5.15. The peak value and duty ratio of the pulse current were regulated by neuron self-learning PSD controller during welding process. The Fig. 5.16 is photos of controlled work-piece.

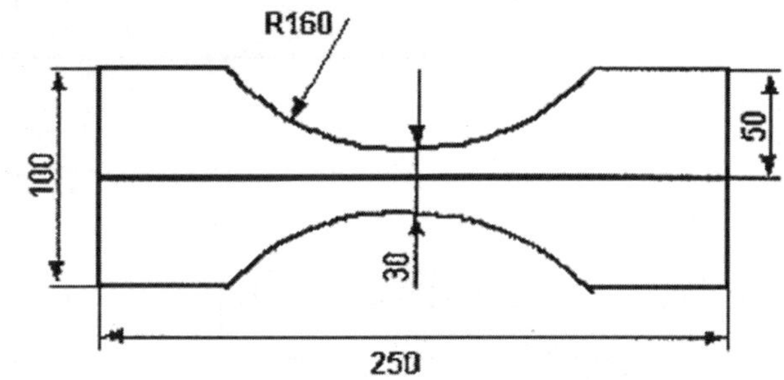

Fig. 5.14 Shape and the size of the work-piece

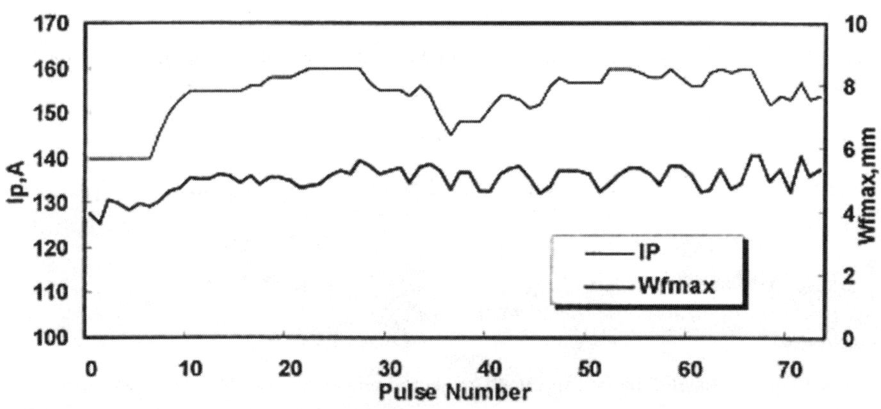

Fig. 5.15 Curves of neuron self-learning PSD control during pulsed GTAW

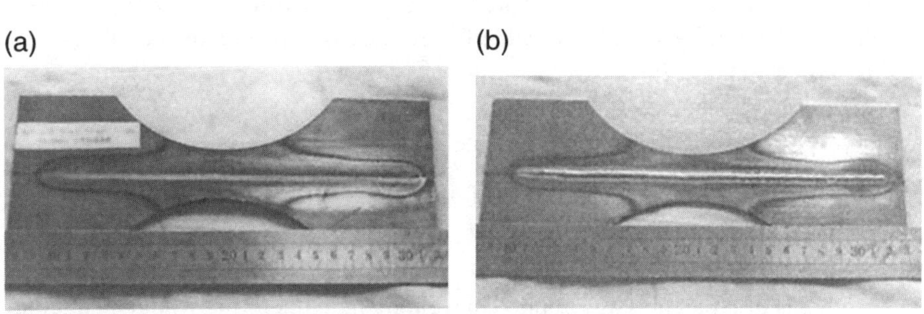

Fig. 5.16 Photographs of the PSD control of weld work-piece (**a**) Topside (**b**) Backside

5.3.2.2 PSD Controller for Low Carbon Steel During Bead-On-Plate Welding

To test the feasibility of the self-learning PSD control schematic for practical use, experiment during the pulsed GTAW is also conducted. The controlled effect is better than the PID controller, shown in Fig. 5.17. The maximum error between BNNM output and desired value is 0.26 mm. Test data is compared with the desired value, the maximum error is 0.30 mm, the average error is 0.10 mm, and the root-mean-square deviation is 0.08 mm. The perfect effect of topside and backside photographs is also shown in Fig. 5.18.

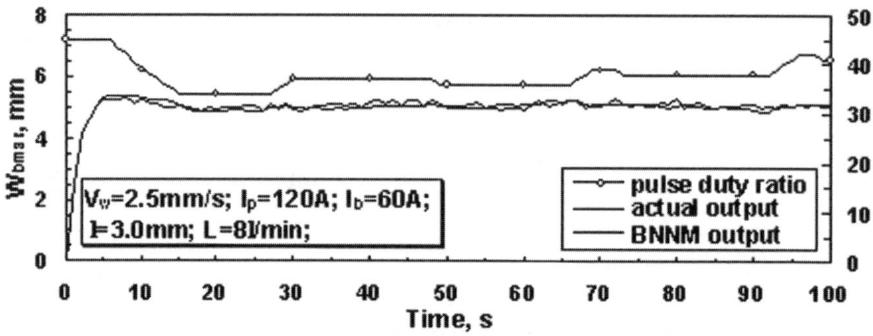

Fig. 5.17 The neuron self-learning PSD closed-loop control curves of dummy bell specimen during pulsed GTAW

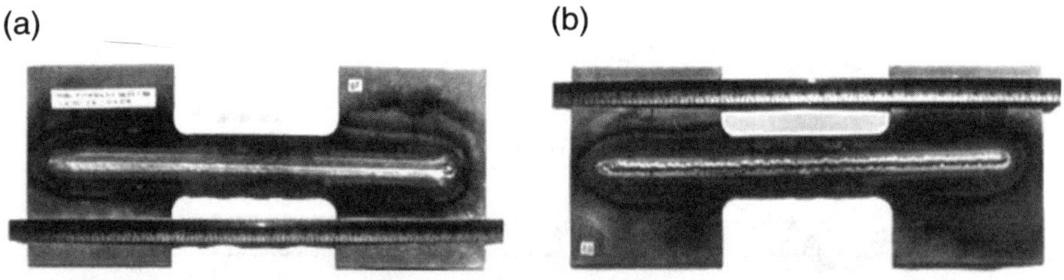

Fig. 5.18 Photographs of dumbbell specimen by neuron self-learning PSD control (**a**) Topside (**b**) backside

Experiment results show that the PSD controller can be successfully implemented for pulse GTAW process control with perfect control effect for different set values and conditions, and is suitable for the varied structure and coefficients of welding process specially.

5.4 NN Self-Learning Controller for Dynamical Weld Pool During Pulsed GTAW

Developments of AI technology applied in welding manufacturing would inaugurate an inspired approach of solving difficult problems for many years. Among all the intelligent control methods, artificial neural network (ANN) shows that the method has been used in a modest scale to develop nonlinear models and control systems. For example, recent studies [1–3] and their cited references have demonstrated the utility and flexibility of the concept within the domain of process engineering. And different with conventional control schematic, fuzzy logic control is based not on mathematical model or physical model, but on the skilled worker's experience, so that it is a suitable method for the complex systems. Generally, fuzzy control design includes determining the structure of fuzzy control, designing control rules, establishing fuzzy correlation and calculation of defuzzier. A combination of fuzzy logic and ANN bring into being the self-learning fuzzy neural network controller (FNNC).

5.4.1 FNNC Control Algorithm

Fuzzy control rules express the fuzzy relationship between the input and output of control system. The result of Extraction of fuzzy rules for welding pool dynamics is discussed in Sect. 4.4.1, and it will be omitted here.

5.4.1.1 FNNC Structure

In Fig. 5.19, every layer and every node of the fuzzy neural network corresponds to a part of the fuzzy system. According to the membership function and inference

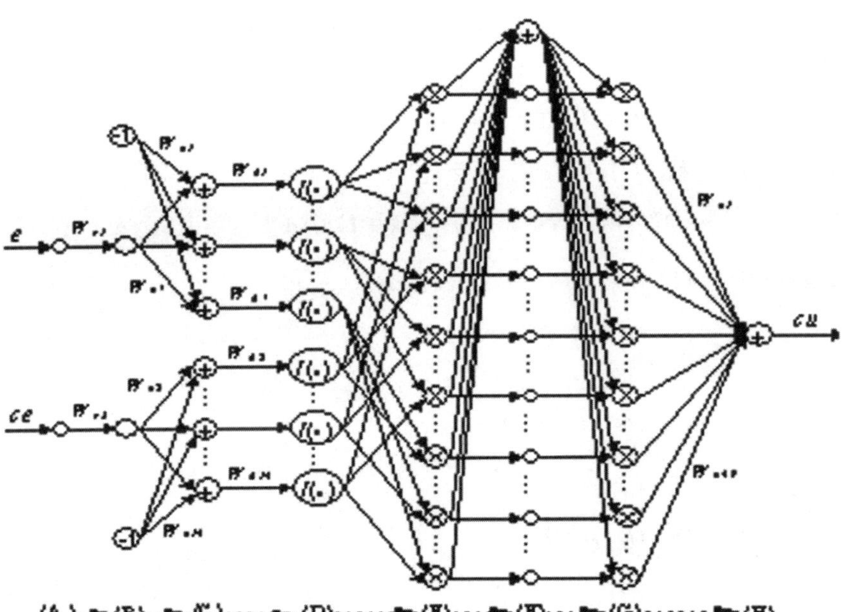

Fig. 5.19 The structure of fuzzy neural network controller

process, its all nodes and weights have certain physical meanings. For the jth fuzzy control rule of the welding process, fuzzy control system is as follows:

Rule$_j$: IF (E is E^j) and (CE is CE^j) THEN (CU is CU^j)

$$j = 1, 2, \ldots \mathrm{m} \tag{5.12}$$

Where $E^j \in \{E_1, E_2, \cdots E_7\}$, $CE^j \in \{CE_1, CE_2, \cdots CE_7\}$, $CU^j \in \{CU_1, CU_2, \cdots CU_7\}$,

The output of the fuzzy system is calculated as follows:

$$cu = \frac{\oplus_{j=1}^{m} h_j CU^j}{\oplus_{j=1}^{m} h_j} \tag{5.13}$$

Where cu is the crisp value of the output variable of the control system, "$\oplus$" means algebraic sum, and h_j is the matched degree between the current input and the jth fuzzy control rule, described as:

$$h_j = \mu_{E^j}(e) \cdot \mu_{CE^j}(ce) \tag{5.14}$$

Where $\mu_{E^j}(e)$ is the membership of e belonging to the fuzzy subset E^j, and $\mu_{CE^j}(ce)$ is the membership of ce belonging to the fuzzy subset CE^j.

Above fuzzy system can be expressed by the fuzzy neural network FNNC showed in Fig. 5.19, the unmarked weights are one. The network includes eight layers, where A $\sim$ D layers transform the input crisp value to the membership of the fuzzy subsets of E and CE, E $\sim$ F layers carry out fuzzy inference, G layer fulfills fuzzy synthesis, H layer acts as fuzzy judging , then the crisp output value is acquired.

Supposing the input of jth node of in the lth layer of the FNNC is I_j^l, the output is O_j^l, the relationship between input and output of each layer of FNNC is described as follows:

$$\text{A layer: } I_j^a = x_j,\ O_j^a = I_j^a,\ j = 1, 2,\ x_1 = e,\ x_2 = ce \tag{5.15}$$

$$\text{B layer: } I_j^b = W_{sj} \cdot O_j^a,\ O_j^b = I_j^b,\ j = 1, 2 \tag{5.16}$$

$$\text{C layer: } I_k^c = (O_j^b - W_{ck}),\ O_k^c = I_k^c, j = [(k-1)/7] + 1,\ k = 1, 2, \cdots 14 \tag{5.17}$$

$$\text{D layer: } I_i^d = O_i^c \cdot W_{di},\ O_i^d = e^{-(I_i^d)^2},\ i = 1, 2, \cdots, 14 \tag{5.18}$$

$$\text{E layer: } I_k^e = O_i^d \cdot O_j^d,\ O_k^e = I_k^e,\ i = 1, 2, \cdots, 7,\ j = 8, 9, \cdots 14,\ k = 7i + j - 14 \tag{5.19}$$

$$\text{F layer: } I_1^f = \oplus_{k=1}^{49} O_k^e,\ O_1^f = (1/I_1^f) \tag{5.20}$$

$$I_k^f = O_{k-1}^e,\ O_k^f = I_k^f,\ k = 2, 3, \cdots, 50 \tag{5.21}$$

$$\text{G layer: } I_k^g = O_1^f \cdot O_{k+1}^f,\ O_k^g = I_k^g,\ k = 1, 2, \cdots, 49 \tag{5.22}$$

$$\text{H layer: } I_1^h = \oplus_{k=1}^{49} O_k^g \cdot W_{bk}, O_1^h = I_1^h \tag{5.23}$$

5.4.1.2 Learning of the FNNC

The learning process of the FNNC consists both off-line and on-line learning. The off-line learning of the FNNC is to get the initial membership of the fuzzy subsets of input fuzzy variables E and CE, to determinate the initial weights between the nodes in the G $\sim$ H layer, and to reduce the on-line learning time of the FNNC. The on-line learning of the FNNC is to modify the membership of the fuzzy subsets and fuzzy control rules during welding, so that the FNNC is adaptive to the fluctuation of the welding.

The learning sample of the FNNC is provided with the fuzzy control rules showed as Table 4.4. We define that the initial quantizing factor W_{si} of e and ce is 6, the initial center W_{ci} of the membership function of the fuzzy subsets in E is $\{-4.80, -3.24, -1.72, 0.00, 1.24, 3.00, 4.80\}$, the initial distributing parameters W_{di} of the membership function of the fuzzy subsets in E is $\{1.068, 1.082, 1.029, 1.126, 1.111, 0.936, 0.926\}$, the initial centers W_{ci} of the membership function of the fuzzy subsets in CE are $\{-5.07, -3.45, -1.90, 0.00, 1.79, 3.28, 4.93\}$, and initial distributing parameters W_{di} of the membership function of the fuzzy subsets in CE is $\{1.029, 1.052, 0.966, 0.903, 1.016, 1.062, 1.010\}$. The initial value of the weights W_{bi} can be acquired by the off-line learning of the FNNC. Training the FNNC for 500 times using back-propagation algorithm, the initial membership of the fuzzy subsets of E and CE are obtained, shown in Fig. 5.20, and e and ce are normalized.

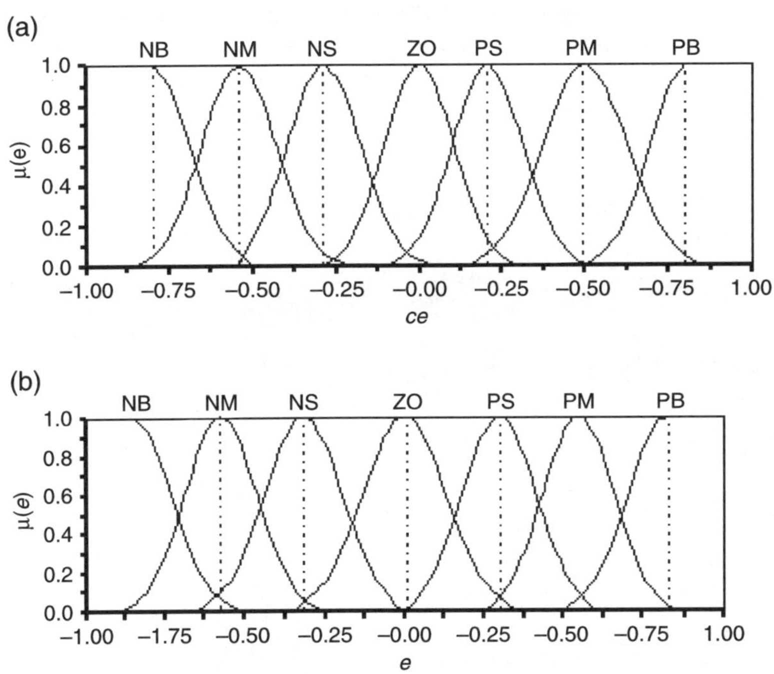

Fig. 5.20 Initial membership function of fuzzy subsets (**a**) Error (**b**) change in error

Fig. 5.21 Initial relationship surface between input and output of FNNC

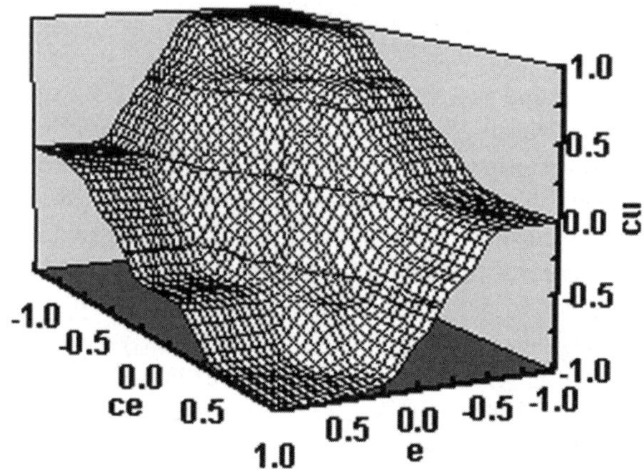

The relationship surface between the input and output of the FNNC after off-line learning is described in Fig. 5.21.

The goal of the on-line learning of the FNNC is to minimize the output error of the controlled object. The error function is defined as follows:

$$E^* = \frac{1}{2}\sum_{i=1}^{n}(y_i - y_d)^2 \tag{5.24}$$

Where n is the number of learning sample, y_i denotes the output of the controlled object, and y_d denotes The desired output of the controlled object. Using back-propagation learning algorithm, the back-propagating error of each layer is calculated and the weights of the FNNC is adjusted as follows:

H layer:

$$\delta_k^h = -\frac{\partial E^*}{\partial I_k^h} = \sum_{i=1}^{n}(y_i - y_d)\cdot\frac{\partial y_i}{\partial u}\cdot f'(I_k^h),\ f'(I_k^h) = \frac{\partial O_1^h}{\partial I_k^h} = 1,$$
$$k = 1,2,\cdots,49 \tag{5.25}$$

$$W_{bk}(t+1) = W_{bk}(t) + \eta\cdot\delta_k^h\cdot O_k^g + \alpha\cdot[W_{bk}(t) - W_{bk}(t-1)] \tag{5.26}$$

G layer:

$$\delta_k^g = -\frac{\partial E^*}{\partial I_k^g} = \delta_k^h\cdot W_{bk}(t)\cdot f'(I_k^g),\ f'(I_k^g) = \frac{\partial O_k^g}{\partial I_k^g} = 1,\quad k = 1,2,\cdots,49 \tag{5.27}$$

F layer:

$$\delta_1^f = -\frac{\partial E^*}{\partial I_1^f} = \sum_{l=1}^{49}\left(\delta_l^g\cdot O_{l+1}^f\right)\cdot f'(I_1^f),\ f'(I_1^f) = \frac{\partial O_1^f}{\partial I_1^f} = -\left(O_1^f\right)^2 \tag{5.28}$$

$$\delta_k^f = -\frac{\partial E^*}{\partial I_k^f} = \delta_l^g \cdot O_1^f \cdot f'(I_k^f),\ f'(I_k^f) = \frac{\partial O_k^f}{\partial I_k^f} = 1, \quad l = k-1, \quad k = 2,3,\cdots,50 \tag{5.29}$$

E layer:

$$\delta_k^e = -\frac{\partial E^*}{\partial I_k^e} = \left(\delta_1^f + \delta_l^f\right) \cdot f'(I_k^e),\ f'(I_k^e) = \frac{\partial O_k^e}{\partial I_k^e} = 1, \quad l = k+1, \quad k = 1,2,\cdots,49 \tag{5.30}$$

D layer:

$$\delta_k^d = -\frac{\partial E^*}{\partial I_k^d} = \sum_{l=1}^{7} \left(\delta_{7(k-1)+l}^e \cdot O_{l+7}^d\right) \cdot f'(I_k^d),\ f'(I_k^d) = \frac{\partial O_k^d}{\partial I_k^d} = -2I_k^d \cdot O_k^d, \quad k = 1,\cdots,7 \tag{5.31}$$

$$\delta_k^d = -\frac{\partial E^*}{\partial I_k^d} = \sum_{l=1}^{7} \left(\delta_{7(l-1)+k-7}^e \cdot O_l^d\right) \cdot f'(I_k^d),\ f'(I_k^d) = \frac{\partial O_k^d}{\partial I_k^d} = -2I_k^d \cdot O_k^d, \quad k = 8,\cdots,14 \tag{5.32}$$

$$W_{dk}(t+1) = W_{dk}(t) + \eta \cdot \delta_k^d \cdot O_k^c + \alpha \cdot [W_{dk}(t) - W_{dk}(t-1)], \quad k = 1,2,\cdots,14 \tag{5.33}$$

C layer:

$$\delta_k^c = -\frac{\partial E^*}{\partial I_k^c} = \delta_k^d \cdot W_{dk}(t) \cdot f'(I_k^c),\ f'(I_k^c) = \frac{\partial O_k^c}{\partial I_k^c} = 1, \quad k = 1,2,\cdots,14 \tag{5.34}$$

$$W_{ck}(t+1) = W_{ck}(t) + \eta \cdot \delta_k^c \cdot O_k^b + \alpha \cdot [W_{ck}(t) - W_{ck}(t-1)] \tag{5.35}$$

B layer:

$$\delta_k^b = -\frac{\partial E^*}{\partial I_k^b} = \sum_{l=1}^{7} \delta_{7(k-1)+l}^c \cdot f'(I_k^b),\ f'(I_k^b) = \frac{\partial O_k^b}{\partial I_k^b} = 1, \quad k = 1,2 \tag{5.36}$$

$$W_{sk}(t+1) = W_{sk}(t) + \eta \cdot \delta_k^b \cdot O_k^a + \alpha \cdot [W_{sk}(t) - W_{sk}(t-1)] \tag{5.37}$$

Where δ_j^l denotes the back propagating error of the jth node in the lth layer of the FNNC, η is learning coefficient, and α is momentum factor. The regulation of the FNNC weights can be realized with the formulas (5.25), (5.26), (5.27), (5.28), (5.29), (5.30), (5.31), (5.32), (5.33), (5.34), (5.35), (5.36), (5.37), which is one-step learning algorithm of the FNNC.

5.4.2 Experiment of FNNC Control Scheme

5.4.2.1 FNNC Controller for Low Carbon Steel During Butt Welding Without Wire Feeding

In order to verify the above FNNC algorithm for control of the pulsed GTAW butt welding, the closed-loop system with FNNC control scheme is designed as Fig. 5.22. The experiment condition is shown in Sect. 4.3.4.

Given the backside maximum width of weld pool is 6.0 mm, the FNNC simulation is accomplished. The learning coefficient η is 0.45, and the momentum factor α is 0.20. We define that the error $e \in [-2.0mm, +2.0mm]$, the change in error $ce \in [-1.5mm, +1.5mm]$, and the change in pulse duty ratio $cu \in [-12\%, +12\%]$. The simulating result is shown in Fig. 5.23, the override of the backside maximum width of weld pool is 3.26%, the ascending time of the backside maximum width of weld pool is 3 s.

To validate the effectiveness of the FNNC control schematic, welding experiments are conducted. The specimen is 2 mm-thick-low carbon steel plate, shown in Fig. 5.24, to imitate the gradual changes in heat transfer during welding.

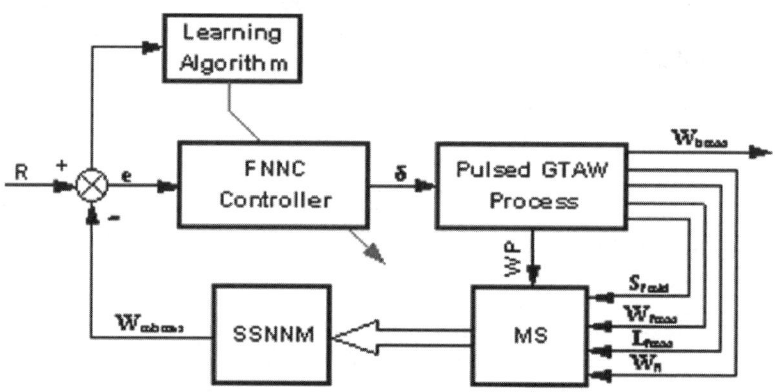

Fig. 5.22 Schematic diagram of FNNC closed-loop control system

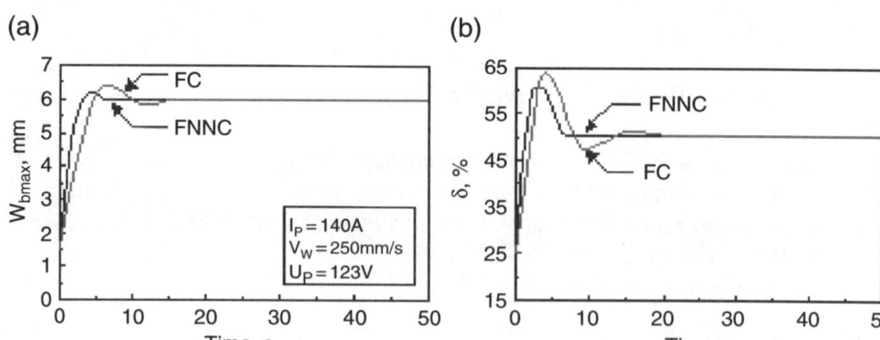

Fig. 5.23 Simulating curve of FNNC **(a)** $W_{b\max} = 6.0$mm **(b)** $W_{b\max} = 5.0$mm

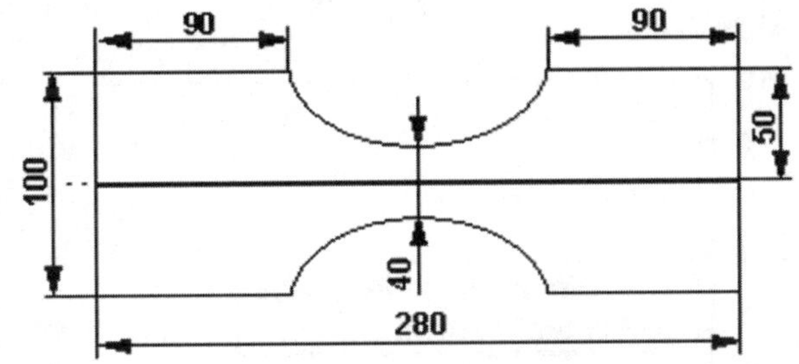

Fig. 5.24 Geometry of arc specimen

Given $W_{b\max}$ is 6.0 mm, FNNC closed-loop control experiments are carried on the specimens, with the minimum regulating unit of pulse duty ration is 1%. The schematic of the control system is shown in Fig. 5.22. The output variable is pulse duty ratio. The output parameters of the welding process are topside size and shape parameters, and backside size parameters. $W_{b\max}$ is the controlled variable. MS is to sense topside size and shape parameters, such as $W_{f\max}$, $L_{f\max}$, S_{fmid}, and ten W_{fi}, and welding parameters. Weights are corrected based on the above algorithm for self-learning on-line.

Figure 5.25 shows the variations of the backside size parameters with constant welding parameters. The backside size parameters fluctuated greatly when the heat transfer becomes worse at the 35 pulse and 70 pulse.

The variations of backside size parameters and pulse duty ratio are shown in Fig. 5.26. From Fig. 5.26(b), we can see the regulation according to the variation of heat transfer, similar with the performance of skilled operator. $W_{b\max}$ maintains around the given value on the whole, and the statistics shows that the maximum error is 0.46 mm, the average error is 0.07 mm, the root-mean-square error is 0.20 mm. This indicates that the size parameters on the width direction can be controlled by regulating pulse duty ratio.

Figure 5.27 shows member functions of normalized error (e) and change in error (ce) of FNNC control after on-line self-learning process. Figure 5.28 shows the final relationship surface between input and output of FNNC, and the differences indicate that the control rules are corrected by self during welding process to realize the

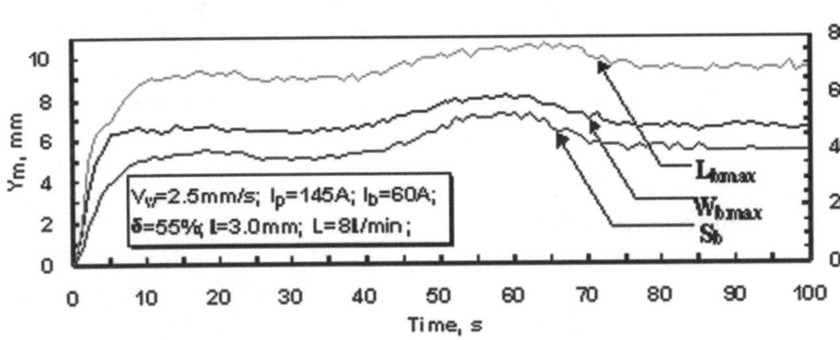

Fig. 5.25 Weld pool sizes in constant welding parameters

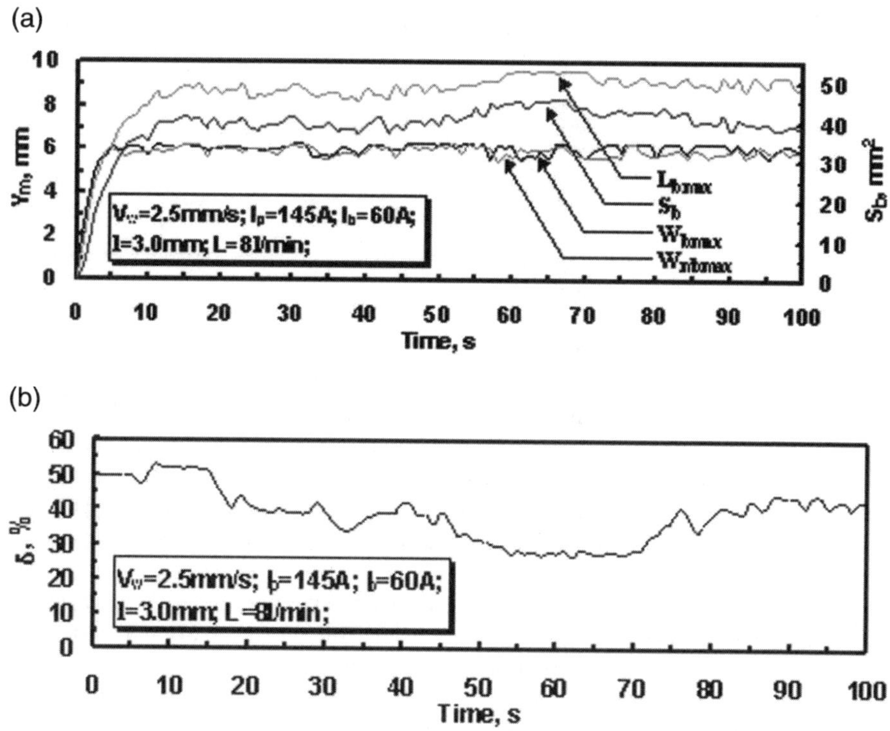

Fig. 5.26 FNNC closed-loop control curves e during pulsed GTAW (a) Backside sizes of weld pool (b) pulse duty ratio

intelligent control function. The relation surface changes more intensely than that of Fig. 5.28, which indicates a large nonlinear correlation during welding process.

5.4.2.2 FNNC Controller for Low Carbon Steel During Butt Welding with Wire Filler and Gap Variation

Compared with no wire filler and constant gap size condition, in this part a self-learning fuzzy neural network controller for pulsed GTAW in butt welding process with wire filler and varied gap is presented. The experiment is as follows, pulse peak current 120 A, base current 60 A, pulse duty ratio 40%, and travel speed 2.5 mm/s. According to imaging principle, image is determined by light source, camera and object shape. Here, imaging current is set as 30 A, and the light source of arc can be supposed as a point light source.

In order to verify the above FNNC algorithm for control of the pulsed GTAW butt welding with gap variations, the closed-loop system with FNNC control scheme is designed as Fig. 5.29. The BWHDNNM model for predicting the backside width of the weld pool is discussed in Sect. 4.3.1.

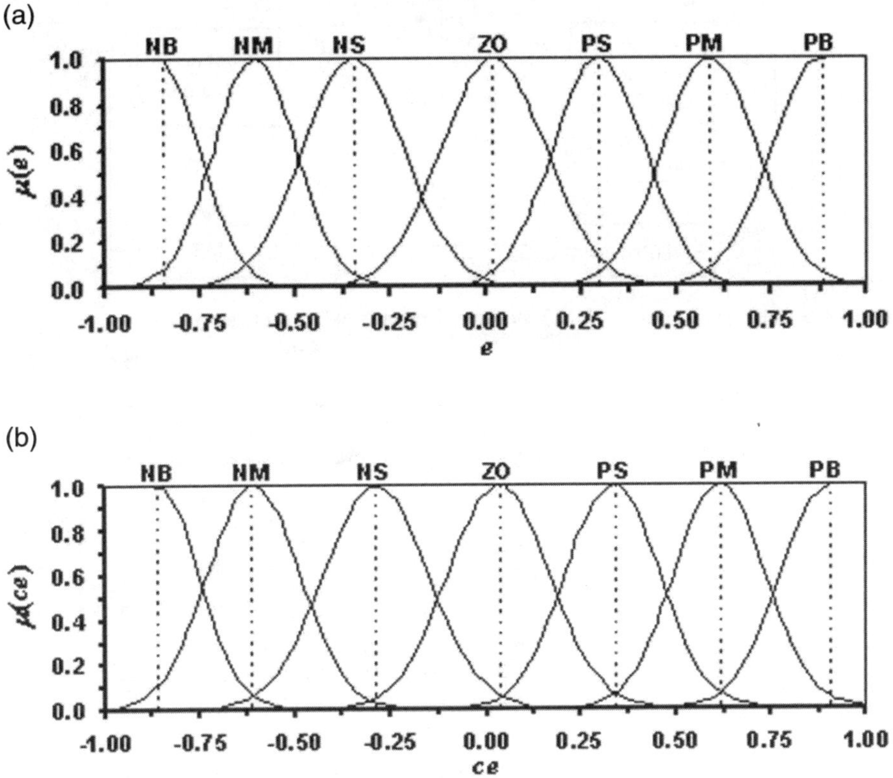

Fig. 5.27 The membership functions of error and error change (**a**) Error (**b**) change in error

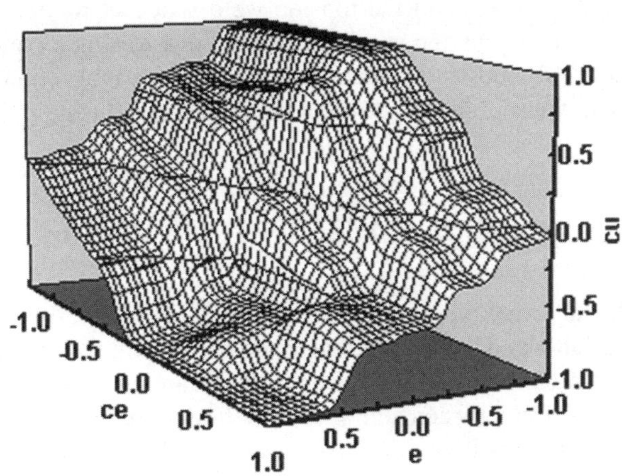

Fig. 5.28 Final relationship surface between input and output of FNNC

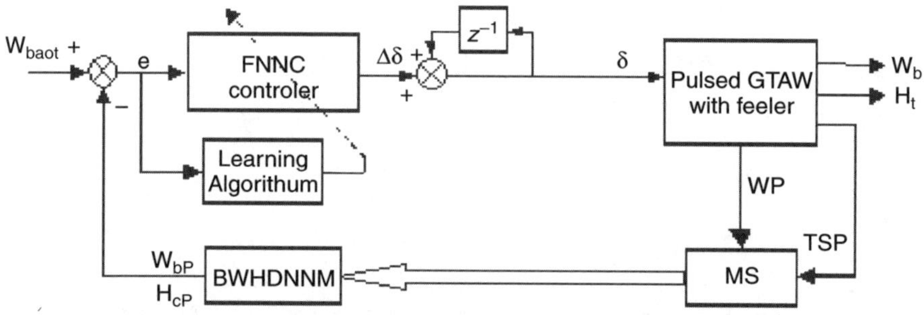

Fig. 5.29 Schematic diagram of FNNC controller for butt welding with gap variation

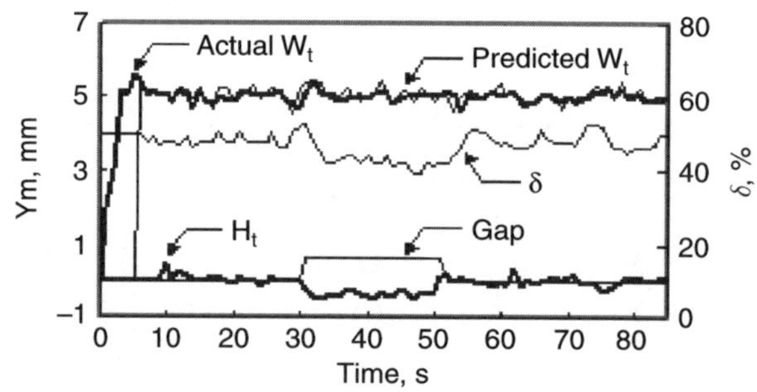

Fig. 5.30 The FNNC closed-loop control curves of varied gap specimen during pulsed GTAW

The FNNC control experiments on pulsed GTAW are conducted. The backside maximum width is given as 5.0 mm, and controlled results as following: the maximum error 0.39 mm, the average error 0.014 mm, and the root-mean-square deviation 0.14 mm. The controlled curves are shown in Fig. 5.30, the topside and backside photos of the weld work-piece are shown in Fig. 5.31.

Experiment results show that for different welding specimen, the visible fluctuations of the controlled performances are obviously depressed due to the robustness of the FNNC. But the topside weld shape of the work-piece is still not desirable, particularly in gap varying segment, due to single control input to the welding process. Obviously, the advanced and multi-variable controller should be developed for both appropriate penetration and fine shape during pulsed GTAW process.

5.5 Model-Free Adaptive Controller for Arc Welding Dynamics

Because arc welding is characterized as inherently nonlinear, time varying, multivariable and having a strong coupling among welding parameters, it is very difficult to find a reliable mathematical model and to design an effective control scheme for arc welding by conventional modeling and control methods [4–9]. To design control system only based on information from the I/O data of the GTAW process will be

(a)

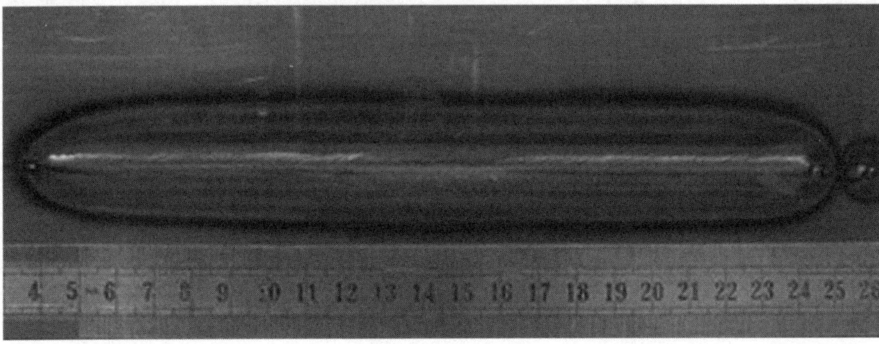

(b)

Fig. 5.31 Photographs of varied gap specimen with FNNC closed-loop control

of great significance. Recently, some kinds of model-free adaptive control scheme have been explored, which based on information from the I/O data of the controlled plant. The controller is constructed through use of a function approximator (FA) such as a neural network or polynomial (no FA is used for the unmodeled system equations). A convergence result for stochastic approximation algorithms with time-varying objective functions and feedback is established [10–14]. A PD-type cross-coupled controller is developed to asymptotically stabilize multi-axis motions while synchronizing positions of all axes in the set-point position control. Experimental results verify the effectiveness of the proposed approaches. Based on a new concept of partial derivative called "pseudo-partial-derivative" (PPD), the model-free learning adaptive control (MFLAC) of a class of nonlinear systems is presented. The BIBO stability of the regulator is proved [8–13]. In fact, this model-free adaptive control is the approach of unity of modeling and control. The modeling and real time feedback control are united in the identification approach, and the pattern of parameter adaptive is broken up [8–14]. The model free control method has excellent performance in oil refining, chemical industry, power, light industry. The application results of the STEC composition control have verified the feasibility and effectiveness of the model-free adaptive control [8–17]. Because the model-free adaptive control can deal well with the control problem of the discrete time nonlinear plant whose structure, parameters and order are time-varying, the attempt to apply model-free adaptive control for the shape of the weld pool is significant and novel.

This section addresses the model-free adaptive controller (MFC) for arc welding dynamics to achieve full penetration and fine formation of the weld seam. The MFC controller only needs the observed input output data, but can improve the performance of the controller.

5.5.1 Preliminary of Model-Free Adaptive Control (MFC)

5.5.1.1 Universal Process Model

The following general discrete SISO nonlinear systems is considered

$$y(k+1) = f\left(Y_k^{k-n}, u(k), U_{k-1}^{k-m}, k+1\right) \tag{5.38}$$

Where

$k = 0, 1, \cdots$, stands for discrete time
$y(k+1)$ represents a one-dimensional state output
$u(k)$ is an input variable
$Y_k^{k-n} = \{y(k), \cdots, y(k-n)\}$ are the sets of system outputs
$U_k^{k-m} = \{u(k), \cdots, u(k-m)\}$ are the sets of system inputs
n and m are the orders of output $y(k)$ and input $u(k)$
$f(\cdots)$ is a general nonlinear function

Following assumptions are made about the controlled plant:

When the system is in the steady state, it satisfies the condition that if $u(k) = u(k-1)$, then $y(k+1) = y(k)$.

The nonlinear function $f(\cdots)$ has a continuous gradient with respect to control input $u(k)$.

From the assumptions above and using the mean value theorem in the Calculus, we have

$$\begin{aligned} & f\left(Y_k^{k-n}, u(k), U_{k-1}^{k-m}, k+1\right) - f\left(Y_k^{k-n}, u(k-1), U_{k-2}^{k-m}, k+1\right) \\ & = \nabla f\left(Y_k^{k-n}, \overline{u(k-1)}, U_{k-1}^{k-m}, k+1\right)^{\mathrm{T}} [u(k) - u(k-1)] \end{aligned} \tag{5.39}$$

Where $\overline{u(k-1)} = u(k-1) + \theta\,(u(k) - u(k-1))$, θ satisfies $0 \leq \theta \leq 1$. Therefore, we have

$$y(k+1) - y(k) = \nabla_{u(k-1)} f\left(\overline{u(k-1)}, k+1\right)[u(k) - u(k-1)] + \xi(k+1) \quad (5.40)$$

If $\|u(k) - u(k-1)\| \neq 0$, let

$$\varphi(k) = \nabla_{u(k-1)} f\left(\overline{u(k-1)}, k+1\right) + \left[(u(k) - u(k-1)) \Big/ \|u(k) - u(k-1)\|^2\right]$$

Then equation (5.40) can be written as

$$y(k+1) - y(k) = \varphi(k)\,[u(k) - u(k-1)] \quad (5.41)$$

Where $\varphi(k)$ can be considered a pseudo gradient of model (5.41). Note that when the system is in a steady state, because of $\|u(k) - u(k-1)\| = 0$, we have $y(k+1) = y(k)$, so in this case, Eq. (5.41) is a valid expression, which is called universal model.

5.5.1.2 Model-Free Adaptive Control Basic Algorithm (MFCB)

Estimation of the Pseudo Gradient $\varphi(k)$

It is clear that the necessary condition that the universal model (4) can be used in practice is that the estimation of $\varphi(k)$, denoted as $\hat{\varphi}(k)$, is available in real-time, and is sufficiently accurate. Considering the control action is known, define the cost function

$$J(\varphi(k)) = \left[y^*(k+1) - y(k) - \varphi(k)^{\mathrm{T}} \Delta u(k-1)\right]^2 + \mu \|\varphi(k) - \varphi(k-1)\|^2 \quad (5.42)$$

where $\Delta u(k-1) = u(k-1) - u(k-2)$, because at the moment $\Delta u(k)$ is unavailable we substitute $\Delta u(k-1)$ for it, $y^*(k+1)$is the desired output of the controlled plant, μ is positive weighting constant which constrains the change of the pseudo gradient $\varphi(k) - \varphi(k-1)$. By using (5.41), the minimization of the cost function (5.42), gives estimation

$$\widehat{\varphi}(k) = \widehat{\varphi}(k-1) + \left(\eta \Delta u(k-1) / \mu + \Delta u^2(k-1)\right) \cdot (\Delta y(k) - \widehat{\varphi}(k-1) \cdot \Delta u(k-1)) \quad (5.43)$$

Where η is a suitable small positive number.

Design of Model-Free Adaptive Control

At k+1, the controller needs to determine the control action ($u(k)$) based on the feedback ($y(k)$) to drive the welding process to reach the desired output ($y^*(k+1)$). The model-free adaptive control is then described as follows: Assume that the observed data $\{u(k-1), y(k)\}$ $(k = 1, \ldots)$ are known, and the expected output $y^*(k+1)$ at $(k+1)$th time is given. Find a controller $u(k)$, such that the output of the system $y(k+1)$ matches $y^*(k+1)$. In order to achieve a robust control, it is required that the following cost function is minimized:

$$J(u(k)) = |y^*(k+1) - y(k+1)|^2 \tag{5.44}$$

It is known that fluctuations in welding parameters will generate unsmooth weld appearance, which is not acceptable.

Also, the large changes of the control actions could make the closed-loop system unstable. Hence, the following modified cost function is used:

$$J(u(k)) = |y^*(k+1) - y(k+1)|^2 + \lambda \, \|u(k) - u(k-1)\|^2 \tag{5.45}$$

Where λ is the weight. The analytic solution is

$$u(k) = u(k-1) + \left(\rho\varphi(k) / \left(\lambda + \varphi^2(k)\right)\right) \cdot (y^*(k+1) - y(k)) \tag{5.46}$$

Where ρ is called a control parameter, which selection is closely related to the convergence of the control law. Equation (5.46) is called the model-free adaptive control basic algorithm. And its convergence has been proved [6, 7].

5.5.2 *The Improved Model-Free Adaptive Control with G Function Fuzzy Reasoning Regulation*

5.5.2.1 General Algorithm of Model-Free Adaptive Control with G Function (MFCG)

The model-free adaptive control general algorithm with G function regulation is described as following.

$$u(k) = u(k-1) + \left(\rho\varphi(k) / \left(\lambda + \varphi^2(k)\right)\right) \cdot \{[y^*(k+1) - y(k)] + G\} \tag{5.47}$$

Where G is a feasible function and represents the part of functions combination of controller. The necessary condition that the model-free adaptive control with G function can be sued is that this algorithm is convergent.

Let

$$\eta(k) = \begin{cases} 1, & y^*(k+1) - y(k) = 0 \\ 1 + G / (y^*(k+1) - y(k)), & y^*(k+1) - y(k) \neq 0 \end{cases} \quad k = 2, 3, \cdots \tag{5.48}$$

Then if $y^*(k+1) - y(k) \neq 0$ we have

$$\begin{aligned} y^*(k+1) - y(k) + G &= [1 + G / (y^*(k+1) - y(k))] \cdot [y^*(k+1) - y(k)] \\ &= \eta(k) \cdot [y^*(k+1) - y(k)] \end{aligned} \tag{5.49}$$

If $y^*(k+1) - y(k) = 0$ we have

$$y^*(k+1) - y(k) + G = 0$$
$$= \eta(k) \cdot [y^*(k+1) - y(k)] \tag{5.50}$$

So the MFCG algorithm (5.47) can be transformed to the basic form (5.46), which has been proved convergence [4–7, 18, 19]. In application, we can design the G function according to demand.

5.5.2.2 MFC with G Function Fuzzy Reasoning Regulation (MFCGF)

Here, we define the G function with fuzzy reasoning regulation. Let the error vector be $e = y^*(k+1) - y(k)$, where $y^*(k+1)$ denotes the desired contolled variable. The input of fuzzy reasoning regulation are error and rate of error. The main steps of the fuzzy reasoning regulation are to build the membership functions, the fuzzy reasoning rules, and the fuzzy transducer. The structure diagram of fuzzy reasoning regulation and the membership functions of input are as shown in Figs. 5.32 and 5.33 respectively.

Having analyzed the arc welding process, we define the reasoning rules. In general, the weld pool increases as the current increases. So the reasoning rules is:

1) if E is NB and $\dot{E}$ is N then B is NB;
2) if E is NB and $\dot{E}$ is Z then B is NS;

⋮

15) if E is PB and $\dot{E}$ is P then B is PB.

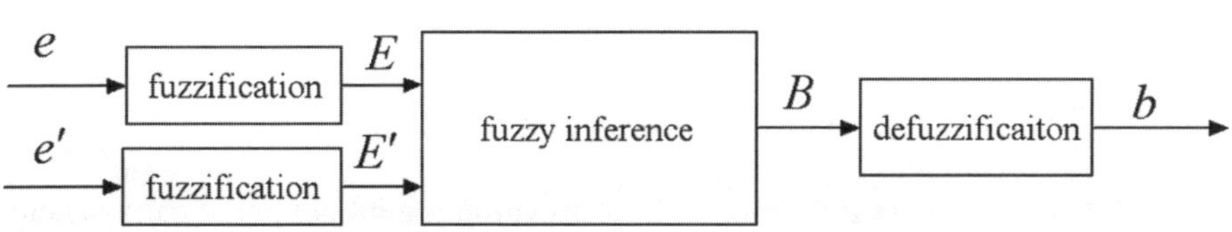

Fig. 5.32 The structure diagram of fuzzy reasoning regulation

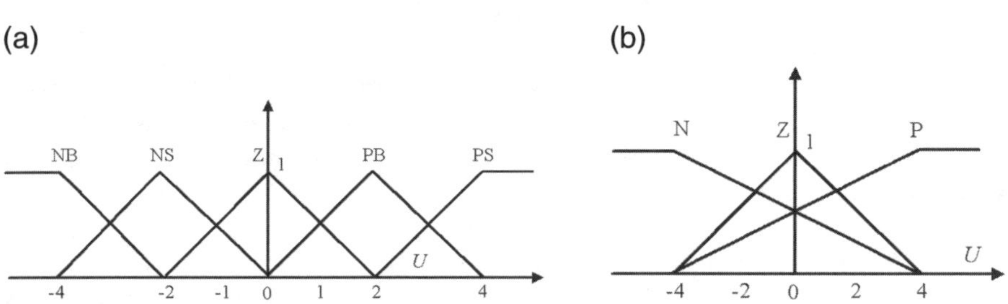

Fig. 5.33 The membership functions of input (**a**) $E(B)$ membership function (**b**) $\dot{E}$ membership function

5.5.3 Realization and Simulation of Improved Control Algorithm

5.5.3.1 Arc Welding Process Modeling

GTAW is influenced by a number of parameters, including the welding current, arc length, and welding speed. In general, the weld pool increases as the current increases and the welding speed decreases. When the arc length increases, the arc voltage increases so that the arc power increases, but the distribution of the arc energy is decentralized so that the efficiency of the arc decreases. As a result, the correlation between the weld pool and arc length may not be straightforward.

Compared with the arc length, the roles of the welding current and welding speed in determining the weld pool and weld fusion geometry are much more significant and definite. For the case of full penetration, the state of the weld penetration is specified by W_b. Here, we selected I_p as the control variable. The controlled process can therefore be defined as a GTAW process in which the welding current adjusted on-line to achieve the desired backside width of the weld pool.

A polynomial Auto-regressive with exogenous input or ARX model representation [8–16] is selected as the model representation. Consider the ARX model below:

$$A(q^{-1})y(k) = B(q^{-1})u(k) + e(k) \tag{5.51}$$

The model belongs to linear-in-the-parameter model, therefore the parameter estimation can be performed using least square method. The model in equation (5.51) can be represented

$$y(k) = \boldsymbol{\Phi}^{\mathrm{T}}(k)\boldsymbol{\theta} + e(k) \tag{5.52}$$

Where

$$\theta^{\mathrm{T}}(k) = \begin{bmatrix} a_1 & \cdots & a_n & b_1 & \cdots & b_n \end{bmatrix}$$

$$\boldsymbol{\Phi}^{\mathrm{T}}(k) = \begin{bmatrix} y(k-1) & \cdots & y(k-n) & u(k-d-1) & \cdots & u(k-d-n) \end{bmatrix}.$$

The identification is thus simplified by estimation the model parameters. There are n+n parameters to be identified, and u is welding current, y is backside width of the weld pool. Based on the input output data $\{u(k), y(k)\}, k = 1, 2, \ldots, N$, and the variation of square sum of residuals, we can determine approximately that the evaluation of n is 5.

Also through experiment data, ARX model of backside weld width (W_b) with welding parameters (I_p) is derived using the least square method developed with the Matlab program. The model can be derived using statistic hypothesis testing method as follows:

$$\begin{aligned} y(k) = {} & a_1 y(k-1) + a_2 y(k-2) + a_3 y(k-3) + a_4 y(k-4) + a_5 y(k-5) \\ & + b_1 u(k-1) + b_2 u(k-2) + b_3 u(k-3) + b_4 u(k-4) + b_5 u(k-5) \end{aligned} \tag{5.53}$$

Where

$$A = \left[a_1\ a_2\ a_3\ a_4\ a_5\right] = \left[1.2245\ {-0.7935}\ 0.45269\ {-0.23124}\ 0.11518\right]$$

$$B = \left[b_1\ b_2\ b_3\ b_4\ b_5\right] = \left[0.0085696\ {-0.3748}\ 0.0039714\ {-0.16826}\ 0.0023674\right]$$

The feasibility of this model is verified by comparing the simulation results with the Matlab program and actual outputs. The square sum of residuals is 0.0303437.

5.5.3.2 Simulation

Following are several steps that describe how the model-free adaptive algorithm works.

(1) For the observed input output data $\{u(k-1), y(k)\}$ and the developed universal model (5.41), we can obtain $\widehat{\varphi}(k)$ which is the estimation of pseudo partial derivative $\varphi(k)$, using least squares algorithm.
(2) For the desired output $y^*(k+1)$, we have

$$u(k) = u(k-1) + \left(\rho\varphi(k) / \left(\lambda + \varphi^2(k)\right)\right) \cdot \{[y^*(k+1) - y(k)] + G\}$$

where G is the fuzzy reasoning regulation.Then a new set of data $\{u(k-1), y(k)\}$ can be achieved, applying the control action $u(k)$ to the GTAW process (here we use the model (5.53)).
(3) Repeat steps (1) and (2) above to generate a serial of data $\{u(k-1), y(k)\}$. The desired backside weld width was chosen as $y^* = 6\,\text{mm}$. Figures 5.38 and 5.39 show the response of the system under the controller with $\eta = 0.5$, $\mu = 15$, $\rho = 4.2$ and $\lambda = 0.5$.

Simulation of the model-free adaptive control with fuzzy reasoning regulation performance was conducted. The comparison simulation results between MFCGF and MFC were shown in Fig. 5.34. It could be seen that the model-free adaptive control had small overshoot, quick adjusting speed.

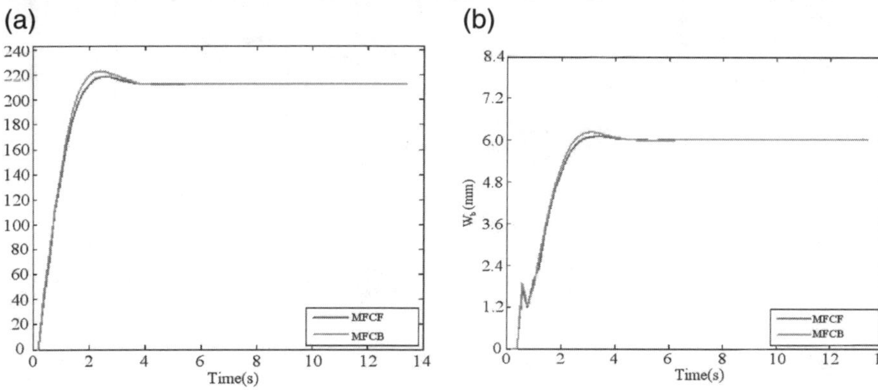

Fig. 5.34 Simulation results of MFC with G function fuzzy reasoning regulation and MFC controller ($W_b = 6mm$) (a) Control actions (**b**) Outputs

5.5.4 Controlled Experiments on Pulsed GTAW Process

5.5.4.1 Sensing and Control System for Arc Welding Process

In practical welding, the main aim of the control of the welding process is to ensure a stable weld of desirable appearance, which is very important for the weld quality. For this purpose, a mouding method using top and back information of the weld pool is necessary for improving the weld quality. The experimental set-up includes a welding power source, wire feeder, image sensor and other equipment. The welding current, wire feeding speed, and welding speed was controlled by a computer. A diagram of the system is shown in Fig. 5.35.

The system took a PC computer as control center, included two parts: welding parameters control part and image sensing part. The former part controlled welding power source, movable workplate, and sampled welding current, welding voltage etc. The latter part realized simultaneous image sensing of front topside, back topside and backside images of weld pool in a frame. In this study, experiments with groove welds in a butt joint were done on 4-mm-thick Ld10 plate. The experiment conditions were tabulated in Table 5.1.

The weld pool geometry is measured by a vision system. A complete weld pool image in a frame is shown in Fig. 5.36, in which the top left is the back topside image, the top right is the front topside image, and the bottom is the backside image. The back topside image of the pool can be divided into the following parts: nozzle,

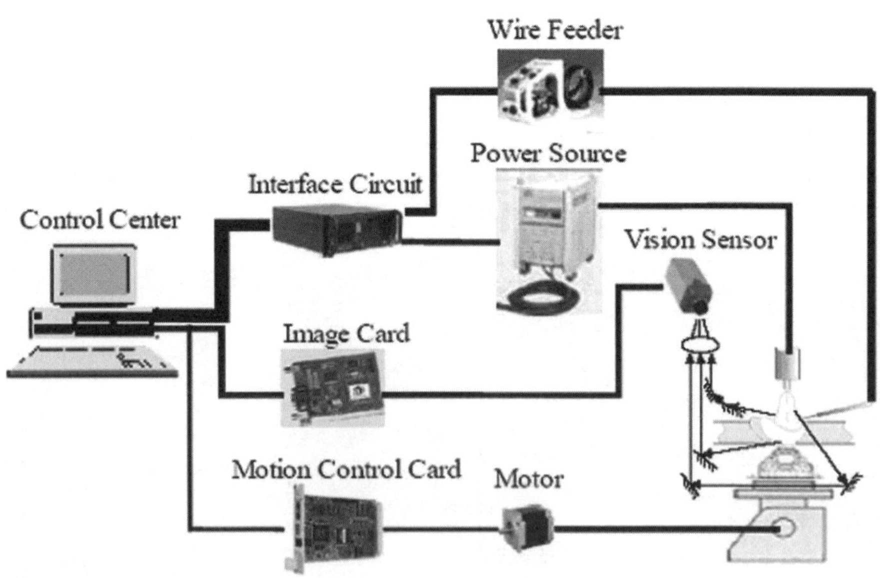

Fig. 5.35 The structure diagram of experimental system

Table 5.1 The experimental condition for image sensing

Pulse Frequency f(Hz)	Peak Current I_P(A)	Base Current I_b(A)	Welding Speed V_w(mm/s)	Filling Rate V_f(mm/s)	Arc Length L(mm)	Electrode Diameter Φ(mm)	Angle of Tip θ(deg)	Arc flow rate (l/min)	Pulse duty ratio (%)
2	215	50	2.8	10	4	3.2	90	13	1

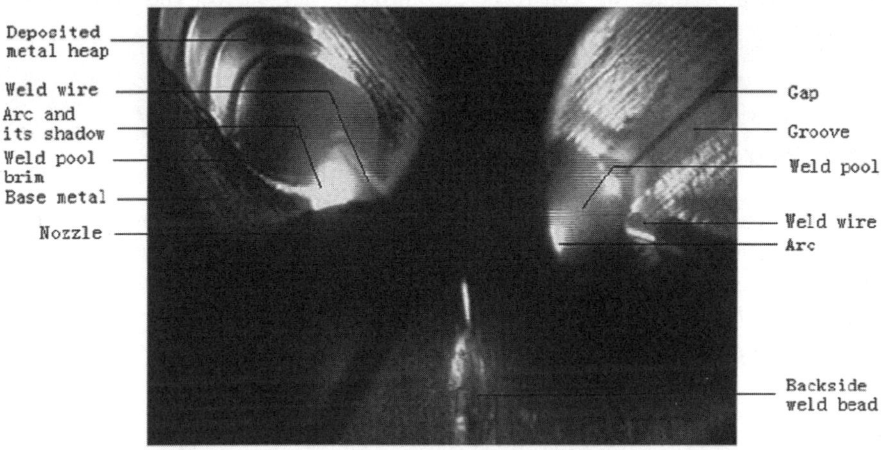

Fig. 5.36 The front topside, the back topside and the backside synchronous image

deposited area of metal heap, weld pool brim, base metal, arc & its shadow and weld wire etc. In the front topside image, the gap, groove, weld pool and weld wire are clear.

The weld pool contains abundant information about the welding process. The geometry parameters of the weld pool are important character for weld quality. As shown in Fig. 5.37, the topside length L_t, maximum width W_t, and half-length ratio

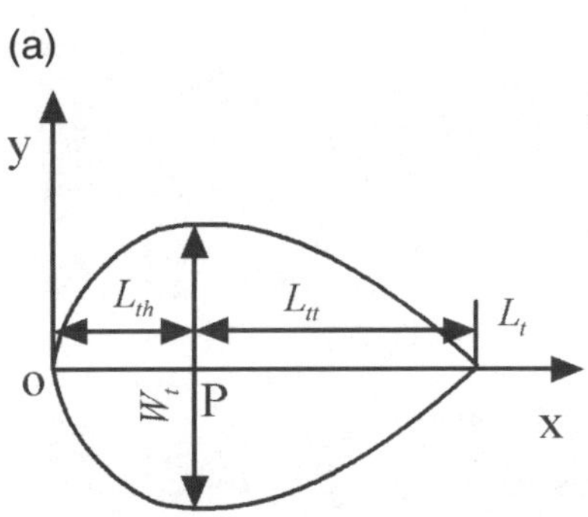

Fig. 5.37 The definition of the geometry features of weld pool (**a**) Topside weld pool (**b**) Backside weld pool

R_{hl} can be used as the characteristic parameters to describe the shape and size of the topside weld pool, the half-length ratio R_{hl} was defined by the following formula: $R_{hl} = \frac{L_{tt}}{L_{th}+L_{tt}} \times 100\%$. The backside pool is specified by the maximum width W_b. A real-time image processed algorithm has been developed to extract these parameters [4–11, 18, 19]. The process of image processing include scale-multiplication-based edge detection, noise removal, calibration, and piecewise curve fitting. By using of this procedure, we can receive controlled variable W_b and other useful parameters.

5.5.4.2 Controlled Experiments Results

To test the feasibility of the model-free adaptive with fuzzy reasoning regulation control, experiments during pulsed GTAW are conducted. The specimens for the test were 4-mm-thick Ld10 plate with root opening. The three kinds of shape including trapezia-shaped specimen, graded dumbbell-shaped specimen and mutant

Fig. 5.38 The trapezia-shaped workpiece with constant welding parameters (**a**) Topside (**b**) Backside

(a)

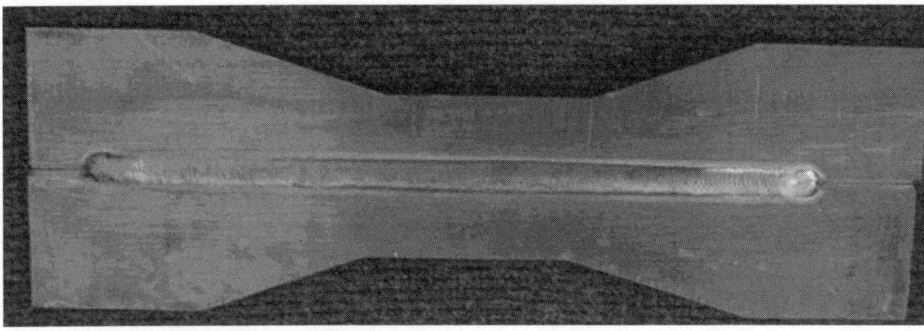

(b)

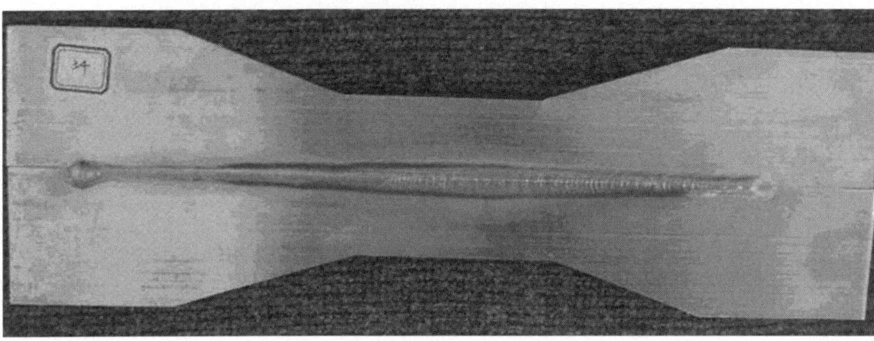

Fig. 5.39 The graded dumbbell-shaped workpiece with constant welding parameters (**a**) Topside (**b**) Backside

dumbbell-shaped specimen imitated various changes in heat-transfer conditions in the welding. The experiment on controlled objects is to obtain fine formation of W_b of the specimen. Without regulation actions, W_b changes violently beyond the desired value in Figs. 5.38, 5.39 and 5.40. That is the weld sizes became larger when the size of the workpiece became narrower.

Figures 5.41, 5.42 and 5.43 shows the model-free adaptive control with G function fuzzy reasoning regulation effect with the desired backside width $W_d = 6.0mm$. Figures 5.44, 5.45 and 5.46 shows the variation in shape parameters. The control variable is the peak pulse current, and it decreased with the heat-sinking conditions turning poor and increased with the condition reversing. Apparently with manipulation of the process by the model-free adaptive control scheme, the controlled results of weld bead width are satisfactory. As is seen from the control results, the model-free adaptive controller can guarantee the backside width uniformity. The control results of this controller can satisfy the requirement for weld shaping in most applications.

(a)

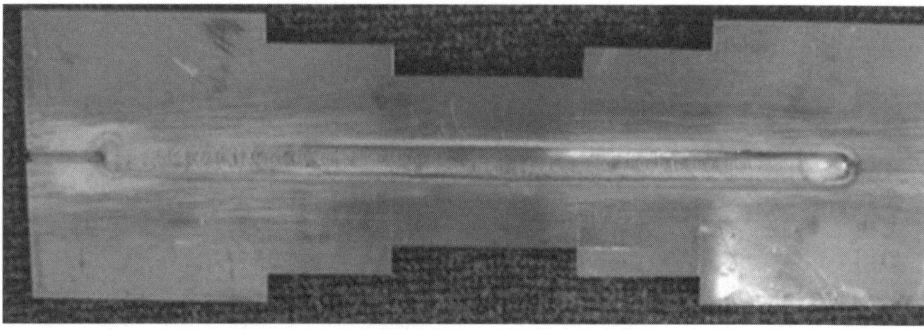

(b)

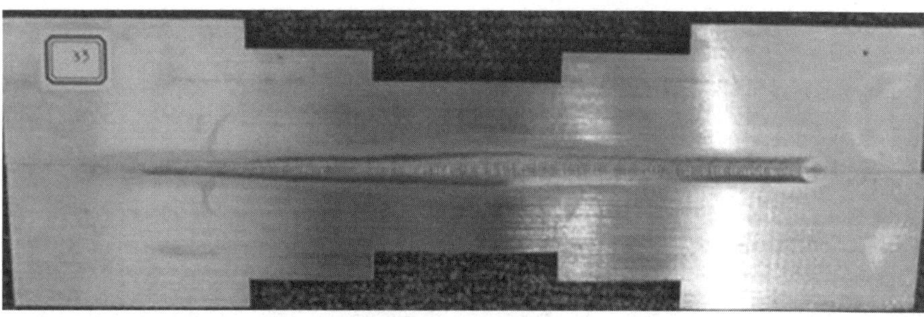

Fig. 5.40 The mutant dumbbell-shaped workpiece with constant welding parameters (**a**) Topside (**b**) Backside

The experiment results of the developed model-free adaptive controller with G function fuzzy reasoning regulation demonstrates that for the welding process with time-varing conditions, the model-free adaptive controller with G function fuzzy reasoning regulation has adequate adaptive capability, and can achieve satisfactory control performance under different conditions. Meanwhile this control method only needs the observed input output data. Thus, the developed model-free adaptive controller with G function fuzzy reasoning regulation provides a promising technology for GTAW quality control.

5.6 Composite Intelligent Controller for Weld Pool Dynamics During Pulsed GTAW

The above-mentioned control strategies are all SISO (Single- Input- Single- Output) methods, which will not be possible for a more precise control for the welding process because the welding is, in nature, a multi-variable coupled process. For a better imitation of the intelligent welder, MIMO (Multiple- Input- Multiple- Output)

(a)

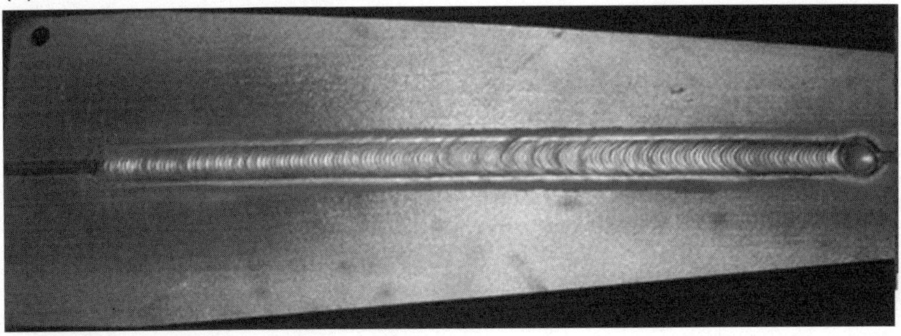

(b)

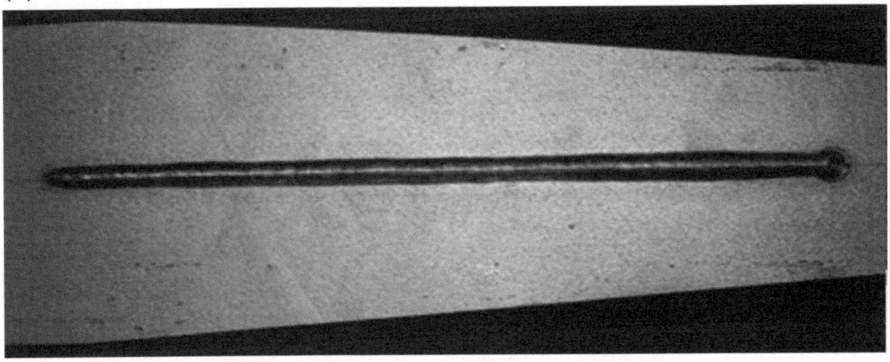

Fig. 5.41 The trapezia-shaped workpiece with model-free adaptive control with G function fuzzy reasoning regulation (**a**) Topside (**b**) Backside

composite intelligent controller is necessary. In the following part, four different applications of composite intelligent controller will be discussed in detail.

5.6.1 FNNC- Expert System Controller for Low Carbon Steel During Butt Welding

5.6.1.1 Double Variables Control Scheme

The backside width can be controlled on the given value with the single variable control FNNC, but the backside shape may not be always satisfied. Through investigations on butt welding experiments, the topside half-length and area of the weld pool are found to increase when the backside geometry is turning bad. This indicates that the regulation of single pulse duty ratio is not capable to guarantee the good weld bead geometry. Meantime, the step impulse on geometry parameters

(a)

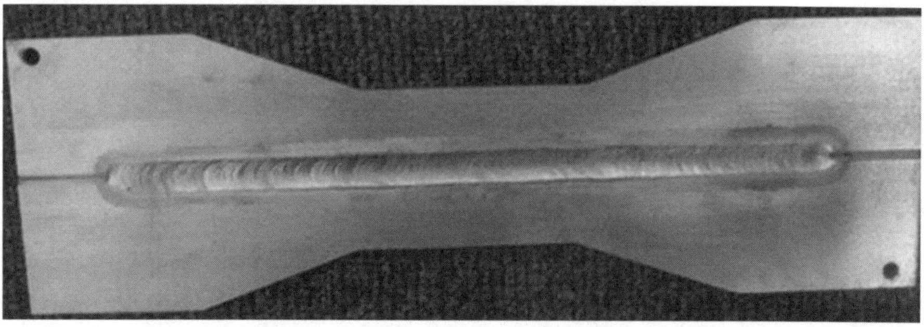

(b)

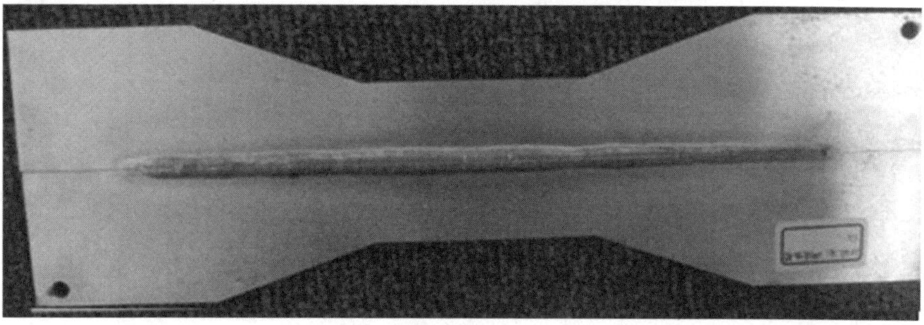

Fig. 5.42 The graded dumbbell-shaped workpiece with model-free adaptive control with G function fuzzy reasoning regulation (**a**) Topside (**b**) Backside

show that the welding speed has an important impact on the size parameters at the welding direction. Therefore, both welding speed and pulse duty ratio should be considered as control variables.

An expert system on weld geometry control is applied to adjust welding speed, along with the single variable fuzzy neural network control, constructs the double variables intelligent control system.

To some extent, expert system is to translate the expert experiences of solving some problems into the generated rules represented by knowledge base. Basically, an expert system consists of knowledge base and ratiocination machine. The knowledge base is to stock the expert knowledge in problem regions, and ratiocination machine is to yield the control strategy from the knowledge base according to the current status. The generated rule is often represented as "*If* A *Then* B".

The schematic diagram of the closed-loop control system is shown in Fig. 5.47, where the above dot line frame represents the fuzzy neural network for control backside width by adjusting pulse duty ratio, and the below dot line frame represents the expert system to adjust welding speed. The inputs of the expert system are $L_{f\max}$ and S_{fmid}, which are combined and converted as a shape quota through signal converter.

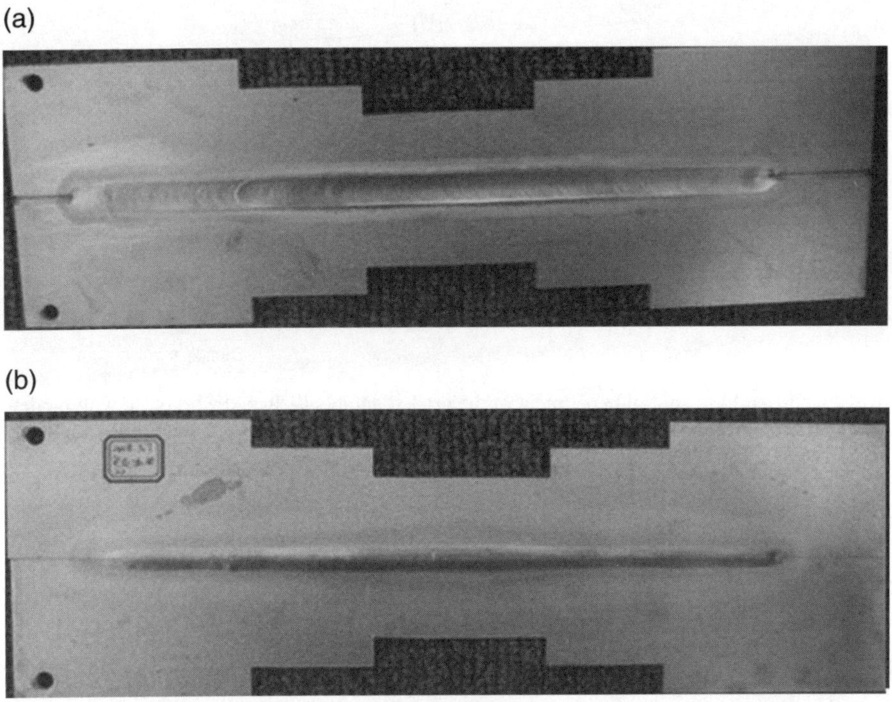

Fig. 5.43 The mutant dumbbell-shaped workpiece with model-free adaptive control with G function fuzzy reasoning regulation (**a**) Topside (**b**) Backside

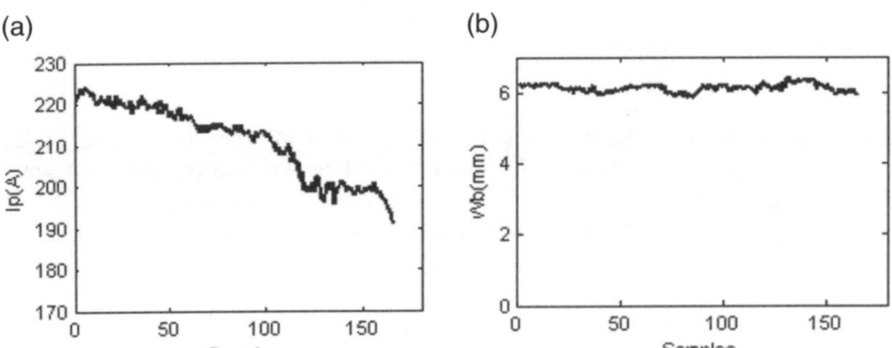

Fig. 5.44 Closed-loop control experiment of the trapezia-shaped workpiece with model-free adaptive control with G function fuzzy reasoning regulation (**a**) Control action (**b**) Output

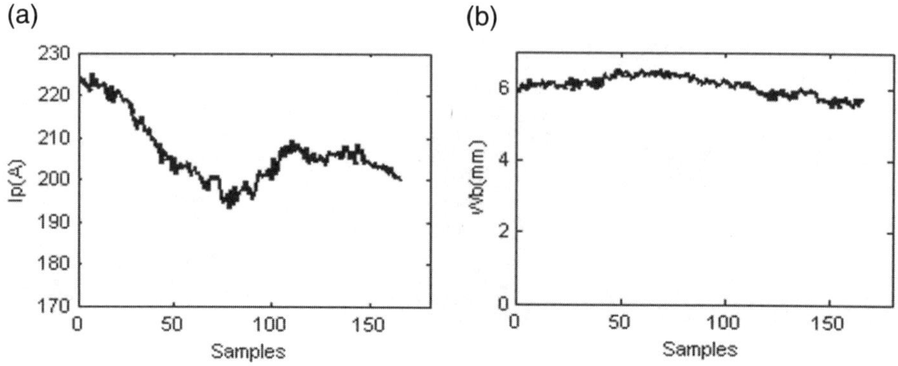

Fig. 5.45 Closed-loop control experiment of the graded dumbbell-shaped workpiece with model-free adaptive control with G function fuzzy reasoning regulation (**a**) Control action (**b**) Output

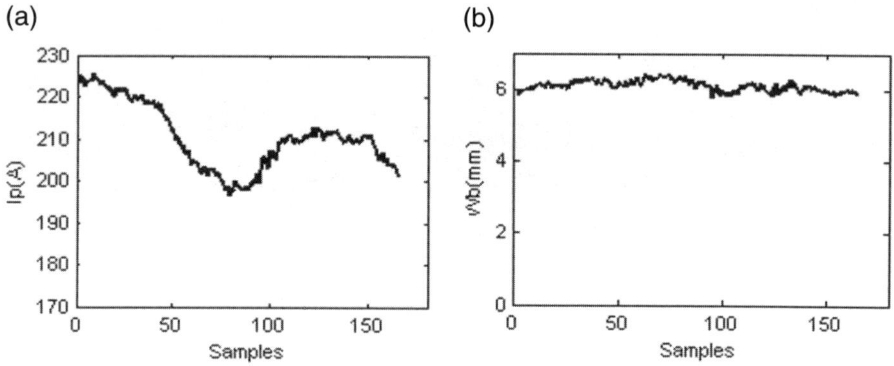

Fig. 5.46 Closed-loop control experiment of the mutant dumbbell-shaped workpiece with model-free adaptive control with G function fuzzy reasoning regulation (**a**) Control action (**b**) Output

The difference between the shape quota and given value matches the generated rules in the knowledge base of the expert system, to yield the welding velocity V_w. V_w and δ are as inputs of the welding process for completing the control. The SSNNM model for predicting the backside width of the weld pool is discussed in Sect. 4.3.4.

When the weld bead shape is perfect, the corresponding geometry parameters can be derived from the welding experiments curves, the topside maximum half-length $L_{f\max}$ is about 7.40 mm, and the topside half-area S_{fmid} is about 37.5 mm^2. While the weld bead shape turns bad, the $L_{f\max}$ reaches 8.95 mm, and S_{fmid} reaches 45.0 mm^2. Based on the above experiences, the variation ranges input to the expert controller are defined from 6.40 mm to 8.40 mm of $L_{f\max}$, from 30.0 mm^2 to 45.0 mm^2 of S_{fmid}. The center values are 7.40 mm and 37.5 mm^2 respectively. To avoid the affect from different quantity and range, the normalization of the values is needed, shown as follows:

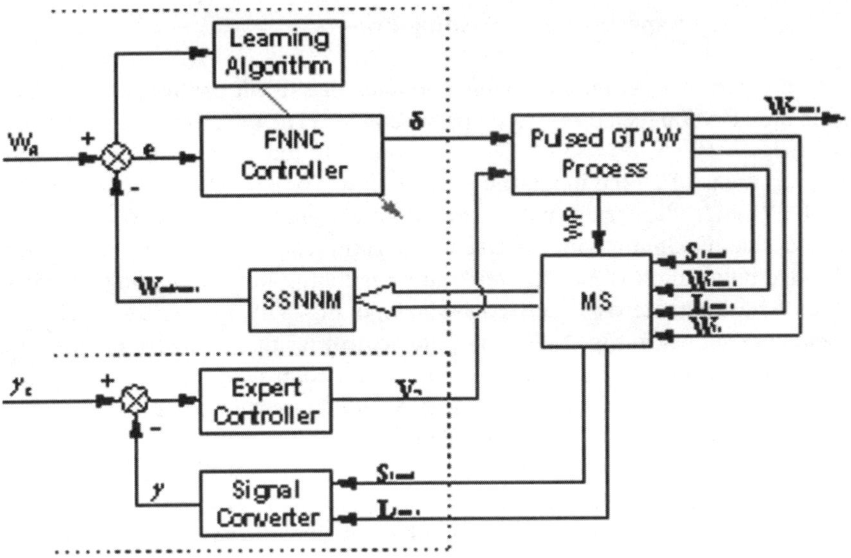

Fig. 5.47 Schematic diagram of double variables closed-loop intelligent control system of pulsed butt GTAW

$$\bar{x}_L = \frac{x_L - x_{L\min}}{x_{L\max} - x_{L\min}} \tag{5.54}$$

$$\bar{x}_S = \frac{x_S - x_{S\min}}{x_{S\max} - x_{S\min}} \tag{5.55}$$

where $\bar{x}_L$ and $\bar{x}_S$ are the normalization results. The shape quota γ is determined by the two factors, and defined as follow:

$$\gamma = 0.5\bar{x}_L + 0.5\bar{x}_S \tag{5.56}$$

The generated rules in the expert controller are the summarization of experiences on some control region of specialist and skilled operators. Aiming at the pulsed GTAW butt welding process, the generated rules can be concluded as follows, which represent the human behavior of adjusting welding speed to ensure the weld bead shape.

R1: IF $\gamma < 0.00$ THEN $\Delta V_w = -0.50\,\text{mm/s}$
R2: IF $0.00 \le \gamma < 0.20$ THEN $\Delta V_w = -0.33\,\text{mm/s}$
R3: IF $0.20 \le \gamma < 0.45$ THEN $\Delta V_w = -0.17\,\text{mm/s}$
R4: IF $0.45 \le \gamma < 0.55$ THEN $\Delta V_w = 0.00\,\text{mm/s}$
R5: IF $0.55 \le \gamma < 0.80$ THEN $\Delta V_w = 0.17\,\text{mm/s}$
R6: IF $0.80 \le \gamma < 1.00$ THEN $\Delta V_w = 0.33\,\text{mm/s}$
R7: IF $\gamma \ge 1.00$ THEN $\Delta V_w = 0.50\,\text{mm/s}$

5.6.1.2 Controlled Experiments of Welding Process

To testify the performance of the double variables intelligent control, welding experiments are also conducted on arc shape specimens. The experiment condition is as follows, $I_{p0} = 140\,\mathrm{A}$, $\delta = 45\%$, $V_{w0} = 2.5\,\mathrm{mm/s}$. To receive a complete welding data, the variations of all the inputs are set as follows: $\Delta Ip = \pm 25\,\mathrm{A}$, $\Delta\delta = \pm 20\%$, $\Delta V_w = \pm 0.67\,\mathrm{mm/s}$, the corresponding step variations are 5 A, 5%, and 0.167 mm/s.

The backside maximum width is given as 6.0 mm, and the minimum regulating unit of pulse duty ration is 1%, the minimum regulating unit of welding speed is 0.17 mm/s. The control curve is derived as shown in Fig. 5.48, the topside and backside photos are shown in Fig. 5.49. From the control results, both size parameters at the width direction and at the length direction are controlled uniformly. The geometry parameters are maintained more stable than the results generated in single variable controller, clearly seen from the photos.

The statistics show that the maximum error of $W_{b\max}$ is 0.34 mm, the average error is 0.09 mm, the root- mean-square error is 0.11 mm. The other statistics of $L_{f\max}$, S_{fmid}, $L_{b\max}$ and S_b are more smaller than the that of single variable control. The results verify the feasibility and accuracy of the double variables intelligent control.

From the above experiment results, we come to the conclusions as follows,

(1) The shape parameters proposed here along with the size parameters developed in bead-on-plate welding process are combined to describe the weld pool geometry. The neural network model with both size and shape parameters can predict the backside width more accurately rather than with the single size parameters.
(2) Single input and single output fuzzy neural network control is developed for control the backside width. From the results of the butt welding experiments conducted on arc shape specimen of 2 mm thickness mild steel plate, the backside width is successfully controlled in the desired range.
(3) Double variables intelligent control incorporated with the single variable fuzzy neural network control and an expert system, is verified the feasibility for control the geometry parameters at length and width direction by regulating pulse duty ratio and welding velocity. The good weld bead shape can be derived with the double variables control system.

5.6.2 FNNC- Forward Feed Controller for Low Carbon Steel During Butt Welding with Gap Variations

5.6.2.1 Composite Intelligent Control Scheme for Penetration and Shape of Pulsed GTAW Butt Welding with Gap Variations

A multi-variable composite intelligent control scheme is presented, which includes a forward-feed controller and the FNNC. The schematic diagram of the closed-loop

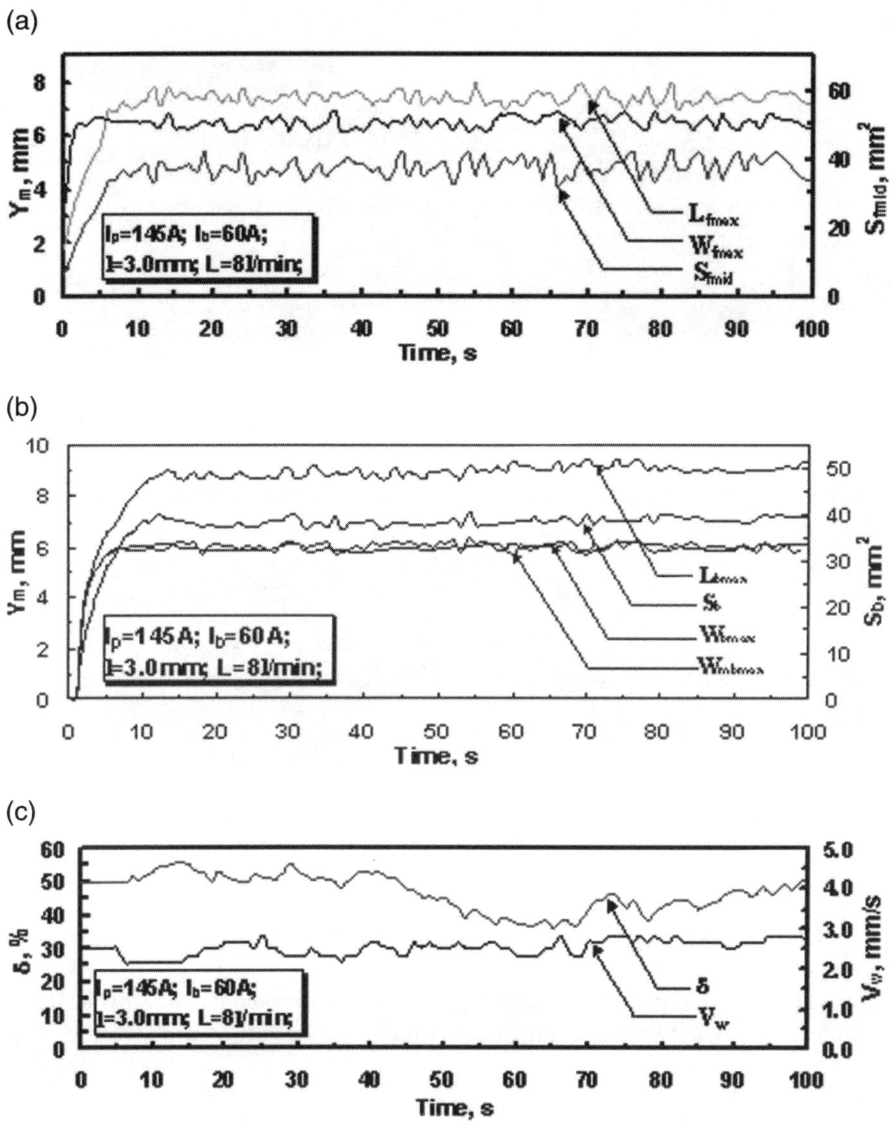

Fig. 5.48 The double variables intelligent control curves of arc specimen during pulsed GTAW (**a**) Topside sizes of weld pool (**b**) backside sizes of weld pool (**c**) controlling variables

control system is shown in Fig. 5.50, where the FNNC is mainly used to regulate pulse duty ratio for control of topside and backside width of dynamical weld pool during pulsed GTAW, the ST is for signal transformer from gap to wire feed speed V_f (corresponding to topside height of weld seam, H_t). The V_{fset} to H_{tset} is setup for forward-feed control, and the forward-feed controller is used to regulate wire filling velocity for compensating gap variations of the work-piece. The

(a)

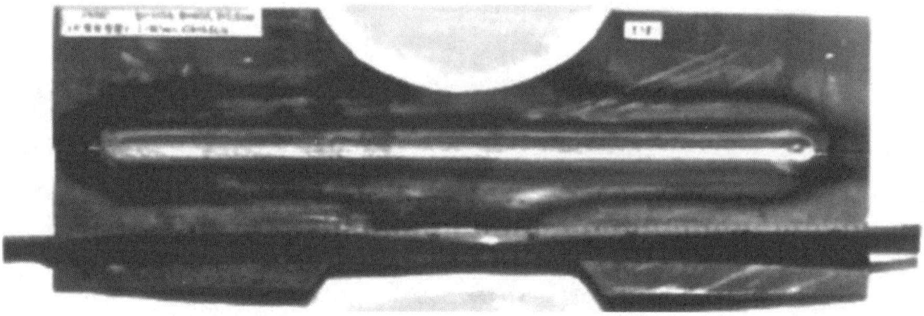

(b)

Fig. 5.49 A photograph of arc specimen with double variables intelligent control (**a**) Topside (**b**) backside

BWHDNNM model for predicting the backside width of the weld pool is discussed in Sect. 4.3.1.

Based on analysis and data processing of welding experiments, the forward-feed controller is designed by knowledge base, which is an expert control scheme, The

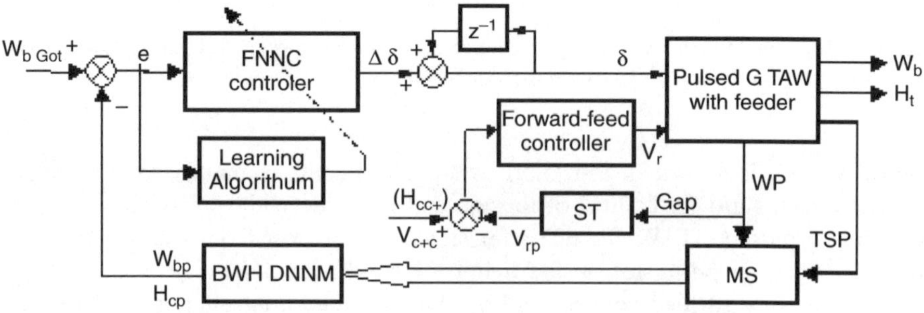

Fig. 5.50 Schematic diagram of composite controller for butt welding with gap variations

Table 5.2 The compensating value of filler rate in different gap

Gap(mm)	0.1	0.2	0.3	0.4	0.5	0.6	0.7	0.8	0.9	1
ΔV_f(cm/s)	0.2	0.4	0.7	1	1.5	2	2.4	2.8	3.1	3.5
V_f(cm/s)		≥ 2	≥ 3	≥ 4	≥ 4	≥ 5	≥ 6	≥ 6	≥ 7	≥ 7

generated rules of the forward-feed controller are often represented as "*If* A *Then* B", shown as Table 5.2.

The difference between gap variations quota and given value matches the generated rules of the forward-feed controller, to yield the wire filler velocity V_f. V_f with δ are inputs of the control system. The control period in the welding process is one second, and the response time of the closed system is less than 200 ms, including extraction of size characters from weld pool image, a cycle algorithm in the neural network model and controller.

5.6.2.2 Controlled Experiment

To testify the performance of the composite intelligent control system for the welding process, the pulsed GTAW experiments are conducted on specimens shown in Fig. 5.51. The experiment condition is shown in Table 5.3. The backside maximum width is given as 5.0 mm, and controlled results as following: the maximum error 0.32 mm, the average error 0.02 mm, and the root-mean-square deviation 0.18 mm. The average value of the topside height is 0.018 mm,the maximum error 0.09 mm, and the root-mean-square deviation 0.08 mm. The controlled curves are shown in Fig. 5.52, the topside and backside photos of the weld work-piece are shown in Fig. 5.53.

From the controlled results, the topside, backside width and the topside height of the weld seam have been controlled uniformly during pulsed GTAW. The geometry parameters are maintained more stable than the results generated in single input controller designs, e.g., PID or FNNC control schemes, clearly seen from the photos of welded work-piece in Fig. 5.53.

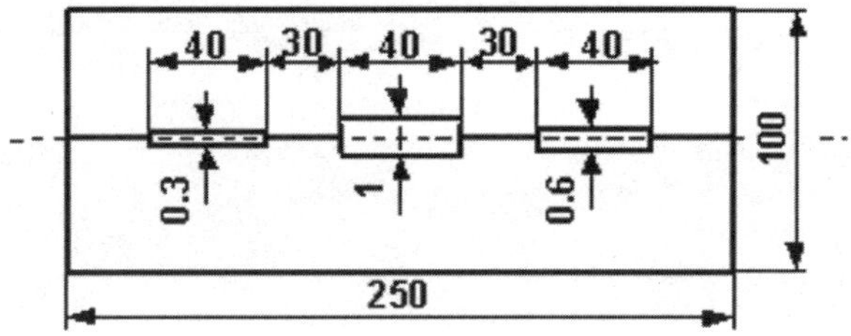

Fig. 5.51 Sketch map of varied gap specimen

Table 5.3 Experimental conditions of pulsed GTAW

pulsed period T, s	1	Arc Length l, mm	2.0
Pulse duty ratio δ, %	40–60	Electrode diameter ϕ, mm	3.0
Peak current I_p, A,	120–150	Electrode tip included angle θ	30
Base current I_b, A	30	Flux of argon L, l/min	8.0
Traveling speed, mm/s	2.5	Workpiece size, mm×mm×mm	250×50×2
Feeding velocity (cm/s)	5.0		

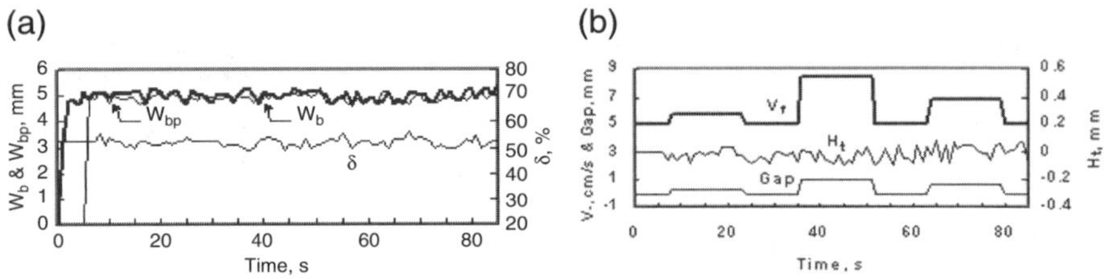

Fig. 5.52 Controlled curve of composite intelligent controlled welding

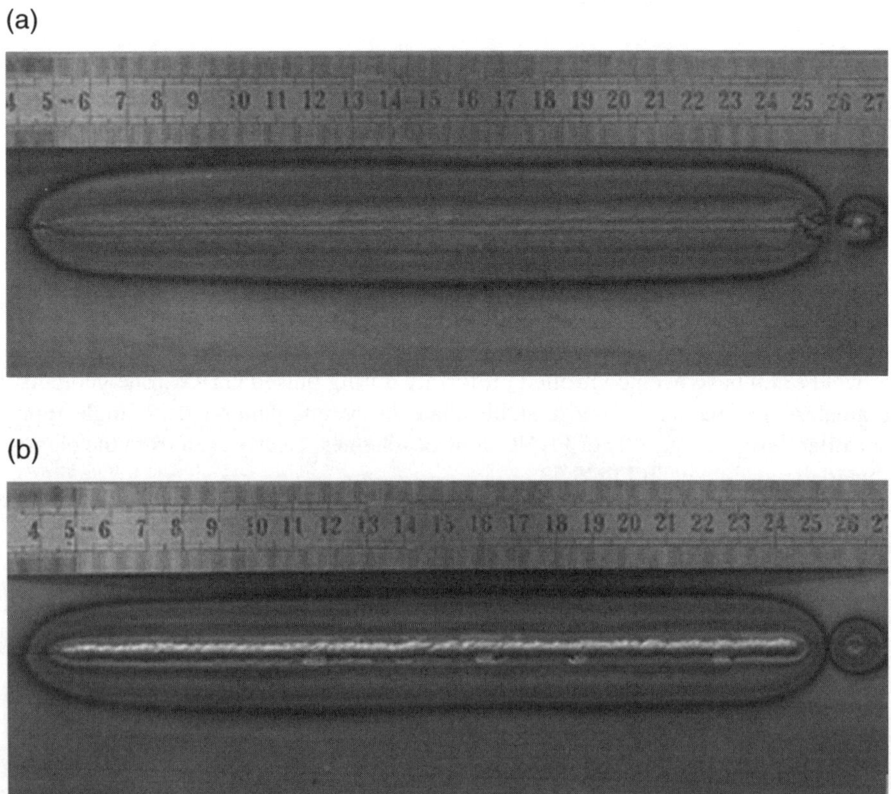

Fig. 5.53 Photograph of varied gap specimen by composite intelligent control scheme

5.6.3 *Compensated Adaptive- Fuzzy Controller for Aluminium Alloy During Butt Welding*

An adaptive- Fuzzy composite controller with the adaptive controller as its main control strategy is presented for aluminium alloy weld pool dynamics.

Although the single variable adaptive controller by regulating welding current reveals various effect for control of weld pool width during Al pulsed GTAW, it is difficult for one to insure the shaped weld seam equably asymmetric and smooth, specially in the case of strong interference and large rang of welding parameter changes. The large number of welding experiments show that welding current is a main variable to dominate the width of the weld pool, and the regulation of the wire feeding speed is more effective to shaping of weld seam, e.g., the sunk quantity of weld pool surface, and the reinforced height of weld seam. The experiment process also indicates that combining welding current and wire feeding speed is a better control scheme for proper penetration and fine shape of weld seam.

Combining designed fuzzy monitor with the above adaptive control scheme, the wire feeding speed can be regulated in real time, a multiplex intelligent control system for dynamic process of aluminium alloy pulsed GTAW is structured as Fig. 5.54.

The multiplex control systems in Fig. 5.54 includes two control loop, one is the adaptive control loop of welding current, which is realized by identification of predictive model and minimum square error control algorithm; another is the fuzzy monitor for regulating wire feeding speed. The system can simultaneously dominate welding penetration, i.e., the backside width of weld pool, and weld seam shape, i.e., the reinforced or sunk height of topside weld pool surface. The BWTWC model for predicting the backside width of the weld pool is discussed in Sect. 4.2.1.

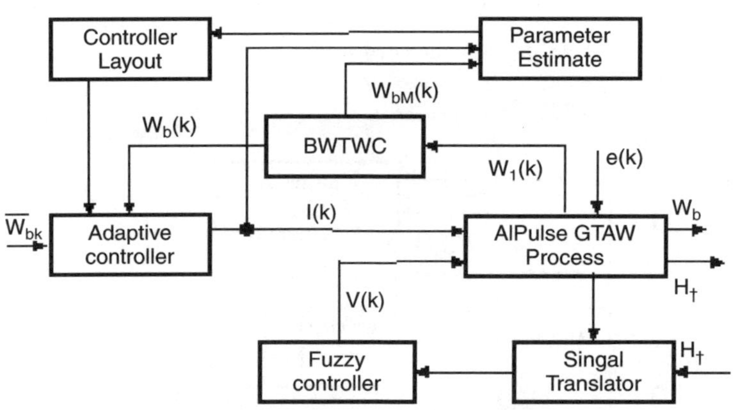

Fig. 5.54 Closed systems with adaptive controller compensated fuzzy monitor during Al alloy pulsed GTAW

5.6.3.1 Minimum-Squared-Error Adaptive Controller

Based on information from visual sensing of topside weld pool of aluminium alloy pulsed GTAW, combining minimum-squared-error control law and the predictive model BWTWC, the adaptive controller of backside dynamic width of aluminium alloy weld pool during pulsed GTAW is designed as Fig. 5.55, the regulating variable and controlled variable in the closed system are $I(k)$ and $W_b(k)$ respectively.

The closed system is composed of the controlled weld pool process, welding parameter estimator, controller layout (modifying parameters), adaptive controller, and model estimator BWTWC of backside weld pool width, which is used to estimate W_b(k) from W_f(k), here the visual sensing part is omitted. The welding parameter estimator is used to identify welding process parameters by the recursive least square algorithm. The adaptive controller is designed as following:

Supposing the common system is

$$A(q^{-1})y(k) = B(q^{-1})u(k-d) + e(k)$$

and contrasting with the model relation between pulse peak current and backside weld pool width

$$\begin{aligned} W_b(k) &= a_1W_b(k-1) + a_2W_b(k-2) + a_3W_b(k-3) + a_4W_b(k-4) \\ &\quad + b_1I(k-1) + b_2I(k-2) + b_3I(k-3) + b_4I(k-4) + e(k) \end{aligned} \tag{5.57}$$

Denoting

$$A(q^{-1}) = 1 - a_1q^{-1} - a_2q^{-2} - a_3q^{-3} - a_4q^{-4},$$
$$B(q^{-1}) = b_0 + b_1q^{-1} + b_2q^{-2} + b_3q^{-3},\ C(q^{-1}) = 1$$

and the structure parameters are determined as $n = 4$, $m = 4$, $d = 1$, and let

$$E(q^{-1}) = 1, \quad G(q^{-1}) = g_0 + g_1q^{-1} + g_2q^{-2}g_3q^{-3}$$

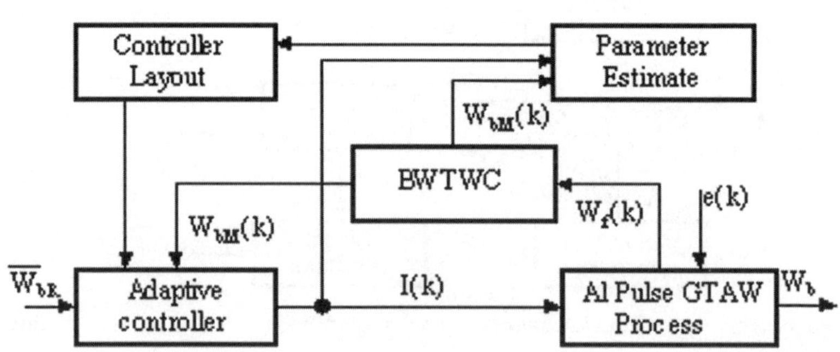

Fig. 5.55 The minimum-squared-error adaptive controller with adjusting welding current

Solving the Diophantine equation for designing minimum-squared-error regulator,

$$C(q^{-1}) = A(q^{-1})E(q^{-1}) + q^{-d}G(q^{-1})$$

we have

$$g_0 = a_1, \quad g_1 = a_2, \quad g_2 = a_3, \quad g_3 = a_4^{\circ}$$

And then the minimum-squared-error control law for backside weld pool width by regulating welding current is following as:

$$\begin{aligned} I(k) &= -\frac{G(q^{-1})}{B(q^{-1})E(q^{-1})}W_b(k) \\ &= -\frac{a_1 + a_2q^{-1} + a_3q^{-2} + a_4q^{-3}}{b_0 + b_1q^{-1} + b_2q^{-2} + b_3q^{-3}}W_b(k) \\ &= -1/b_0(a_1W_b(k) + a_2W_b(k-1) + a_3W_b(k-2) + a_4W_b(k-3) + b_1I(k-1) \\ &\quad + b_2I(k-2) + b_3I(k-3)) \end{aligned} \tag{5.58}$$

In the adaptive control system, denoting $W_b(k)$ as $y(k)$, $I(k)$ as $u(k)$, the system as

$$y(k) = \phi^T(k)\theta + e(k),$$

and observed vector as $\phi^T = [-y(k-1), \ldots -y(k-n), u(k-d), \cdots, u(k-m-d)]$ estimated parameter vector as

$$\theta = [a_1, a_2, \ldots a_n, b_0, b_1, \ldots b_m]^T$$

under the index:

$$J = \varepsilon^T\varepsilon = (y - \phi\bar{\theta})^T(y - \phi\bar{\theta})$$

and the recursive least square algorithm is used to estimate the following:

$$\bar{\theta}(N+1) = \bar{\theta}(N) + K(N+1)[y(N+1) - \phi^T(N+1)\bar{\theta}(N)] \tag{5.59}$$

$$K(N+1) = \frac{P(N)\phi(N+1)}{\rho + \phi^T(N+1)P(N)\phi(N+1)} \tag{5.60}$$

$$P(N+1) = \frac{1}{\rho}[I - K(N+1)\phi^T(N+1)]P(N) \tag{5.61}$$

Here, ρ is the declining factor, $0 < \rho < 1$, and determined as $\rho = 0.93$, $\bar{\theta}$ is the estimation of $\theta = [a_1, a_2, \ldots a_n, b_0, b_1, \ldots b_m]^T$, $P(N) = [\phi^T(N)W_N\phi(N)]^{-1}$, $W_N = diag\{\rho^{N-1}, \rho^{N-2}, \cdots \rho, 1\}$.

In the closed system, the minimum-squared-error adaptive control law is realized by the following algorithms:

1) Setup the initial value $\bar{\theta}(0)$, $P(0)$, $u(0)$ and $y(0)$, here $\bar{\theta}(0)$ is the parameter vector of the identified model of the aluminium alloy welding process. $P(0)$ is the matrix in the recursive identification, and $u(0)$ is welding current, $y(0)$ is backside weld pool width;

2) Take a new topside weld pool width $W_f(k)$ and estimate the backside weld pool width $W_{bM}(k)$ by the model BWTWC;
3) Makeup the measured vector with the present and past data of welding process;
4) Calculate renewed parameter vector $\bar{\theta}(k)$ and $P(k)$ by the recursive least square algorithm (5.59), (5.60) and (5.61);
5) Calculate the regulated welding current $u(k)$ by (5.58);
6) Output $u(k)$;
7) Return step (2) and continue.

In order to verify the above the control scheme, the designed controller is applied to actual welding closed system. The experiment on the adaptive control of dynamic changes of backside weld pool width during aluminium alloy pulsed GTAW is conducted, shown as the next section.

5.6.3.2 Fuzzy Logic Controller

A fuzzy inferential model is developed to imitate functions of welder's experiential operations. This fuzzy model as Table 5.4 is realized as a fuzzy monitor in Fig. 5.54. The fuzzy monitor is used to dominate weld seam shape, e.g., the reinforced or sunk height of topside weld seam, by regulating wire feeding speed.

The inputs of the fuzzy monitor are the error and change of error of the sunk height of topside weld pool, which can be estimated by topside weld pool width or by off-line calculation of visual image of weld pool surface. The fuzzy rule is described in the form as

$$\text{if } e \text{ is A and } ce \text{ is B, then } \Delta V(k) \text{ is C.}$$

Based on the welding experiments, here the change range of wire feeding speed V is determined as [5.-5 mm/s, 5 mm/s], the change range of the sunk height error e is [5.-1.5 mm, 1.5 mm], the change range of change of the error, ce, [5.-1 mm/s, 1 mm/s]. The fuzzy sets of e and ce are defined as negative big (NB), negative small (NS), zero (Z), positive big (PS) and positive big (PB). The fuzzy rules are shown as Table 5.4.

Table 5.4 The control rules of fuzzy monitor

E \ CE	NB	NS	Z	PS	PB
NB	NB	NB	NS	Z	PS
NS	NB	NS	Z	Z	PS
Z	NS	Z	Z	PS	PS
PS	Z	Z	PS	PS	PB
PB	PS	PS	PB	PB	PB

5.6.3.3 Controlled Experiment

The welding experiment on the systems with adaptive controller compensated fuzzy monitor is conducted to control weld pool dynamics during aluminium alloy pulsed GTAW. The experiment condition is shown in Table 2.2.

The desired width of backside weld pool is setup as 8 mm, and the desired reinforced height of topside weld seam is 0.8 mm, other conditions are shown as Table 2.3.1. The experimental results on the systems are shown in Figs. 5.56 and 5.57.

From the controlled curves in Fig. 5.57, and contrasting the experiment results by single variable adaptive control of backside weld pool width, the multiplex system with fuzzy monitor and adaptive control scheme has a faster regulating speed of backside weld pool width by the welding current, and realize more symmetrical weld width and shape, e.g., the reinforced height of the topside weld seam. In the controlled process, the multiplex control system with double variables is obviously superior to performance of the single variable adaptive system in dominating weld pool dynamics and seam shaping of aluminium alloy pulsed GTAW simultaneously.

Summing up on the above investigation of real time control strategies for aluminium alloy weld pool dynamics during pulsed GTAW, main conclusions are as follows:

1) The dynamic properties of aluminium alloy pulsed GTAW process are investigated and some characteristics different with that of the low carbon weld pool dynamics in the previous studies are discovered;
2) Aiming at the randomicity of the disturbances during aluminium alloy pulsed GTAW, the linear stochastic models between weld pool characteristics and technical parameters, such as topside and backside width of the weld pool, welding current, wire feeding speed, have been identified for predicting dynamical changes of the shaped parameters of the weld pool.
3) The minimum-squared-error adaptive controller with single variable regulation of welding current has been developed to control the backside width of the weld pool during aluminium alloy pulsed GTAW, the automatic regulating effects on the penetration and backside shape of the weld pool have been obtained.
4) In order to simultaneously dominate proper weld penetration, backside shape of weld pool and topside forced height of weld seam, the welding current adaptive

(a) (b)

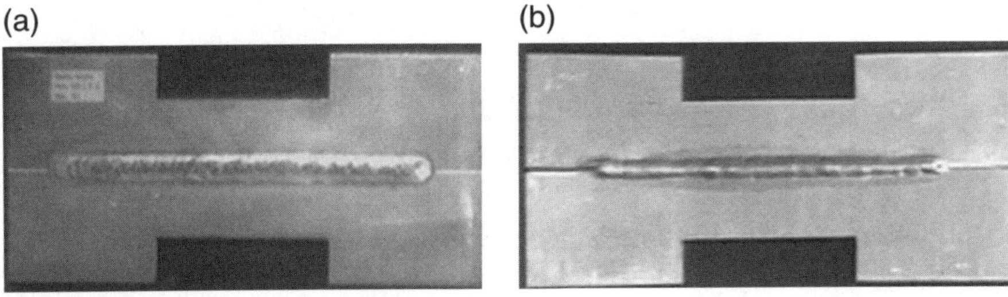

Fig. 5.56 Photographs of dumbbell-shaped specimen using multiplex compensated controller (**a**) Topside (**b**) Backside

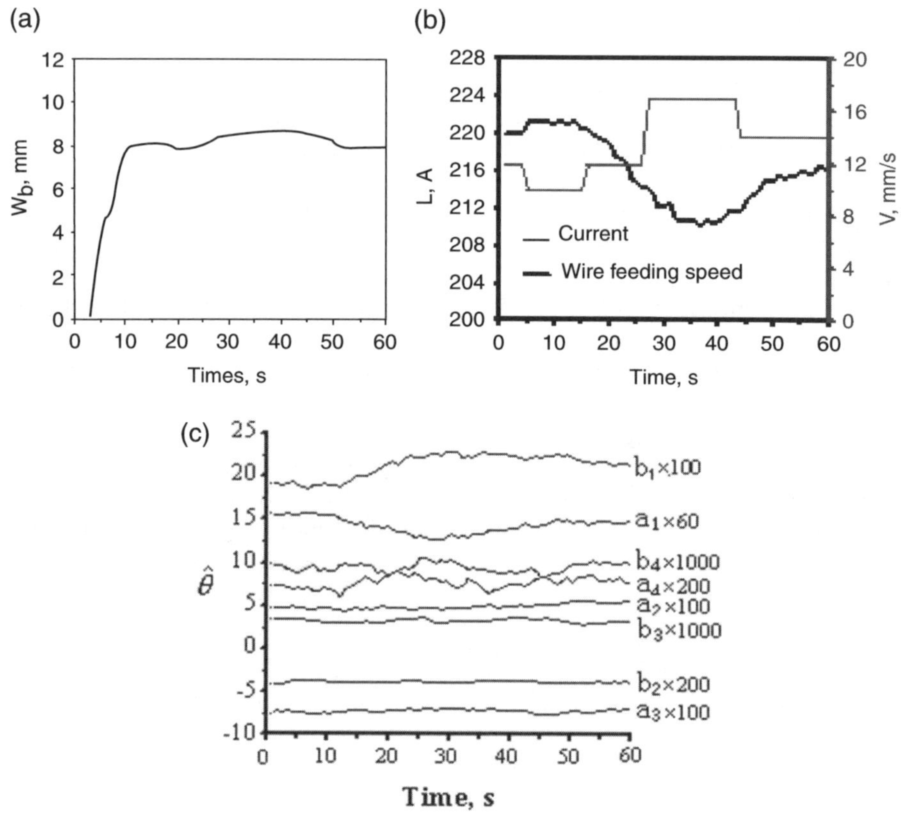

Fig. 5.57 Control curves of dumbbell-shaped specimen using multiplex compensated controller

controller compensated with fuzzy monitor of wire feeding speed has been developed. The stable dynamics of the weld pool and fine shape of the weld seam are obtained on the composed control systems during aluminium alloy pulsed GTAW.

The results show that adaptive control compensated fuzzy monitoring scheme is an effective real-time control strategy for weld pool dynamics and forming quality in aluminium alloy pulsed GTAW.

5.6.4 Adaptive-Fuzzy Controller Based on Nonlinear Model for Low Carbon Steel During Butt Welding with Wire Filler

5.6.4.1 Self-Tuning Fuzzy Control Scheme

A schematic of the self-tuning fuzzy control system [19] is shown in Fig. 5.58 where the self-tuning controller maps the regulated W_b into control action I_p based on the

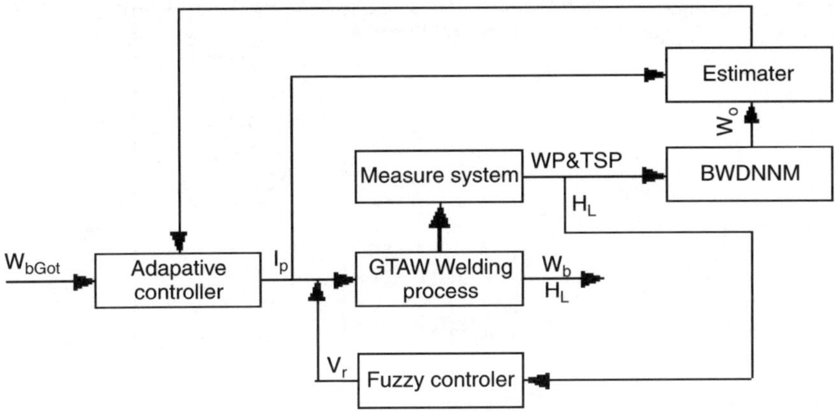

Fig. 5.58 Block diagram of the intelligent self-tuning fuzzy control system

Hammerstein model, while the fuzzy controller maps the regulated H_t into control action V_f; the output signals (I_p, V_w, V_f, L_{tt}, W_t, L_{tt}, D_t, W_t) of the welding process are detected by the measure system; BWDNNM denotes the neutral network model to map W_b, which is not measurable during practical welding process, into output signals of measure system; estimator serves to establish a Hammerstein Model mapping W_b into I_p. The structure and algorithms of each component, respectively, are explained as follows. The BWDNNM model for predicting the backside width of the weld pool is discussed in Sect. 4.2.2.

Because the direct measurement of backside width of weld pool (W_b) is not available in practical welding condition, a Backside Width Neural Network Model (BWDNNM) model is built to correlate W_b with topside size parameters of weld pool, the welding parameters and their history values. Previous researches show that the BP network with an input layer, a hidden layer of sigmoidal function and an output layer can approximate any rational function.

Therefore, the architecture of BP neural network is determined with 22 nodes in the input layer, 1 node in the output layer and 6 nodes in the hidden layer, as shown in Fig. 5.59.

$$
\text{Inputs:}\quad
\begin{array}{lll}
x(1)=I_p(t) & x(10)=L_{tt}(t) & \\
x(2)=V_w(t) & x(11)=W_t(t) & \\
x(3)=V_f(t) & x(12)=L_{th}(t) & \\
x(4)=I_p(t-1) & x(13)=L_{tt}(t-1) & x(19)=D_t(t) \\
x(5)=V_w(t-1) & x(14)=W_t(t-1) & x(20)=H_t(t) \\
x(6)=V_f(t-1) & x(15)=L_{th}(t-1) & x(21)=D_t(t-1) \\
x(7)=I_p(t-2) & x(16)=L_{tt}(t-2) & x(22)=H_t(t-1) \\
x(8)=V_w(t-2) & x(17)=W_t(t-2) & \\
x(9)=V_f(t-2) & x(18)=L_{th}(t-2) &
\end{array}
$$

$$
\text{Output:}\quad y(1)=W_b(t)
$$

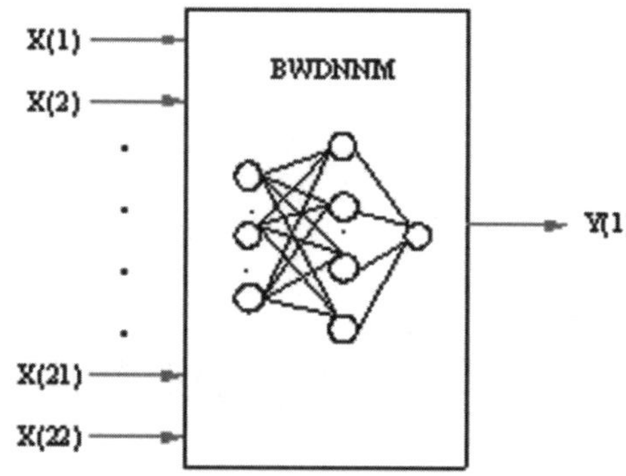

Fig. 5.59 The architecture of neural network model for W_b prediction

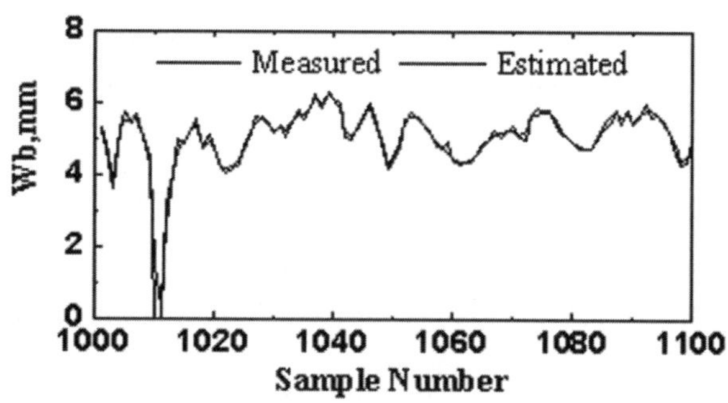

Fig. 5.60 Testing results of BWDNNM

The main welding parameters and the topside size parameters of weld pool have been used as the inputs of the network for modeling the backside width of weld pool. The training is performed using Matlab software off-line. In the experiment, totally 1000 points are used for network learning and 100 points are used for verification. The result of verification of the network learning process is shown in Fig. 5.60, with the average error being 0.0032 mm and mean square deviation being 0.0493.

5.6.4.2 Self-Tuning Controller

According to the Hammerstein model built above, the welding process is described as follows,

$$\begin{aligned} A(q^{-1})y(k) &= B(q^{-1})\sum_{i=1}^{p} r_i u^i(k-d) + C(q^{-1})e(k) \\ &= q^{-d}B(q^{-1})x(k) + C(q^{-1})e(k) \end{aligned} \tag{5.62}$$

Where $x(k) = \sum_{i=1}^{p} r_i u^i(k-d)$ is an intermediate variable.

$$A(q^{-1}) = 1 + a_1 q^{-1} + \ldots + a_n q^{-n}$$
$$B(q^{-1}) = b_0 + b_1 q^{-1} + \ldots + b_m q^{-m}$$
$$C(q^{-1}) = 1 + c_1 q^{-1} + \ldots + c_l q^{-l}$$

According to (5.62), we get

$$e(k) = \frac{A(q^{-1})}{C(q^{-1})} y(k) - \frac{q^{-d} B(q^{-1})}{C(q^{-1})} x(k) \tag{5.63}$$

$$y(k+d) = \frac{B(q^{-1})}{A(q^{-1})} x(k) + \frac{C(q^{-1})}{A(q^{-1})} e(k+d) \tag{5.64}$$

In (5.65), $\frac{C(q^{-1})}{A(q^{-1})}$ is transformed into

$$\frac{C(q^{-1})}{A(q^{-1})} = E(q^{-1}) + \frac{q^{-d} G(q^{-1})}{A(q^{-1})} \tag{5.65}$$

where $E(q^{-1})$ and $G(q^{-1})$ are unknown vectors.

$$E(q^{-1}) = 1 + e_1 q^{-1} + \ldots + e_l q^{-l}$$
$$G(q^{-1}) = g_0 + g_1 q^{-1} + \ldots + b_i q^{-i}$$

Equation (5.65) is the Diophantine Equation in its transformation Diophantine Equation is shown in (5.66)

$$C(q^{-1}) = A(q^{-1}) E(q^{-1}) + q^{-d} G(q^{-1}) \tag{5.66}$$

By substituting (5.65) into (5.64), we get (5.67)

$$y(k+d) = \frac{B(q^{-1})}{A(q^{-1})} x(k) + E(q^{-1}) e(k+d) + \frac{G(q^{-1})}{A(q^{-1})} e(k) \tag{5.67}$$

By substituting (5.63) into (5.67), we get (5.68)

$$\begin{aligned} y(k+d) &= E(q^{-1}) e(k+d) + \left[\frac{B(q^{-1})}{A(q^{-1})} - \frac{q^{-d} B(q^{-1}) G(q^{-1})}{A(q^{-1}) C(q^{-1})} \right] x(k) + \frac{G(q^{-1})}{C(q^{-1})} y(k) \\ &= E(q^{-1}) e(k+d) + \frac{B(q^{-1}) E(q^{-1})}{C(q^{-1})} x(k) + \frac{G(q^{-1})}{C(q^{-1})} y(k) \end{aligned} \tag{5.68}$$

At the time of $(k+d)$, the predicted error is marked as $\tilde{y}(k+d|k)$, then

$$\tilde{y}(k+d|k) = y(k+d) - y(k+d|k) \tag{5.69}$$

Where $y(k+d)$ is the measured output and $y(k+d|k)$ is the predicted output.

The prediction variance is

$$J = E\{\tilde{y}^2(k+d|k)\} \tag{5.70}$$

Substitute (5.70) into the prediction variance and we get

$$\begin{aligned} J &= E\{\tilde{y}^2(k+d|k)\} \\ &= E\{[E(q^{-1})e(k+d)]^2\} + E\left\{\left[\frac{G(q^{-1})}{C(q^{-1})}y(k) + \frac{B(q^{-1})E(q^{-1})}{C(q^{-1})}x(k)\right]^2\right\} \end{aligned} \tag{5.71}$$

In (5.71), the first item is uncontrollable. In order to minimize the prediction variance, the second item should equals to zero. Therefore, the minimum variance control law is as follows,

$$x(k) = -\frac{G(q^{-1})}{B(q^{-1})E(q^{-1})}y(k) \tag{5.72}$$

The controller is designed with I_p as its output and W_b as its input. According to the established BWHM model,

$$\begin{aligned} W(k) &= -a_1W(k-1) - a_2W(k-2) - a_3W(k-3) \\ &\quad + b_0I(k-1) + b_1I(k-2) + b_2I(k-3) + b_3I(k-4) \\ &\quad + r_2(b_0I^2(k-1) + b_1I^2(k-2) + b_2I^2(k-3) + b_3I^2(k-4)) + e(k) \\ &= -a_1W(k-1) - a_2W(k-2) - a_3W(k-3) \\ &\quad + b_0x(k-1) + b_1x(k-2) + b_2x(k-3) + b_3x(k-4) + e(k) \end{aligned} \tag{5.73}$$

Where $x(k) = r_1I(k) + r_2I^2(k)$
We get the parameters in the Diophantine Equation

$$\begin{aligned} &A(q^{-1}) = 1 + a_1q^{-1} + a_2q^{-2} + a_3q^{-3} \\ &B(q^{-1}) = b_0 + b_1q^{-1} + b_2q^{-2} + b_3q^{-3} \\ &C(q^{-1}) = 1 \\ &m = n = 3, \ d = 1, \ p = 2 \\ &E(q^{-1}) = 1 \\ &G(q^{-1}) = g_0 + g_1q^{-1} + g_2q^{-2} \end{aligned}$$

Where only $G(q^{-1})$ is the unknown parameter.
By solving the Diophantine Equation

$$\begin{aligned} &C(q^{-1}) = A(q^{-1})E(q^{-1}) + q^{-d}G(q^{-1}) \\ &1 = 1 + (g_0 + a_1)q^{-1} + (g_1 + a_2)q^{-2} + (g_2 + a_3)q^{-3} \\ &g_0 = -a_1, \quad g_1 = -a_2, \quad g_2 = -a_3 \end{aligned}$$

According to the minimum variance control law, we get the intermediate variable

$$\begin{aligned} x(k) &= -\frac{G(q^{-1})}{B(q^{-1})E(q^{-1})}W(k) \\ &= \frac{a_1 + a_2 q^{-1} + a_3 q^{-2}}{b_0 + b_1 q^{-1} + b_2 q^{-2} + b_3 q^{-3}} W(k) \\ &= \frac{1}{b_0}(a_1 W(k) + a_2 W(k-1) + a_3 W(k-2) \\ &\quad - b_1 x(k-1) - b_2 x(k-2) - b_3 x(k-3)) \end{aligned} \tag{5.74}$$

According to the equation $x(k) = r_1 I(k) + r_2 I^2(k)$
And $r_1^2 + 4r_2 x(k) \geq 0$
We get the controlling intput u(k)

$$\begin{aligned} I(k) &= \min \left| \frac{-r_1 \pm \sqrt{r_1^2 + 4r_2 x(k)}}{2r_2} \right| \\ &= \min \left| \frac{-r_1 \pm \sqrt{r_1^2 + 4\frac{r_2}{b_0}\left(\sum_{i=1}^{3} a_i q^{-(i-1)} W(k) - r_1 \sum_{i=1}^{i=3} b_i q^{-i} I(k) - r_2 \sum_{i=1}^{i=3} b_i q^{-i} I^2(k)\right)}}{2r_2} \right| \end{aligned} \tag{5.75}$$

The pulse peak current (I_p) regulating process is described as follows,

1) Set the initial value of $\bar{\theta}(0)$, $P(0)$ and $I_p(0)$. The values of $\bar{\theta}(0)$,$P(0)$ have been discussed above;
2) Get the new data of pulse peak current (I_p), welding velocity (V_w) , and feeding velocity (V_f) to calculate the backside width $W_b(k)$ by BWDNNM;
3) Construct the measured vector of $W_b(k)$ and $I_p(k)$;
4) Calculate the BWHM by RLS on line;
5) Calculate the self-tuning controlled variable $I_p(k)$;
6) Return to step (2) until the welding process ends.

5.6.4.3 Fuzzy Controller

The V_f in Fig. 5.58 is used to regulate H_t, which varies in so small a range that it is difficult to be controlled precisely. A fuzzy controller is, therefore, designed to ensure uniform topside height of the welding pool. With the deviation of topside height (e) and its variation (ce) as its inputs and the feeding velocity (V_f) as its output, the fuzzy controller uses the control rule described as follows, if e is A and ce is B, then u is C.

V_f varies in the domain of [5.-4 mm/s, 4 mm/s]; e in [5.-0.5 mm, 0.5 mm] and ec in [5.-0.8 mm/s, 0.8 mm/s], among which e and ec are described as corresponding fuzzy subsets containing the linguistic variables {PB; PM; PS; O; NS; NM; NB}, respectively, where P;O, and N are meant for positive, zero and negative, respectively, and B;M, and S for big, medium, and small, respectively. The fuzzy control rules are shown in Table 5.5 and the query table of fuzzy controller is shown in Table 5.6.

Table 5.5 Control rules of fuzzy controller

ec \ e	NB	NM	NS	ZE	PS	PM	PB
NB	NB	NB	NM	NM	NS	ZE	ZE
NM	NB	NB	NM	NM	NS	ZE	ZE
NS	NB	NB	NM	NS	PM	PM	PM
ZE	NB	NB	NM	ZE	PM	PB	PB
PS	NM	NM	ZE	PS	PM	PB	PB
PM	ZE	ZE	PS	PM	PM	PB	PB
PB	ZE	ZE	PS	PM	PM	PB	PB

Table 5.6 Query table of fuzzy controller

ec \ e	–0.5	–0.3	–0.1	0	0.1	0.3	0.5
–0.8	–4	–4	–2	–2	–1	0	0
–0.5	–4	–3	–2	–2	0	0	0
–0.2	–4	–3	–2	–1	2	2	2
0	–3	–3	–2	0	2	3	3
0.2	–2	–2	0	1	2	3	4
0.5	0	0	1	2	2	3	4
0.8	0	0	1	2	2	4	4

5.6.4.4 Controlled Experiments of Welding Process

Based on the system scheme in Fig. 5.58, a series of experiments on control of backside bead width and topside bead height of the pulse GTAW with wire feeder have been performed. The specimens for the experiments are 2 mm-thick low carbon steel plate shown in Fig. 5.61; the dumbbell specimen is used for imitating sudden changes of heat conduction in the welding process and the arced specimen for imitating gradient change of heat conduction in the welding process. The experiments include the process with PID controller and with self-tuning fuzzy controller with the welding parameters shown in Table 5.7.

The experiment on controlled objects is to obtain fine formation of both front and back weld bead of the specimen; here, the desired backside bead width is $D =$ 5.0 mm.

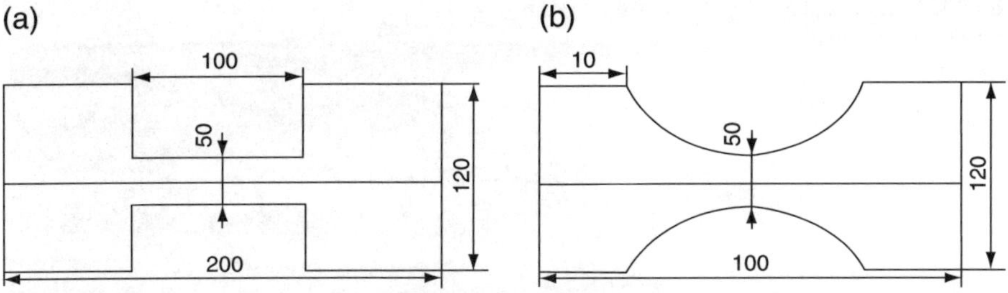

Fig. 5.61 The geometry of specimen for various heat conduction (**a**) Abrupt change (**b**) Gradient change

Table 5.7 Experimental conditions of pulsed GTAW with filler

pulse frequency(f), Hz	1	arc length(l), mm	3.0
duty ratio (δ), %	50	tungsten electrode diameter(φ), mm	3.2
base current (I_b), A	30	electrode angle(θ), °	30
material	Q235B	argon flow rate (L), l/min	8.0
	0.8 mm		
wire diameter	H08Mn2Si	specimen dimension, mm×mm×mm	280×80×2

With the same welding parameters, self-tuning fuzzy control experiments are carried out on heat conduction condition abrupt change and gradient change specimens. Specimen photographs with self-tuning fuzzy control are shown in Figs. 5.62 and 5.64, while their control process curves are shown in Fig. 5.63 and Fig. 5.65. According to the control process curves, the backside width fluctuates around 5 mm and the topside height fluctuates around 0.15 mm even when the heat conduction condition changes violently, which demonstrates the efficiency of self-tuning fuzzy controller in regulating both the backside width and topside height to obtain uniform appearance and full penetration of welding pool under disturbances.

With the self-tuning control and fuzzy logic control, we have proposed a self-tuning fuzzy controller scheme for the uncertain process of pulsed GTAW with wire filler characterized as time delay, nonlinear time variation and strong disturbance. The self-tuning controller is designed to regulate pulsed peak current based on the nonlinear Hamerstein model which adjusts its parameters on line. The fuzzy controller is designed to regulate the wire feeding velocity. Welding results have demonstrated that the proposed control scheme can monitor full penetration and fine formation of the weld seam with adequate accuracy despite of various disturbances during the process.

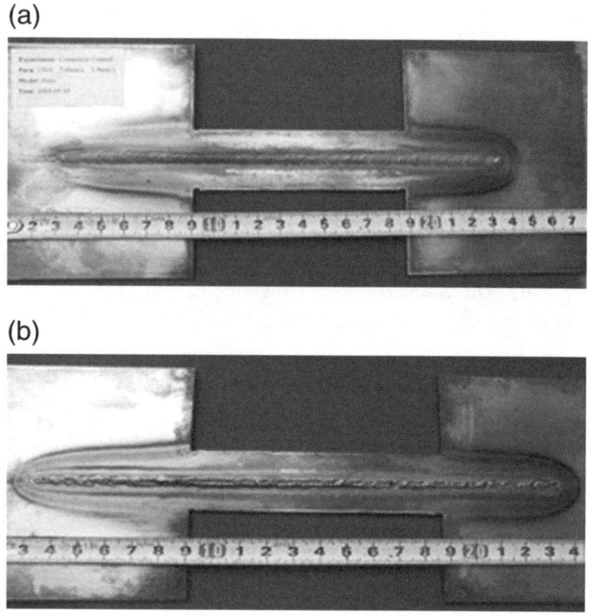

Fig. 5.62 Photographs of heat abrupr-changed specimen with self-tuning fuzzy control (**a**) Topside (**b**) Backside

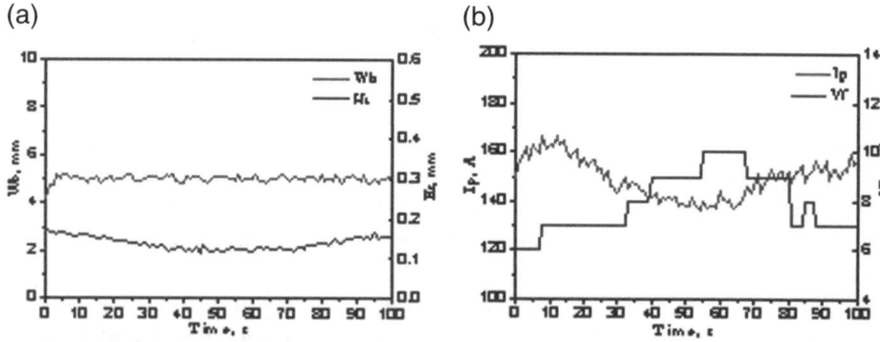

Fig. 5.63 The control process curves of heat abrupt-changed specimen with self-tuning fuzzy controller (**a**) Weld pool shape parameters (**b**) Welding current

5.7 The Chapter Conclusion Remarks

GTAW is a complicated process with non-linearity, time varying and uncertainties, therefore it is very difficult to model and control welding dynamics by classical linear system theory. In this chapter, several different intelligent controllers, including PSD, Neural Network, model-free controller and composite intelligent controller are discussed and they are also compared with the open-loop experiment results.

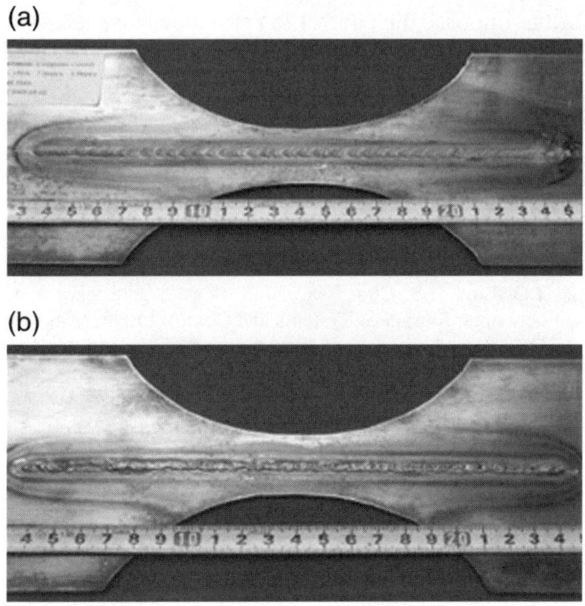

Fig. 5.64 Photographs of heat gradient-changed specimen with self-tuning fuzzy control (**a**) Topside (**b**) Backside

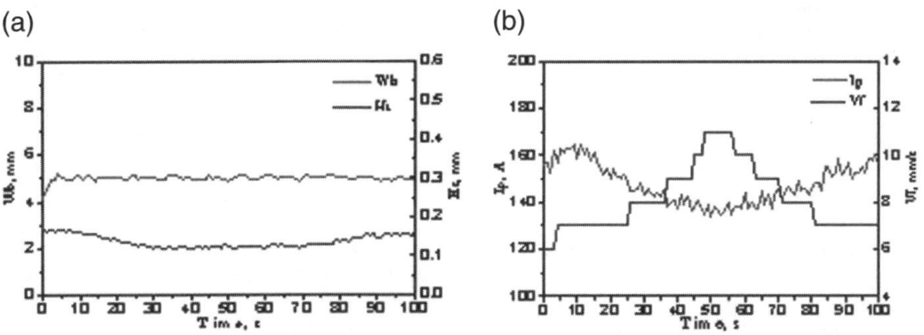

Fig. 5.65 The control process curves of heat gradient-changed specimen with self-tuning fuzzy control (**a**) Weld pool shape parameters (**b**) Welding current

By comparing different control strategy, conclusions can be draw. First, conventional PID controller is fast in response but large in overshoot, thus leading to instability in system. Therefore, backside width in the welding seam fluctuates greatly with PID controller. Fuzzy control can simulate the worker's experience, but does not possess the adaptive regulation with the varied conditions. While the self-learning PSD control can attain perfect control effect with different set values and conditions, and is suitable for the varied structure and coefficients of welding process specially. Model-free controller is a promising control strategy because it needs only the observed input output data. The use of composite controller can further improve the controlling result because it can improve stablize the process on one hand and decrease response time on the other hand.

References

1. K. Andersen, G.E. Cook. Artificial neural networks applied to arc welding process modeling and Control. IEEE Transaction Industry Application. 1990, 26(9):824–830
2. T.G. Lim, H.S. Cho. Estimation of weld pool sizes in Gma welding process using neural networks. Journal of Systems and Control Engineering. 1993, 207(1):15.26
3. D.T. Ham, D. Karaboga. Self-tuning fuzzy controller design using genetic optimization and neural network modeling, Artificial Intelligence in Engineering. 13(2):119–130
4. E. George, R.J.B. Cook, A. Kristinn, A.M. Strauss. Weld modeling and control using artificial neural networks. IEEE Transactions on Industry Applications. 1995, 31(6):1484–1491
5. M. Yu, R.K. Zhang. Neurofuzzy model-based predictive control of weld fusion zone geometry. IEEE Transactions on Fuzzy Systems. 1998, 6(3):389–401
6. L. Chien-Yi, P.-C. Tung, C. Wen-Hou. Adaptive fuzzy sliding mode control for an automatic arc welding system. The International Journal of Advanced Manufacturing Technology. 2006, 29:481–489
7. S.B. Chen, L. Wu, Q.L. Wang. Self-learning fuzzy neural networks for control of uncertain systems with time delays. IEEE Transactions on Systems, Man, and Cybernetics-Part B: Cybernetics. 1997, 27(1):142–148
8. D.B. Zhao, S.B. Chen, L. Wu, M. Dai, Q. Chen. Intelligent control for the shape of the weld pool in pulsed GTAW with filler metal. Welding Journal. 2001, 80(11):253s–260s
9. G.J. Zhang, S.B. Chen, L. Wu. Intelligent control of pulsed GTAW with filter metal. Welding Journal. 2005, 84(1):9s–16s
10. C. James, J.A.C. Spall, Model-free control of nonlinear stochastic systems with discrete-time measurements. IEEE Transactions on Automatic Control. 1998, 43(9):1198–1210
11. D. Sun, X. Shao, G. Feng. A model-free cross-coupled control for position synchronization of multi-axis motions: Theory and experiments. IEEE Transactions on Control Systems Technology. 2007, 15(2):306–314
12. Z. Hou, W. Huang. The model-free learning adaptive control of a class of SISO nonlinear systems. In Proceedings of the American Control Conference. Albuquerque, New Mexico, 1997
13. H. Zhi-Gang, Y. Xinghuo. An adaptive model free control design and its applications. in industrial informatics, 2004. INDIN '04. 2004 2nd IEEE International Conference on. Berlin, 2004
14. H. Zhi-Gang. An integrated approach to modeling and adaptive control. Acta Automatica Sinica. 2004, 30(3):380–389
15. Z. Zhong, I. Yasuyuki, N. Masatoshi. Composition Control for STEC Plant by A Model Free Control Method. in Systems, Man and Cybernetics, 2003. IEEE International Conference on. 2003
16. L. Ljung. System Identification: Theory for the User 2nd Edition. PTR Prentice Hall, 1999
17. C. Fan, L. FL, S.B. Chen. A visual sensing system for seam tracking and welding control in aluminum gas tungsten arc welding. in The 33rd Annual Conference of the IEEE Industrial Electronics Society. 2007
18. Q. Du. Extraction and intelligent control of 3D dynamic weld pool shape information of pulsed GTAW with wire filler, [5.Doctor thesis] Shanghai Jiaotong University, 2005
19. S.B. Chen, J. Wu, Q.Y. Du. Compound nonlinear control of welding pool dynamics during pulsed GTAW with wire filler. IEEE Transactions on Automation Science and Engineering (Submitted, July. 2007, Manuscript ID: T-ASE-2007-155)

Chapter 6
Real-Time Control of Weld Pool Dynamics During Robotic GTAW

Abstract Current teaching play-back welding robot is not with real-time function for sensing and control of weld process. This chapter addresses the real-time vision sensing and intelligentized control techniques for robotic arc welding. Intelligentized robotic system includes a computer visual sensing system with image processing algorithms, weld pool penetration controller and seam tracking controller. Using composite filtering technology, a computer visual sensing system is able to capture clear weld pool images during robotic pulsed GTAW. Corresponding image processing algorithm is described to pick-up characteristic parameters of the weld pool in real time. Furthermore, intelligentized models and real time controller of weld pool dynamics during pulsed GTAW process are discussed in the robotic systems. Seam tracking is another key technology for welding robotic system. Image processing algorithms are presented to extract the seam trajectory and the offset of the torch to the seam in the weld pool images with grooves. An application of intelligentized welding robot systems is also described at the end of this chapter.

At present, more and more welding robots are being applied in the automatic manufacturing process. However, most of them are "teach and playback" robots, which must be taught in advance. Actually many uncertainties, such as errors of pre-machining, fitting work-piece and in-process thermal distortions, during welding process will result in variations of the gap size and seam position. The "teach and playback" robots cannot meet the requirements of quality and diversification. Therefore, an autonomous welding robot is needed to overcome the shortages of "teach and playback" robot.

6.1 Real-Time Control of Low Carbon Steel Weld Pool Dynamics by PID Controller During Robotic Pulsed GTAW

Although artificial intelligent methods have been introduced to extensive fields of advanced manufacturing and robotic systems for many years [1, 2], the teaching play-back welding robots are still the main method for welding production. In recent years, much research has been carried out on robotic arc welding, but current

S.-B. Chen, J. Wu, *Intelligentized Methodology for Arc Welding Dynamical Processes*, Lecture Notes in Electrical Engineering 29,

teaching play-back robotic welding still is open-looped manufacturing process with constant welding parameters, thus resulting in pool welding quality when the change of welding conditions and disturbances occurs. Effective detecting and real time control of weld pool dynamic features are puzzle problems and, at the same time, key technologies for intelligentlized welding robots in the future.

Control of weld penetration is, at all time, a puzzle problem in automatic or robotic welding [3,4]. Backside weld width and penetration are direct factors determining the weld quality. Although the backside bead width can be directly detected from the backside work-piece, sensing from the backside is often inaccessible in most cases in practice. Therefore, control of weld quality by sensing the topside has become a highly challenging problem in robot welding. A lot of researches is aimed at sensing, modeling and control of welding dynamical process for improving robotic welding quality under complex welding conditions [5–14].

Recently, visual sensing technology has been used in some welding practice, owing to its features of no tough or interference with the welding process on one hand and its abundant information provided by the visual images of weld pool on the other. References [5, 6] investigated sensing and control of weld pool based on visual sensing in GTAW by a high-shutter-speed camera assisted with a pulsed laser for capturing weld poor images. References [7, 8] developed a coaxial arc weld pool viewing system, which could get clear weld pools images through blocking the bright arc core from overpowering exposure on the CCD target. Suga Y. [9, 10], et al. investigated vision-based sensing and control of GTAW pool. Some studies on welding path planning and seam tracking by vision sensing are reported in Refs. [15–17].

In our previous research, a double-side visual sensing system was established [11] for monitoring weld pool dynamical changes. Through a composed filter system including a narrow band filter and a neutral density filter, clear weld pool image of pulsed GTAW were captured. And the backside weld pool sizes could be predicted by a backside neural network model. The intelligent control of backside weld width was realized [11, 12].

Modeling and control of welding process has been a perplex difficulty for many years because the arc welding process is inherently variable, nonlinear, time-varying and strong coupling among welding parameters. And conventional models and control methods cannot describe and control the arc welding process precisely. Lim et al. [13] proposed an artificial neural network model to predict weld penetration depth, topside and backside width on-line from the detected surface temperature during GMAW process. Tzafestas et al. [14] presented a hierarchical MIMO predictive control scheme for regulation of GMA welding thermal characteristics. Using the fuzzy logic control and artificial network methodologies, we presented a self-learning fuzzy neural control scheme for general uncertain processes to deal with the problem of uncertain systems with time delays in the arc welding process [18]. In Suga Y.'s research [10], an adaptive control system using artificial neural network was proposed in the TIG welding control. The ANN model inputs were weld pools' characteristics parameters and welding parameters and the output was the change of welding current.

It must be emphasized that sensing and control of weld penetration above mainly focus on the traditional automatic welding process, and welding conditions, such as welding direction and pose of arc torch. And few studies aimes at multi-joint welding robots for sensing and control of weld penetration. Weld quality control in multi-joint robot systems is greatly different to traditional automatic welding due to shaking and position error effects of robot movement, which brings some difficulties for real-time welding processing control.

In this section, based on a platform of arc welding robot with visual sensing system, welding pool characteristics are picked up and analyzed by image processing for real-time control of welding dynamics, which associates with weld penetration and quality. The welding parameters in the robot systems are regulated to ensure a good weld quality under some uncertain welding condition during robotic welding.

6.1.1 Welding Robot Systems with Vision Sensing and Real-Time Control of Arc Weld Dynamics

6.1.1.1 The Functional Structure of the Welding Robot Systems

The robotic welding systems for real-time control of weld pool dynamics during pulsed GTAW is established as Fig. 6.1, which included 6.freedom arc weld manipulator, 3-freedom weld-piece positioner, a computer for weld pools sensing and control (SPPC), welding power source(WP), image processing(IP), visual sensor and et al., the central control computer(CCPC) for supervision of whole robotic welding systems, the moving control computer for coordinating motion between the

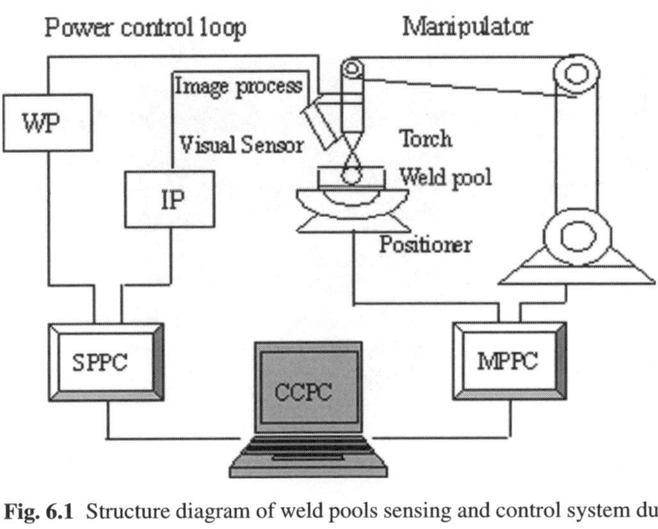

Fig. 6.1 Structure diagram of weld pools sensing and control system during robotic pulsed GTAW

manipulator and positioner. The whole systems is called a weld flexible manufacture cell (WFMC). The SPPC is the core of sub-system for weld pools sensing and control, which realized three functions: communicating with central computer through serial ports, control of welding power with D/A inverter card and interface card, and control of visual sensor through image capturing card. Image signals are transmitted into SPPC and processed. The SPPC set welding current through D/A card, taking image time and the image field size. Those functions are realized through the same controlling computer, which ensured the synchronization between image capturing and welding current setting during robotic arc welding.

6.1.1.2 Vision Sensing Sub-System

The structure diagram and photograph of the visual sensor are shown in Fig. 6.2. The main element in the visual sensor is a charge coupled device (CCD) camera. In order to decreasing disturbances during robot movement, a two-step reflecting light path system is adopted. Through a composed filter system, which consisted of a neutral density filter and a narrow band filter, weld pool images and arc light enters the CCD camera. The center band of the narrow band filter is 661 nm, half width is 10 nm, and the peak attenuation ratio of lens is 83.9%. The sensor is fixed at the rear part of the robot torch so as to observe a wider area. The camera lens center is set in the same plane as reflector center and the axis of the torch.

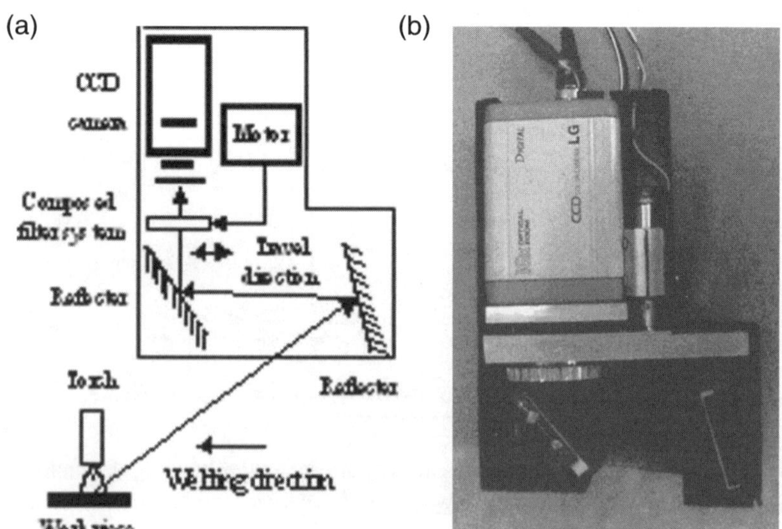

Fig. 6.2 Structure diagram and photograph of robot's image sensor

6.1.2 Weld Pool Image Processing During Robotic Pulsed GTAW

6.1.2.1 Weld Pool Images Capturing During Robotic Pulsed GTAW

Visual sensor parameters are necessary to be selected so as to obtain satisfying weld pool image; such parameters are as narrow band filter and neutral density filter, CCD camera parameters, current and time of capturing images, and et al. Because of strong arc light during a pulse peak current, image capturing time is selected at the beginning of the pulsed base current. Experiment results show that good weld pool images are obtained at the current of 40 A. The images from different taking-time corresponding to the pulsed current time sequence in Fig. 6.3 are shown as Fig. 6.4.

A typical weld pool image of robotic pulsed GTAW is shown in Fig. 6.5, where nozzle of torch, arc, weld pool and solidified metal can be distinguished clearly.

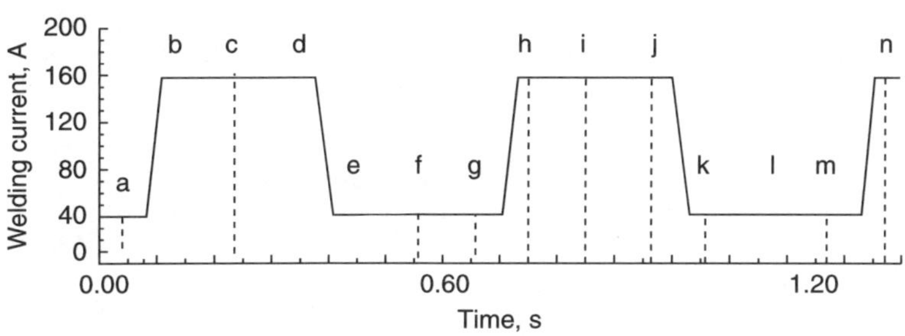

Fig. 6.3 Pulsed current time sequence

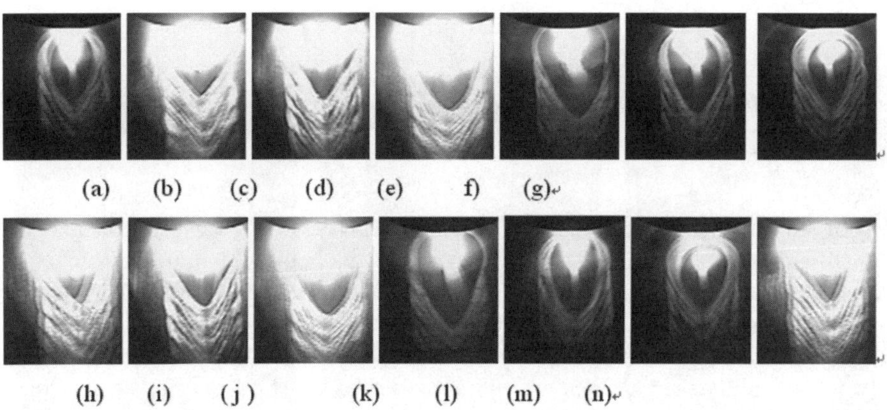

Fig. 6.4 Images from different taking-time

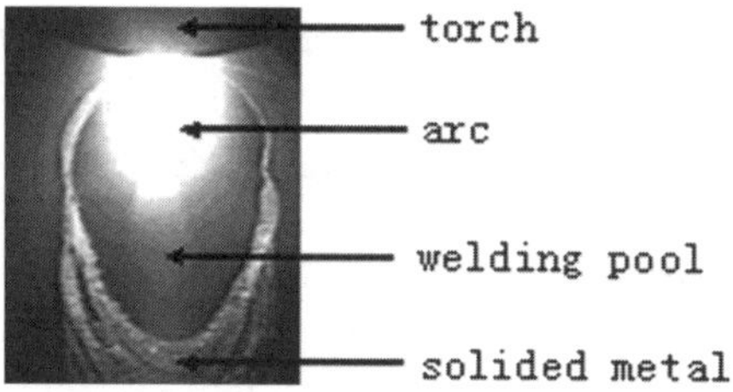

Fig. 6.5 Typical pool image of robotic pulsed GTAW

6.1.2.2 Weld Pool Image Processing

In order to investigate penetration and quality of the weld, two characteristic parameters of the weld pool, topside maximum width W_{fmax} and topside half length L_{fmax} shown as Fig. 6.6, are defined as follows.

The image processing algorithm must deal with the changing of observation direction during robotic arc welding, because the robotic controller do not know the angle between the direction of sensing and the welding direction. Usually, robot weld in the direction of the seam and the torch angle varies during welding, which adds difficulties to the development of image processing algorithm. In Fig. 6.7, three typical weld pool images are captured during robotic welding of the S-shaped seam.

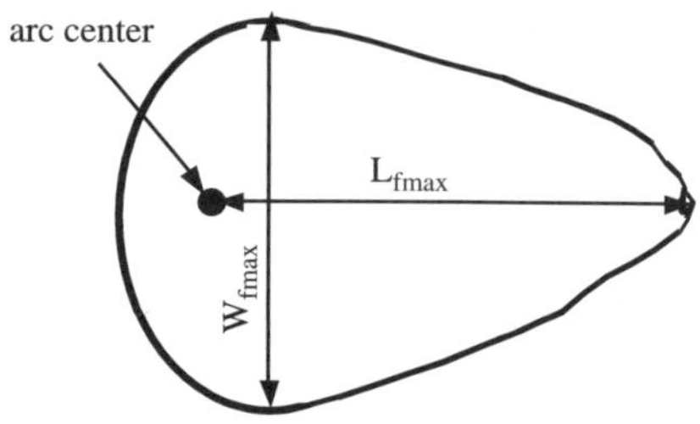

Fig. 6.6 Definition of characteristic parameters of weld pool during robotic pulsed GTAW

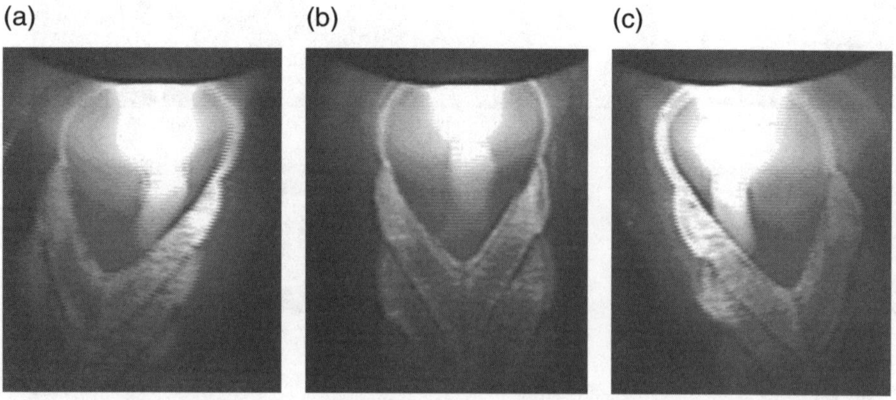

Fig. 6.7 Typical weld pool images during robotic welding of S-shaped seam (**a**) Image from right direction (**b**) Image from backside (**c**) Image from left direction

The steps of processing a typical weld pool image is shown in Fig. 6.8. Through a Gauss fileter with 7×7 window, the original weld pool image in Fig. 6.5 was smoothed (Fig. 6.8a), which decreased the noises from arc light or the path of signal transmitting. The gray distribution property in Fig. 6.9 shows that gray level jumps

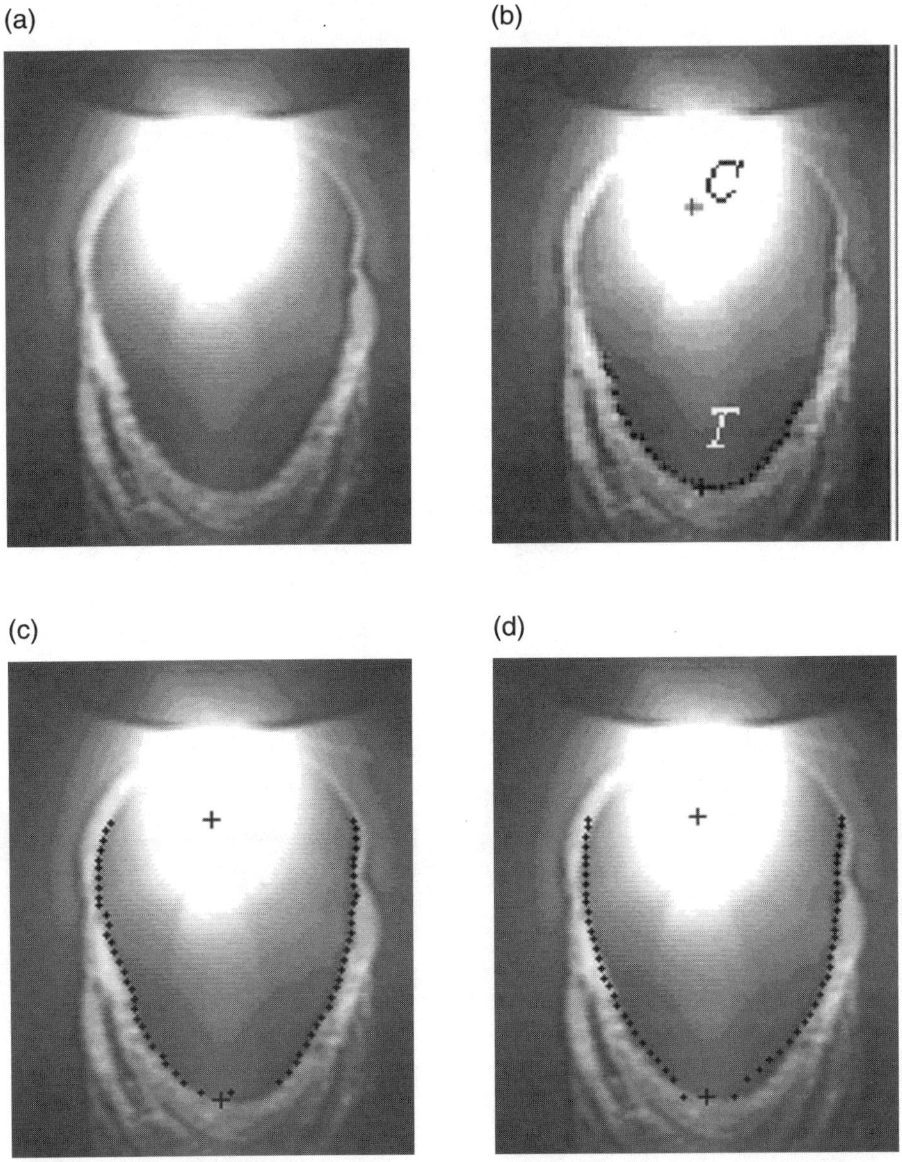

Fig. 6.8 Weld pool images processing for robotic pulsed GTAW (**a**) GAUSS filtering (**b**) Tail point getting (**c**) Original weld pool edge (**d**) Edge points regressing

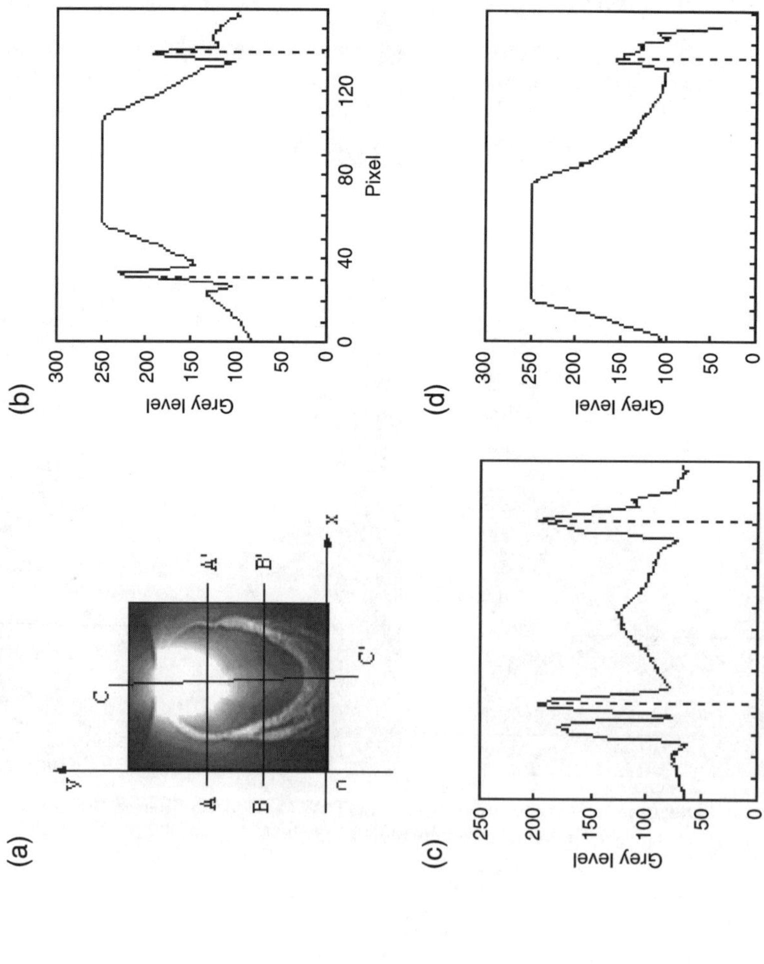

Fig. 6.9 The gray distributions of weld pool image in different directions

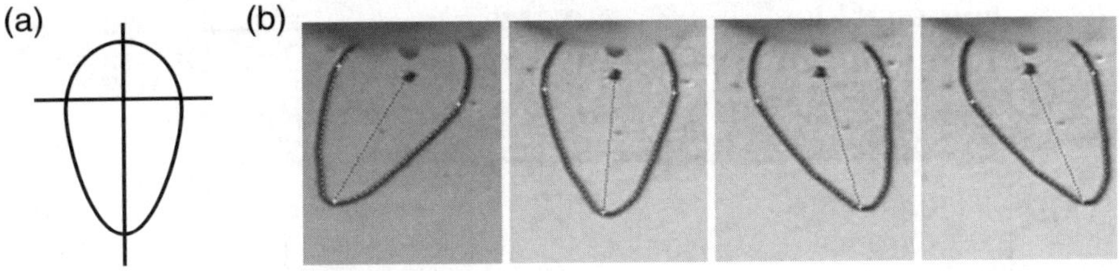

Fig. 6.10 Images of weld pool model observed from various directions during robotic welding (**a**) Real shape and size of weld pool model (**b**) Images of weld pool model from various directions

exist at the weld pool edge points when the image is scaned from the center point to exterior. This gray level jump is used to detect the weld pool edge. The center point of the weld pool is right down the arc center, marking it as C (shown in Fig. 6.8b). The tail point of weld pool is scanned and signed as point T (Fig. 6.8b). The weld pool image is scanned along the spcecial direction vertical to CT in the workpiece coordinate system, and scanning ends when a gray level jump is met with. This point is regarded as the edge point of weld pool shown in Fig. 6.8c. Regressing the edge points reasult is shown in Fig. 6.8d.

Some experiments below are proposed to veritify the adaptablity of the image processing algorithms when the observating dircetion is changing. A weld pool model, whose shape and size are similar to the real weld pools, is put down the torch., and the torch's axes is vertial to the plane of the weld pool model. With turning of the last robot joint, the visual sensor observes the weld pool model from various orientations, shown as Fig. 6.10. And the model's characteristic parameters are measured by the arithmetic above in various observing directions. The actual maximum width of weld pool model is 6.2 mm, and the half length is 8.5 mm (shown as Fig. 6.9a). Measuring results are show in Fig. 6.11, we can find that the maximum error of maximum width of the model is 0.15 mm, the average error is 0.06 mm, and the maximum error of half length is 0.12 mm, the average error is 0.04 mm. Sensing results show that the image processing arithmetic is able to fit for the changing of sensing orientation during robotic welding.

6.1.2.3 Investigation of Torch Height and Pitching Angle Influences on Weld Pool Characters

Besides the change of observing angle of visual sensor, torch height (arc length) and pitching angle also varies inevitably during robotic welding with vision sensing system, thus image quality of weld pool is obviously influenced. It is necessary to determine an effective or robust range for the welding robot systems under disturbances of torch height (arc length) and pitching angle.

In this welding robot system, the torch height is hold by arc sensing and regulation, and the welding experiments are conducted under the following set conditions:

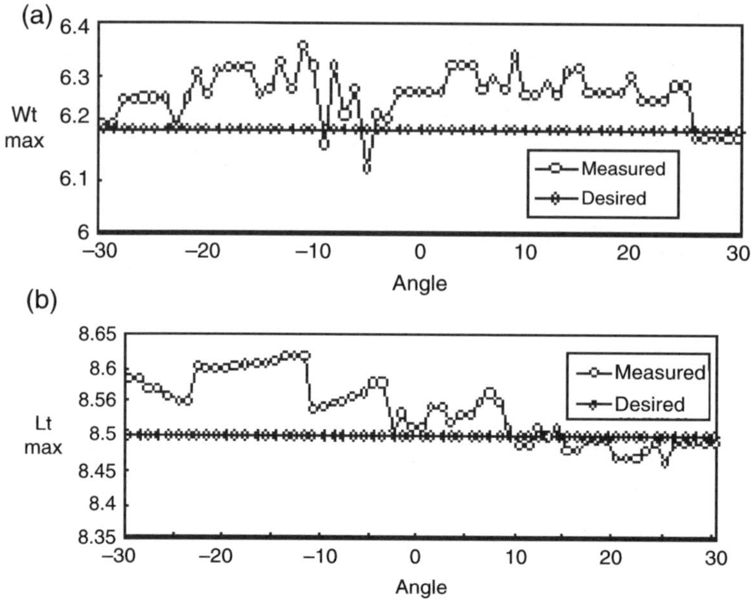

Fig. 6.11 Measuring results of weld pool model in various sensing directions during robotic welding (**a**) Maximum width of weld pool mode (**b**) Half length of weld pool model

(1) The inclined angle range between the torch and normal line of work-piece plane is [6.-10° – 10°];
(2) The arc length can be effectively regulated in [6.1.5 mm-2.5 mm]. The following experiment is to analyze the influences of torch inclined angle and arc length changes on weld pool images characters shown as Fig. 6.12.

Based on weld pool model in Fig. 6.9a, observing from just rear of weld pool, the center line of the torch is upright to weld pool plane and through arc center, the arc length is changing in [6.1.5 mm-2.5 mm] by step-length 0.25 mm, five images of weld pool are obtained.

Using corresponding image processing and character picking-up algorithm, the curve of characteristic changes with arc length is shown as Fig. 6.13. The experiment results indicate that, in limited range of arc length changes, the maximum width error of weld topside pool is 0.03 mm, and the maximum width error of weld topside pool is 0.70 mm; and the measured half-length of weld pool is evidently increased with increase of the arc length.

The Fig. 6.14 is shown the characteristic changes of weld pool during weld torch pitching angle changes. Where the pitching angle changes in [6.-10° – 10°], the step-length is set as 2°. When the pitching angle changed, the maximum width of weld topside pool has a little fluctuating; maximum error of weld pool width is 0.1 mm. The pitching angle changes evidently influenced maximum half-length of weld topside pool, the maximum error is 0.65 mm, and measured value of the maximum half-length of weld topside pool is decreased with pitching angle increasing.

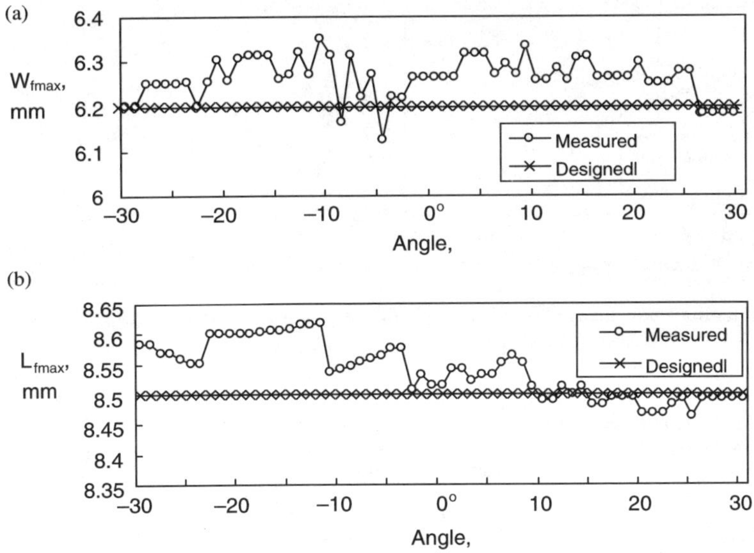

Fig. 6.12 Character parameters of weld pool with different observing angles (**a**) Maximum top width of weld pool (**b**) Maximum top half-length of weld pool

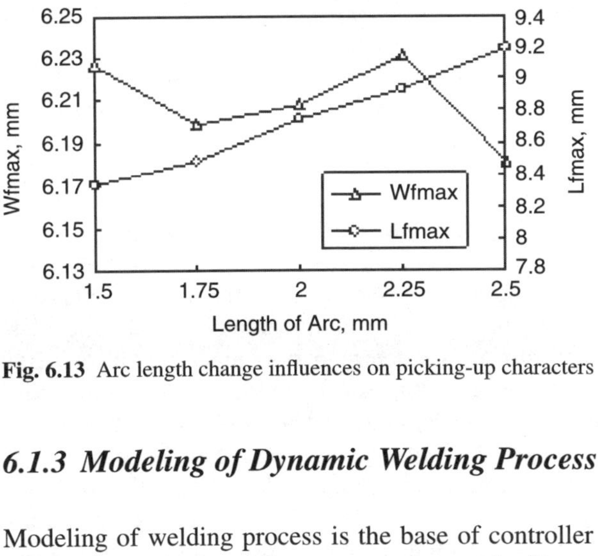

Fig. 6.13 Arc length change influences on picking-up characters of weld pool

6.1.3 Modeling of Dynamic Welding Process

Modeling of welding process is the base of controller design. An effective model must reflect actual welding process accurately. Because of the complexity of the weld pool dynamics, conventional modeling method is not usually fit for welding process. Considering merits of modelling for control system the back-propagation (BP) type of artifical neural nets (ANN) is selected to model weld pool dynamical process. Modeling data are got from robotic welding experiments. Under welding conditions in Table 6.1 and designed pseudorandom sequence inputs, the

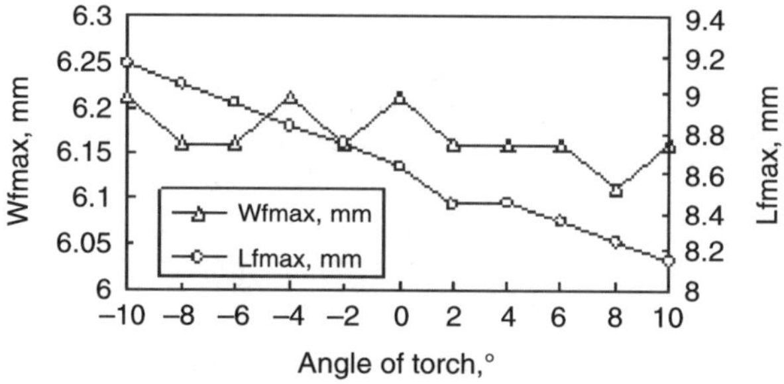

Fig. 6.14 Torch pitching angle change influence on character parameters of weld pool

Table 6.1 Experiment conditions of robotic pulsed GTAW for getting model data

Peak current/A	Varying from 140 to 180 at random
Pulse duty ratio / %	Varying from 35 to 70 at random
Pulse frequency / Hz	1
Traveling speed / mm/s	3.0
Arc Length / mm	2.0
Flux of argon / l / min	8.0
Base current / A	40
Workpiece dimension / mm×mm×mm	250×50×2

dynamic sample data of weld pool characteristic parameters is acquired for trainning ANN model from real-time visual sensing and image processing during robotic welding.

6.1.3.1 Modeling of Dynamical Characteristic of Topside Weld Pool

The model structure of weld pool topside characteristics for robotic pulsed GTAW process are shown in Fig. 6.15. In this model, welding parameters including peak current value I_p and pulse duty ritio δ, are determined as input parameters. Because of heat inertia in the welding process, historical value of welding parameters is also selected as inputs of dynamical ANN model of the weld pool. For example, I_p (t-1) meant the peak current value of last pulse, and I_p (t-2) meant the peak current value of last before last pulse. Character parameters of weld pool, maximum width (W_{fmax}) and half length (L_{fmax}), are determined as outputs of the ANN model. Considering the influence of the process continuity, The historial value of weld pool characteristic is also inputed into the model, e.g., W_{fmax}(t-1), W_{fmax}(t-2) and so on. There are 10 inputs and 2 outputs in the model. The hidden layer in the model contains 7 elements, which is determined by convergence rate of the net model training. The sigmoid function is selected as the nonlinear transfer function of the neurons in the net. The training results of this model are following as: maximum absolute error of maximum width of weld pools is 0.50 mm, average error of maximum width is

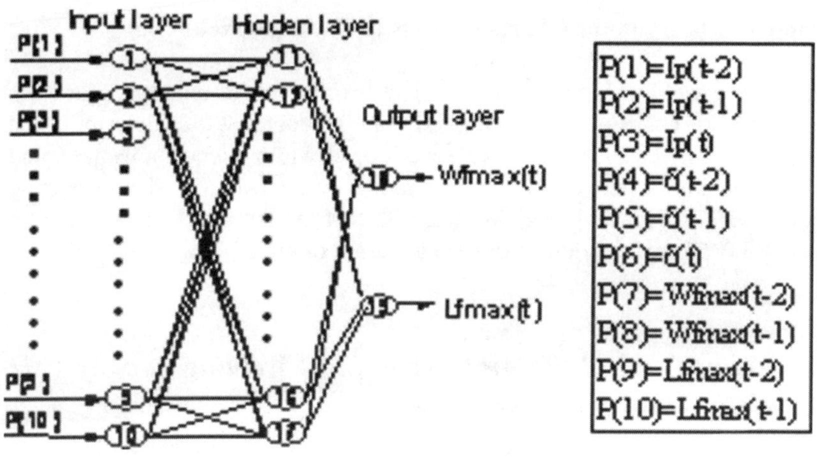

Fig. 6.15 Structure diagram of topside parameters model

0.01 mm, and root-mean-square error (RMS) of maximum width is 0.23 mm. The errors of half length of weld pools are 0.82 mm, −0.02 mm and 0.34 mm respectively. The results are shown as Fig. 6.16, one can see that the identified model can approximate real dynamic welding process.

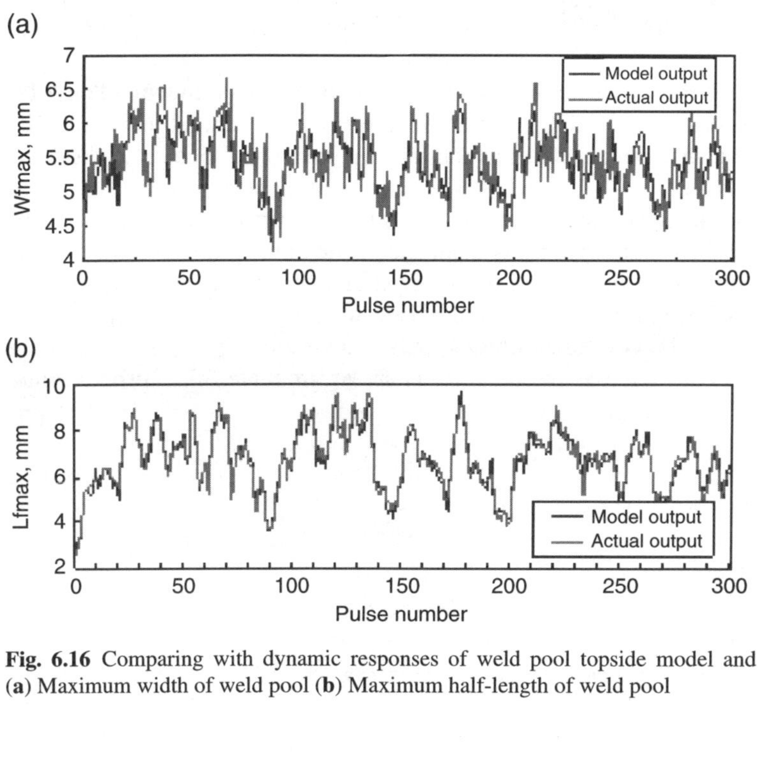

Fig. 6.16 Comparing with dynamic responses of weld pool topside model and actual process (**a**) Maximum width of weld pool (**b**) Maximum half-length of weld pool

6.1.3.2 Modeling of Dynamic Characteristics of Backside Weld Pool

Generally, topside weld pool width cannot always reflect weld quality accurately, but width change of backside weld pool is a main characteristic to determine the weld quality. In many cases, the backside weld bead width cannot be sensed and measured in real time. Therfore, a dynamic model of backside weld pool during robotic pulsed GTAW is established. We use SSNNM model, which is the same as that in Sect. 4.3.4. Therefore, details of the model are omitted here.

6.1.4 Real-Time Control of Low Carbon Steel Welding Pool by PID Regulator During Robotic Pulsed GTAW

As it known to all, the PID controller, is widely used in practice due to its advantages of simplicity and fast regulating speed. In our experiment, the digital PID control algorithm can be first designed and used in the systems. The digital PID control algorithm is following:

$$\Delta u(k) = K_C\left[e(k) - e(k-1)\right] + K_I e(k) + K_D\left[e(k) - 2e(k-1) + e(k-2)\right] \quad (6.1)$$

where proportional constant $K_C = 13.3$, integral constant $K_I = 0.8$, differential constant $K_D = 2.4$, and e means the error of actual output value from expected output value.

The structure of PID controller for topside maximum width is shown in Fig. 6.19. Supposing expected value of W_{fmax} is 5.2 mm, the curves of simulation for the PID controller are shown in Fig. 6.20, in which the maximum overshoot is 2% and the regulating time is 5s.

In order to verify the effectiveness of the PID controller, as shown in Fig. 6.17 robotic pulsed GTAW butt welding is conducted. The dumbbell shape work-piece is low carbon steel plate with 2 mm thickness as shown in Fig. 6.18a for simulating the change of heat transfer condition during welding. The peak current value is regulated by the PID controller, and the pulse duty ratio is 50%. Other welding parameters are as the same as the parameters shown in Table 6.1. During robotic

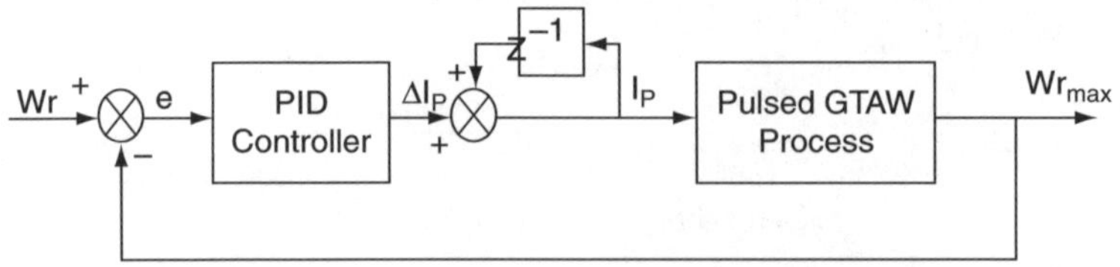

Fig. 6.17 Structure of PID controller for topside maximum width

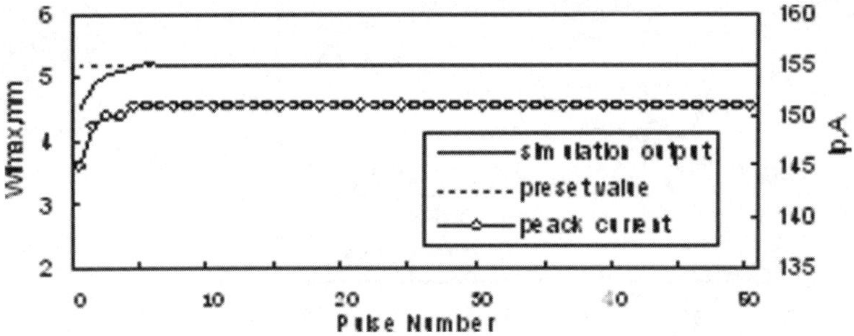

Fig. 6.18 Simulating results of PID controller

welding process, the varying of W_{fmax} and controlling variable, peak current changing with time is shown in Fig. 6.20 during welding. Maximum topside width supposed is 5.2 mm, the maximum absolute error is 0.76 mm, average error is 0.07 mm and RMS error is 0.39 mm. And the photographs of controlled welding work-piece by welding robot are shown as Fig. 6.19b,c. The welding experiment results during robotic welding are shown the effectiveness of closed loop real-time control in welding robot systems.

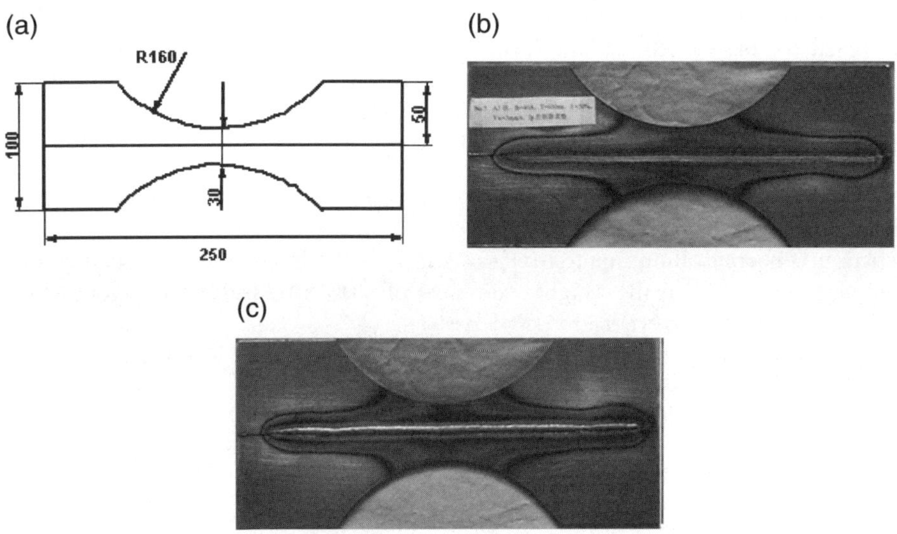

Fig. 6.19 Dumbbell work piece by PID controller during robotic pulsed GTAW (**a**) The shape and the size of the work piece (**b**) Topside (**c**) Backside

Fig. 6.20 The curve of closed-loop PID control for topside maximum width during robotic pulse GTAW

6.2 Real-Time Control of Weld Pool Dynamics and Seam Forming by Neural Self-Learning Controller During Robotic Pulsed GTAW

6.2.1 Neuron Self-Learning PSD Controller for Low Carbon Steel Weld Pool

A neuron self-learning Proportional-Summational-Differential (PSD) controller is designed for the robotic welding process. Main merits of the PSD controller are model-free and adaptability of the complicated process. Measuring errors between the actual outputs and the expected outputs, the PSD controller can realize an adaptive control of the process. A neuron self-learning PSD controller's output follows as

$$\Delta u(t) = \omega_1(t)x_1(t) + \omega_2(t)x_2(t) + \omega_3(t)x_3(t) \tag{6.2}$$

where u(t) is controlling input, $x_i(t)(i = 1,2,3)$ is the input signals of the neuron, and $\omega_i(t)(i = 1,2,3)$ is the weight coefficient of $x_i(t)$. $x_i(t)$ is defined as following. $x_1(t) = e(t)$, $x_2(t) = \Delta e(t)$ and $x_3(t) = \Delta e^2(t)$.

$\omega_i(t)$ is regulated continually during neuron's learning. The learning rule is

$$\omega_i(t+1) = (1-m)\omega_i(t) + dr_i(t) \tag{6.3}$$

where m, d > 0, d is learning ratio and

$$r_i(t) = z(t)u(t)\left[e(t) + \Delta e(t)\right] \tag{6.4}$$

where z(t) is teaching signal.

To ensure the convergence and the robustness of the PSD, the following learning algorithm is adopted:

$$\Delta u(t) = K \sum_{i=0}^{3} \omega'(t) x_i(t) \tag{6.5}$$

where

$$\omega_i'(t) = \sum_{i=1}^{3} |\omega_i(t)| \tag{6.6}$$

and

$$\begin{cases} \omega_1(t+1) = \omega_1(t) + d_I z(t) u(t) \left[e(t) + \Delta e(t) \right] \\ \omega_2(t+1) = \omega_2(t) + d_P z(t) u(t) \left[e(t) + \Delta e(t) \right] \\ \omega_3(t+1) = \omega_3(t) + d_D z(t) u(t) \left[e(t) + \Delta e(t) \right] \end{cases} \tag{6.7}$$

where d_P, d_I and d_D are regulated according to the actual system.

Experiment results showed that control the topside characters of the weld pool during GTAW process can not always ensure an ideal weld penetration due to the variation of gaps of weld seam. Therefore, it is necessary to control the backside characteristics of weld pool, which are the direct parameters for weld penetration. Direct detection of backside characteristics' variation of the weld pool is not accessible in many cases. Using the above neuron self-learning PSD controller, the butt welding experiment during pulsed GTAW process is conducted on the welding robot systems. The control system structure is shown as Fig. 6.21. The output value of PSD controller is the increment of peak current of pulsed GTAW. During welding, topside parameters of weld pool are sensed and measured in real-time. The backside weld width W_{mbmax} can be predicted by the topside pool characters and processing variables. The error and its changes between predicted and measured maximum backside pool width are inputs of the neuron self-learning PSD controller.

The work-piece is dumbbell-shaped low carbon steel plate with 2 mm thickness. The size and the shape of the work-piece are shown in Fig. 6.22a. The peak current value of the pulse is regulated by the neuron self-learning PSD controller during welding process. And the pulse cycle is 0.6s, the pulse duty ratio is 50%. The other welding conditions are shown in Table. 6.2.

During controlling welding process, the varying curves of W_{fmax}, W_{mbmax} and control variable, the peak current value, are shown in Fig. 6.23. Maximum backside width supposed is 4.0 mm. Figure 6.23 shows that the maximum absolute error is 0.81 mm, average error is −0.27 mm and RMS error is 0.47 mm. The photographs of work-piece are shown in Fig. 6.22b,c.

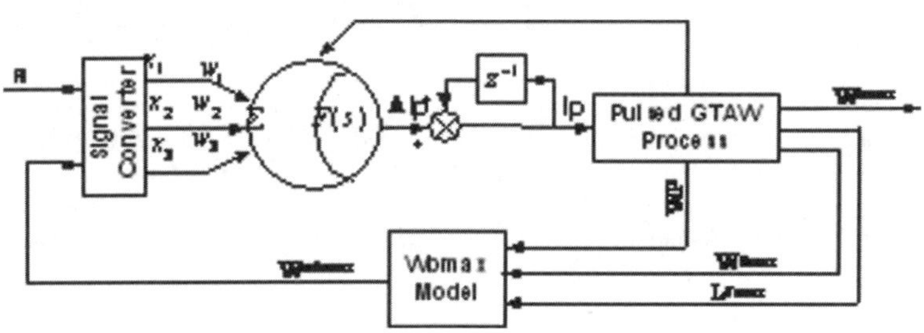

Fig. 6.21 Neuron self-learning PSD control of backside width of pool weld

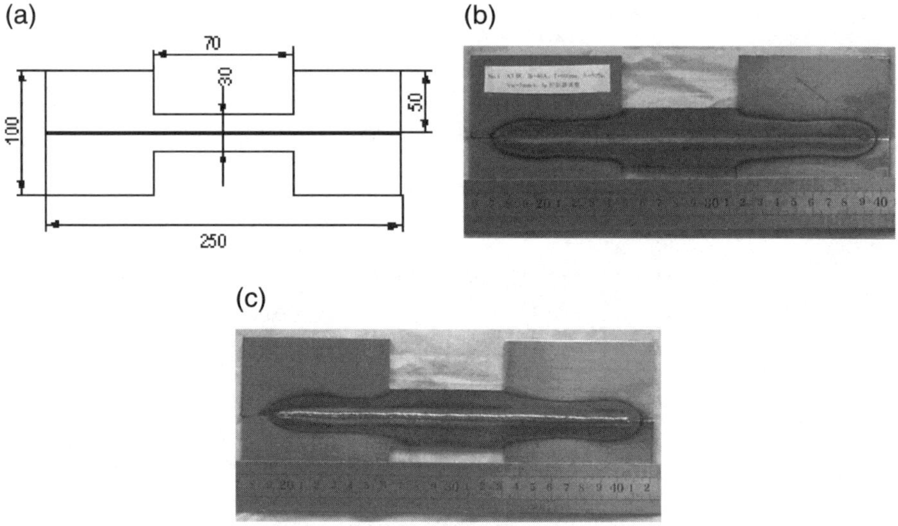

Fig. 6.22 Photographs of neuron self-learning PSD controlling for backside width of dumbbell work-piece

Table 6.2 Controlled welding conditions of aluminium alloy pulse GTAW

Pulse frequency f, Hz	1	Traveling speed V_w, mm/s	2.5
AC frequency f, Hz	50	Arc Length L, mm	2.5
Peak current I_p, A	170	T-pole dia ϕ, mm	3.2
Base current I_b, A	50	Flux of argon L, l/min	10
Wire feeding V_f, mm/s	7	Workpiece size, mm^3	250 × 50 × 2.5

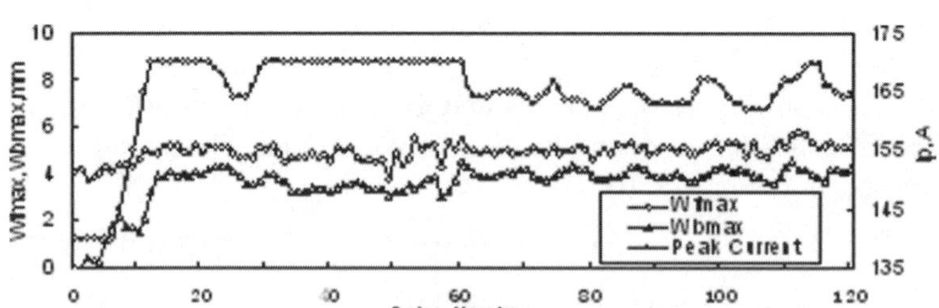

Fig. 6.23 Curve of neuron self-learning PSD control for backside width during robotic pulsed GTAW

6.2.2 Adaptive Neural PID Controller for Aluminium Alloy Welding Pool

In this section, a key intelligentized technologies for the robotic welding will be discussed, such as, computer vision sensing for recognizing weld seam and starting, autonomously guiding in the local circumstance and tracking seam, intelligent control of aluminium alloy welding pool dynamics and seam forming during pulse GTAW. These key technologies are integrated into locally autonomous intelligentized welding robot (LAIWR) systems [19].

6.2.2.1 The Structure and Main Functions of Intelligentized Welding Robot Systems

The principle scheme of an intelligentized welding robot systems is shown as Fig. 6.24, which consists of a 6.freedom manipulator and a freedom visual servo unit (VSU) installed on the sixth axis of the robot for turning a dual camera sensor; a welding seam guiding unit (SGU), a seam tracking unit (STU), a welding penetration control unit (PCU), a knowledge data unit (KDU) and a system simulation

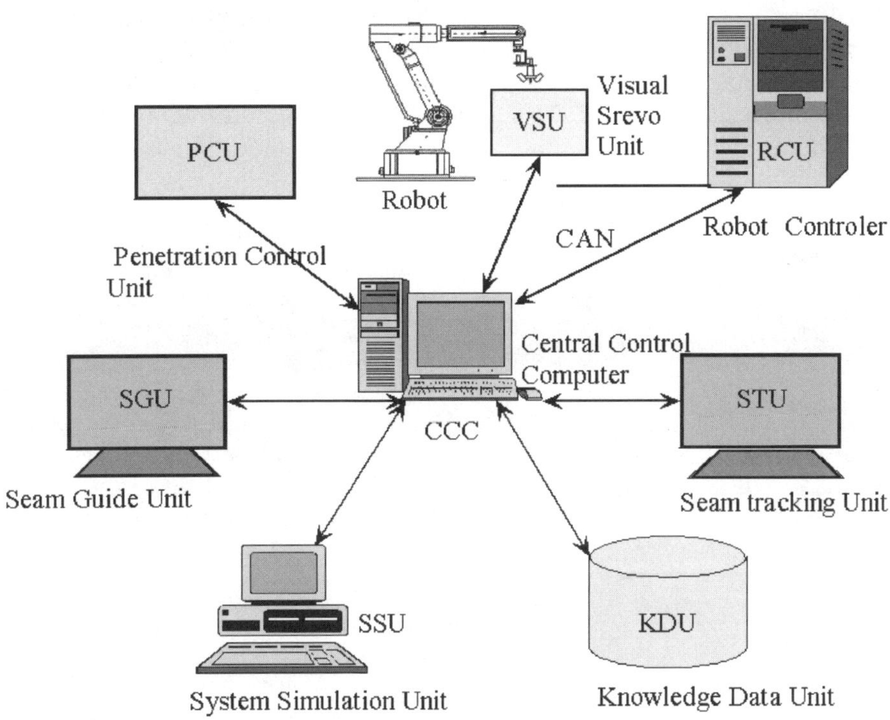

Fig. 6.24 The hardware structure of the LAIWR systems

unit (SSU), all units are dominated by a central control computer (CCC). This combined welding robot system can realize autonomously recognizing weld starting and seam by vision sensing in the local circumstance, guiding robot to the starting, tracking seam, and real-time control of welding pool dynamics and seam forming during pulse GTAW by appropriate intelligentized strategies. It is called the locally autonomous intelligentized welding robot (LAIWR) systems [20].

The software structure of the LAIWR systems is showing as Fig. 6.25. The software systems are divided into different control modules correspond to the task units of the LAIWR systems, which contains a central control module (CCM) for supervising and dominating each unit and whole robot system functions, a seam guiding module (SGM), a seam tracking module (STM), a welding penetration control module (PCM), a knowledge database module (KDM), and a system simulation module (SSM). All modules communicate with the CCC through the Windows Socket. The system also contains a WWW server module for long-distance control of welding robots by the CORBA communication [20].

Real-time control of welding pool dynamics and seam forming is one of most crucial technologies for robotic welding quality. At present, almost teaching playback welding robot is non real-time control of dynamics of welding pool. In the LAIWR system, a real-time control subsystem as Fig. 6.26 is developed for dynamical process of robotic welding.

In the LAIWR systems Fig. 6.26, an adaptive neural PID controller is developed for real-time control of dynamical pool and fore seam during robotic welding. Control, the controller framework is showing as Fig. 6.27, which includes common PID regulator, learning algorithms, neural networks NN1 and NN2 for modeling welding dynamics and modifying PID parameters. The controller algorithms are omitted here [20].

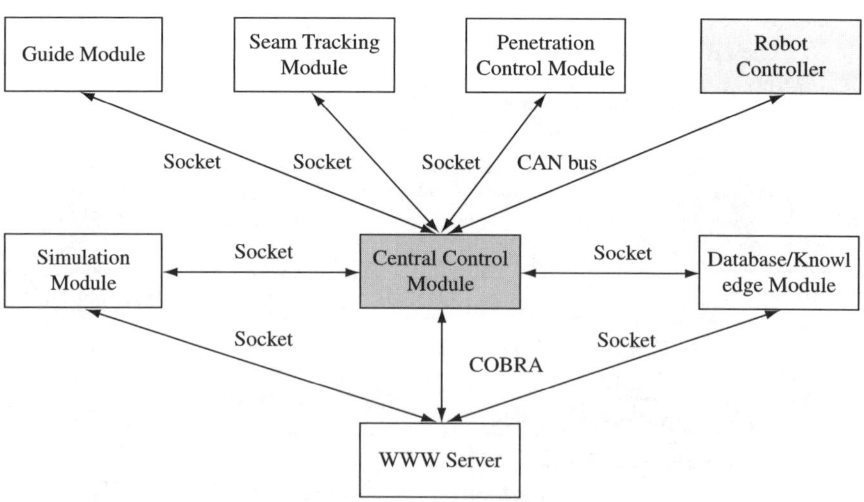

Fig. 6.25 The software structure of the LAIWR systems

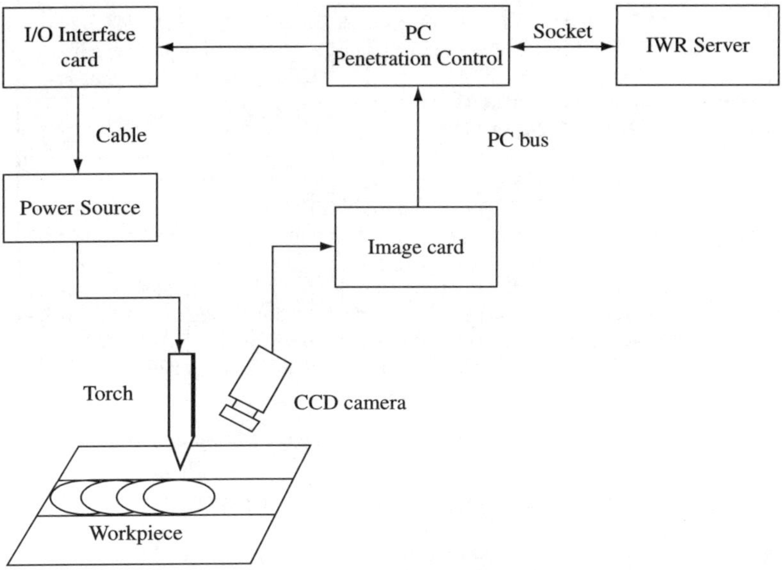

Fig. 6.26 Real-time control subsystem for dynamical process of robotic welding

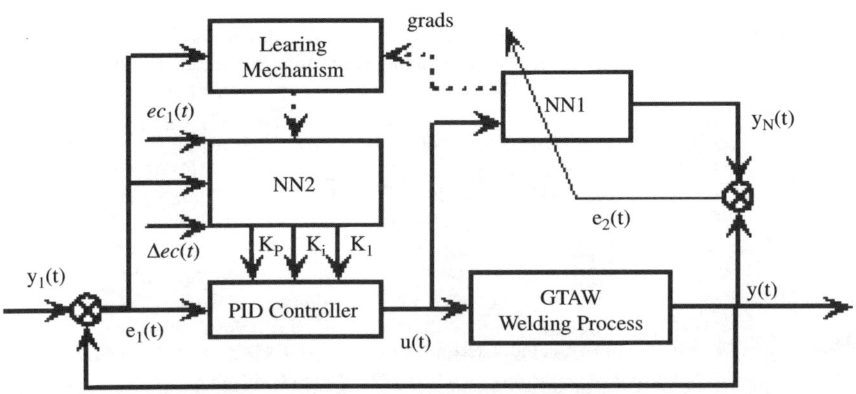

Fig. 6.27 The framework of adaptive neural PID controller for robotic welding process

6.2.2.2 Image Processing and Feature Acquiring of Weld Pool During Robotic Welding

Based on the above controller design and the Table 6.2 welding conditions, the real-time control experiment is completed on the LAIWR systems.

The image processing and results of controlled welding on the LAIWR systems are shown as Figs. 6.28, 6.29 and 6.30, the details are omitted here [21]. The Fig. 6.28 is flow chart of aluminium alloy pool image processing, Fig. 6.29 is showing the image processing results of the pool during robotic welding.

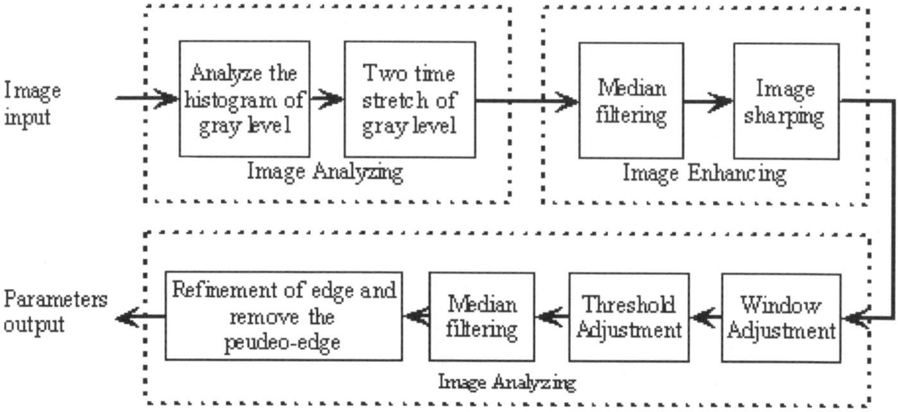

Fig. 6.28 The flow chart of Al alloy pool image processing during robotic welding

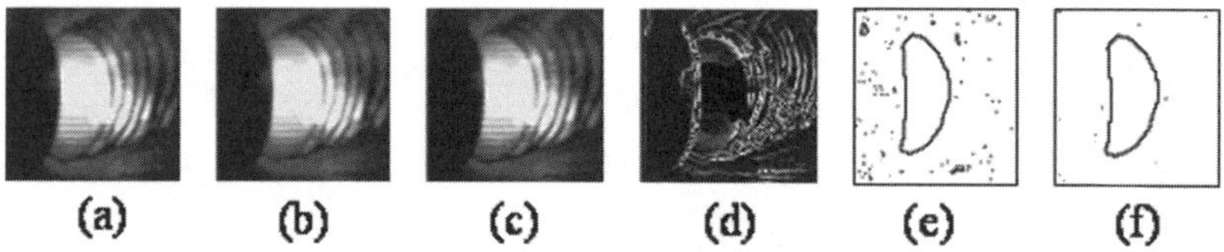

Fig. 6.29 The results of Al alloy pool image processing during robotic welding (**a**) Original (**b**) Median filter (**c**) Image reinforcing (**d**) Edge detecting (**e**) Profile extracting (**f**) Filtering

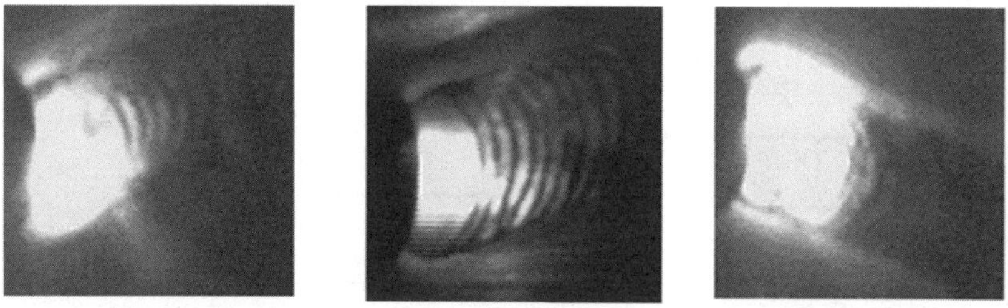

Fig. 6.30 Al alloy pool images in three direction of the S shape seam during robotic welding(**a**) The left rear direction (**b**) The positive rear direction (**c**) The right rear direction

In the robotic welding, the image shape would be changed with seam curve and robot motion direction. The Fig. 6.30 is shown aluminium alloy pool images in three direction of the S shape seam during robotic welding. The corresponding image processing algorithms are developed in LAIWR systems [20], and here it is omitted.

6.2.2.3 Real-Time Control Experiment During Robotic Welding

Using the characteristic information of the welding pool, the closed loop feedback control in LAIWR systems is structured and real-time control of dynamic welding process is realized. The experiments of the constant technical parameters, i.e. without the loop feedback control, and simple PID control scheme are conducted for comparing with the designed adaptive neural PID controller in this paper, the compared results is showing that the adaptive neural PID controller in the LAIWR systems is effective for real-time control of weld pool dynamics and fine seam formation during aluminium alloy pulse GTAW. The Fig. 6.31 and Fig. 6.32 are showing the controlled welding results on the LAIWR systems. The trapezoid and dumbbell workpiece are designed to simulate the different changes of heat conduction and the effectiveness of the controller during robotic welding process. The controlled results are showing that the desired seam width, 7.6 mm for trapezoid workpiece, and 8.0 mm for dumbbell workpiece, are maintained steadily by the peak current regulation during pulse GTAW on the LAIWR systems [20].

A welding robot systems with real-time visual sensing and self-learning neuron control of weld pool dynamics is established in this section to overcome the drawbacks of teaching play-back welding robot without real-time sensing control of weld pool dynamics. Clear pool images of pulsed GTAW during robotic welding are acquired by the visual sensor and composed filter technique. The related image processing algorithm is developed, Dynamic models of topside and backside weld pool of pulsed GTAW are established by artificial neural networks. The PID controller and the neuron self-learning PSD controller are designed for ideal topside maximum width and backside maximum width of weld in real time respectively. The experiment results on the robotic systems show that the visual sensing and controller algorithms designed in welding robot system are effective. The above results will be key prepared technologies for intelligentlized welding robots in our further researches [19, 20, 22, 23].

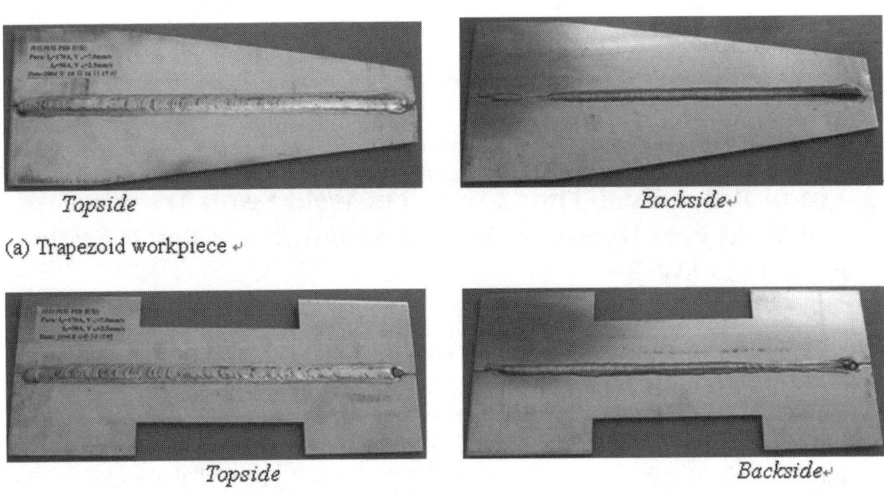

Fig. 6.31 The workpiece pictures of adaptive neural PID controlled welding on the LAIWR

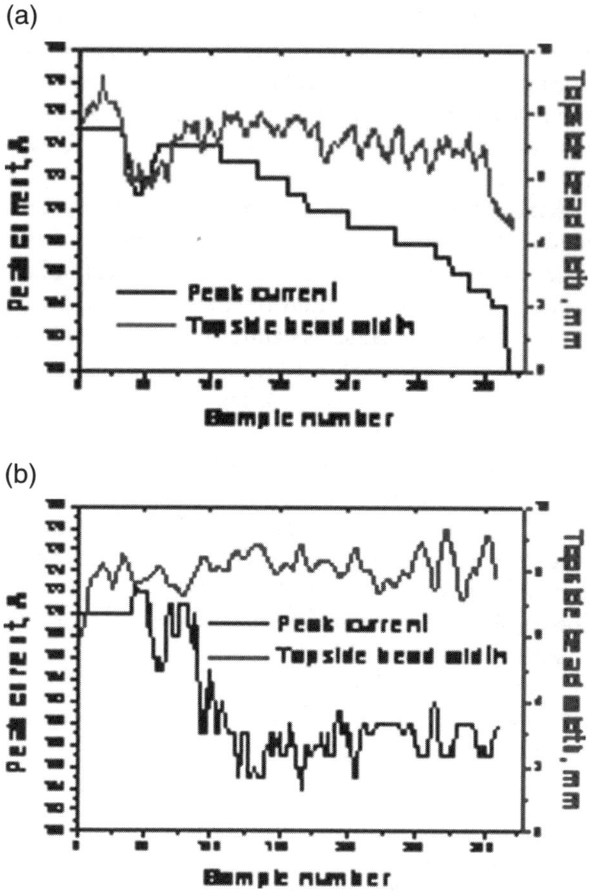

Fig. 6.32 Adaptive neural PID controlled curves of Al alloy welding process on the LAIWR (**a**) Trapezoid workpiece (**b**) Dumbbell workpiece

6.3 Vision-Based Real-Time Control of Weld Seam Tracking and Weld Pool Dynamics During Aluminium Alloy Robotic Pulsed GTAW

Most visual sensors are used in welding seam tracking and weld pool size measuring [21, 24–32]. A number of significant achievements have been made in the field of autonomous welding robot by means of visual sensing [33–36]. In some studies, the camera is directly used to view the weld pool and its vicinity to obtain control information such as the size, the position of the weld pool and the width of the gap [37, 38]. In other studies, the camera is used to view the laser stripe projected by a laser diode to detect the seam position, gap size and the offset etc [39, 40]. But most of the applications need the robot calibration, such as the coordinate systems, the end-effector position and the "hand-eye" calibration. The calibration is professional technology and quite complicated. Few operators can put the achievements into practice. So the seam tracking technology without robot calibration is proposed in this study.

6.3.1 Welding Robotic System

The flange products of the rocket, as shown in Fig. 6.33, will be welded by welding robot in Fig. 6.34. The diameter of the flange is 148 mm and the welding procedure is GTAW with wire filler. The robot must weld around the center of the flange for 400° at least to ensure the welding quality. Because the diameter is too small to keep the robot weld the flange along the seam center exactly unless the operator istes plenty of time to teach the robot at any point, we used the visual sensing technology to develop the seam tracking system to solve the problem.

The vision-based welding robot system consists of a visual sensor, a robot, a rectifying board, a welding power source and a control computer. Figure 6.35 shows a schematic diagram of the vision-based real-time seam tracking welding robot system. The robot is a six-axis industrial robot, made of Motoman Robot Co., Ltd, as shown in Fig. 6.34. It can move in the vertical direction of welding through setting the voltage signal ($-10v$-$10v$) to rectifying board in the robot controller. The visual sensor device and software controller will be presented in the following sections.

Fig. 6.33 The flange product welded by robot

Fig. 6.34 Robot welding system

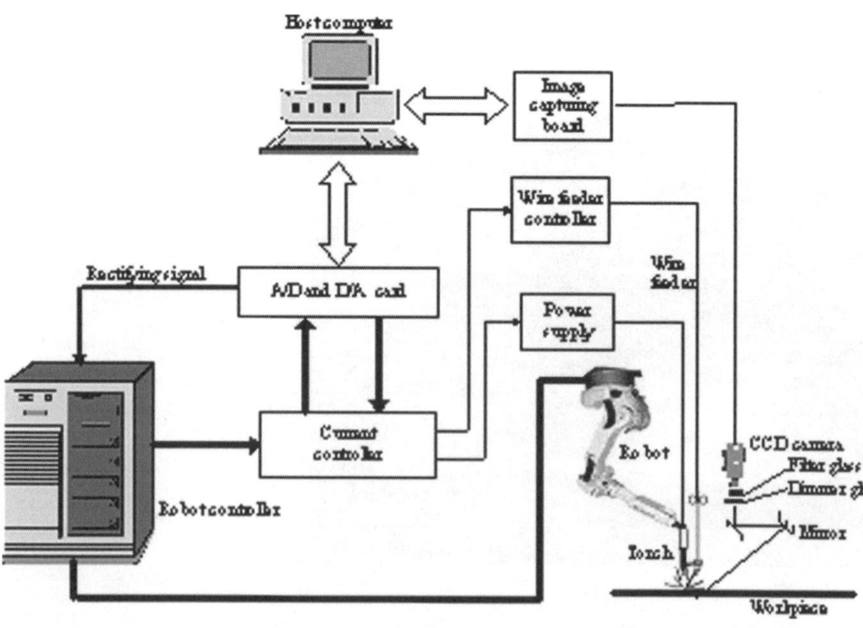

Fig. 6.35 The schematic diagram of the vision-based real-time seam tracking arc welding robot system

6.3.1.1 Visual Sensor

The visual sensor is composed of CCD camera and optical filter. It directly affects the tracking accuracy level as the detecting element.

Structure of the Visual Sensor

Figure 6.36(a) shows the prototype of the visual sensor device. The CCD camera receives the weld pool image after twice reflection, as shown in Fig. 6.36(b).

The double-layer filter system has been designed in this device, because the light intensity of the weld pool is much greater than that of the seam during welding. The bottom layer filter is used to view the seam region and the double layers are used to view the weld pool region. A clear weld pool image is shown in Fig. 6.37(a) using the optical filter adapting to the weld pool, but the seam region is very obscure. If using the optical filter adapting to the seam, the seam region would be very clear, but we can't get any characteristic of the weld pool, as shown in Fig. 6.37(b). Figure 6.37(c) shows the image with the double-layer filter. The left side of the image is captured through the double layers filter; the right side is just through bottom layer filter. The characteristics of the weld pool and the seam are both clear enough. Moreover, the experiments indicate that the computer can capture the good images using this visual sensor when the welding current is in the range of 150–340 A that is appropriate for the medium and thick plate of aluminum alloys.

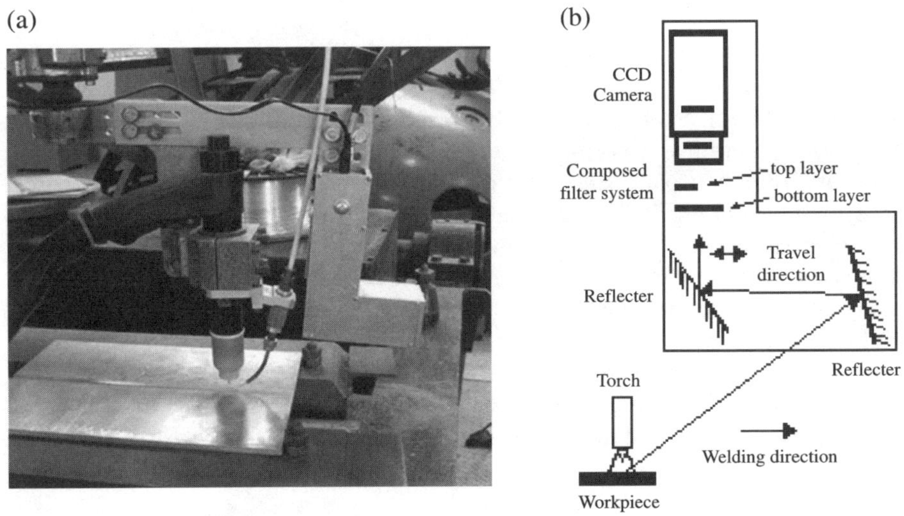

Fig. 6.36 The visual sensor device (**a**) the prototype (**b**) the structure

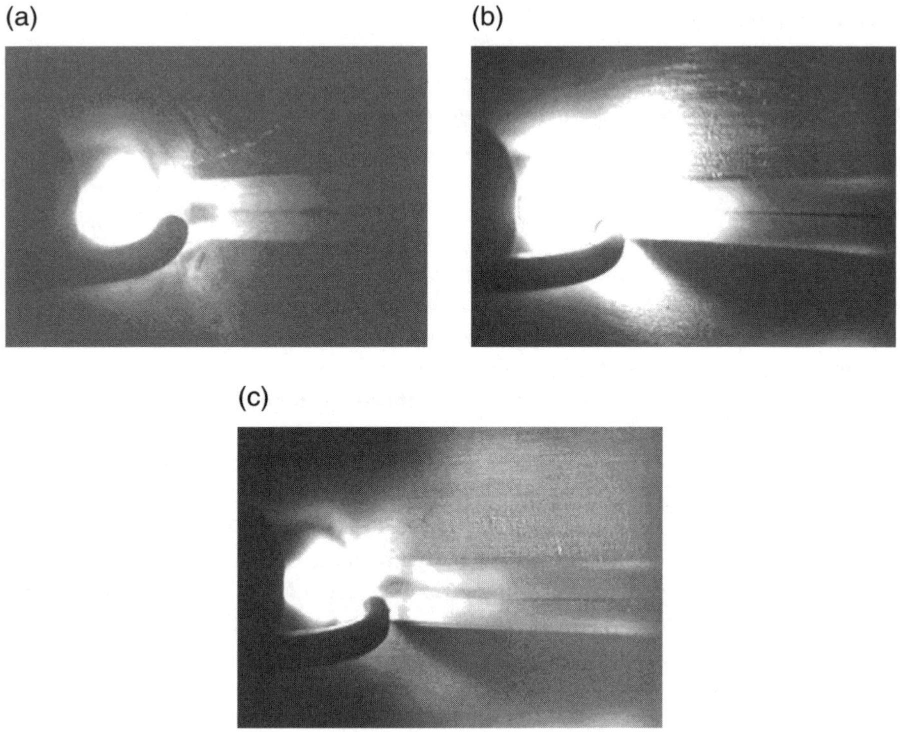

Fig. 6.37 The image with different filter system (**a**) the optical filter adapting to the weld pool (**b**) the optical filter adapting to the seam (**c**) the double-layer filter

Calibration of the Visual Sensor

The deformation inevitably exists in the image plane coordinate system relative to the absolute coordinate system, so the calibration is necessary before any experiment. A calibration plate with a lot of $5\,\text{mm} \times 5\,\text{mm}$ panes simulates the workpiece, as shown in Fig. 6.38. The orientation of the tungsten electrode is normal to the calibration plate. The distance between the tip of the tungsten electrode and the calibration plate is 5 mm, which is appropriate to GTAW. Point $O(0,0)$is the projection point of the tungsten electrode on the calibration plate. And it is defined as the original point of the calibration plate coordinate system. It can be seen that the deformation of the abscissa and the ordinate both increase with the abscissa increasing and the relation between the deformation and the abscissa is basically linear. So the deformations are supposed to be two linear relation represented as follows:

$$\begin{cases} f(n) = k \times n + b \\ f'(n) = k' \times n + b' \end{cases} \qquad n \geq 0 \tag{6.8}$$

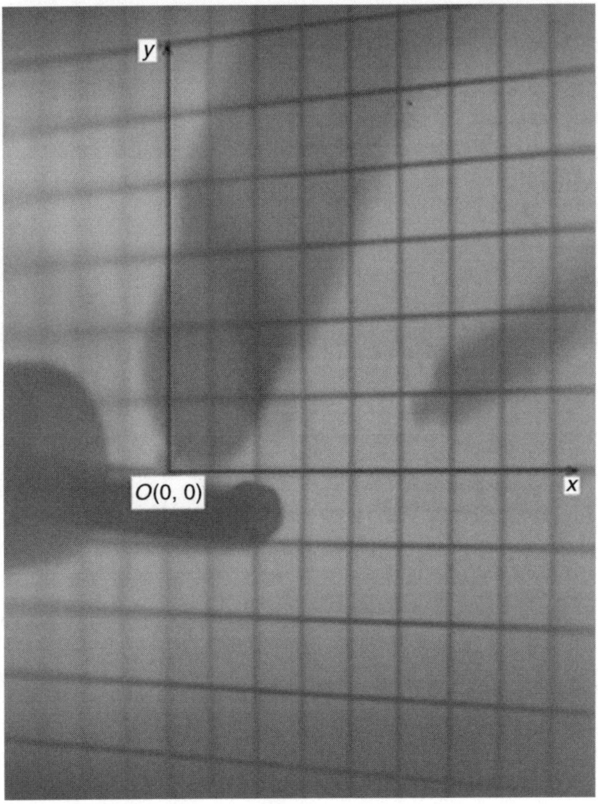

Fig. 6.38 The picture of CCD calibration

Where $f(n)$ and $f'(n)$ are the abscissa and the ordinate deformation with the abscissa increasing in the image plane coordinate system, respectively.

So the relation of the image plane coordinates and the absolute coordinates can be worked out, which is given as follows:

$$\begin{cases} x_{real} = \dfrac{-b+\left[b^2-4k(x_{image\_o}-x_{image})\right]^{1/2}}{2k} \times d_{real} = X(x_{image}) \\ y_{real} = \dfrac{(y_{image\_o}-y_{image}) \times d_{real}^2}{x_{real} \times k' + d_{real} \times b'} = Y(y_{image,X(x_{image})}) \end{cases} \tag{6.9}$$

Where (x_{image}, y_{image}) and (x_{real}, y_{real}) are the image plane coordinates and the absolute coordinates, respectively. $(x_{image\_o}, y_{image\_o})$ is the coordinates of point $O(0,0)$ on image plane coordinate system, d_{real} is the interval of the panes, $X$and Y are the abscissa and the ordinate relation function between the image plane coordinates and the absolute coordinates, respectively.

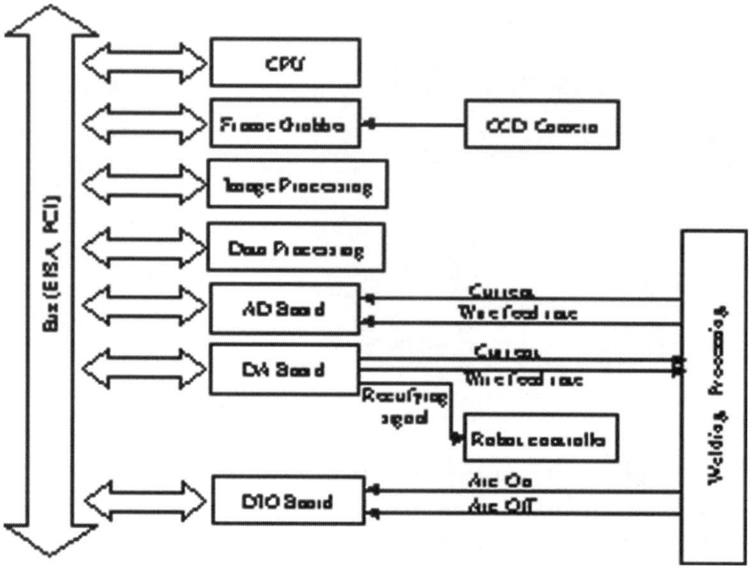

Fig. 6.39 Control system of the robot seam tracking

6.3.1.2 Software Controller for the Robot Seam Tracking

Figure 6.39 shows a block diagram of the software controller of the robot welding system based on a personal computer. The computer acquires images of the weld pool through the CCD camera and the frame grabber, monitors the welding current and wire feed rate through the analog-digital (AD) converter board, adjusts them and sets the rectifying voltage ($-10v$–$10v$) through digital-analog (DA) converter board, detects the arc being or not through the digital input and output (DIO) board.

The computer runs the image processing and data processing programs in respective thread. Figure 6.40 presents the program interface on computer screen during welding process. The period of the tracking control is 400 ms. In a period, the offset of the torch to the seam is extracted one time and the rectifying voltage is updated one time.

6.3.2 Image Processing During the Robot Seam Tracking

The visual sensor is fixed on the robot end joint to capture the images of the weld pool in the topside front direction, as shown in Fig. 6.41. As is known, the sensor moves with the robot during welding process, and the relative position of the torch and sensor is invariable. Also the tip of the tungsten electrode is projected onto the same position of the CCD target.

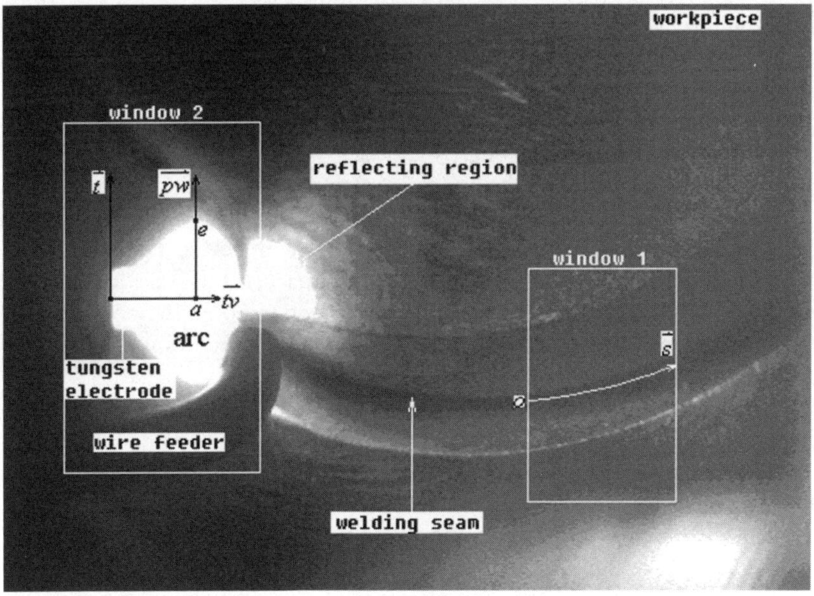

Fig. 6.40 The program interface during welding process

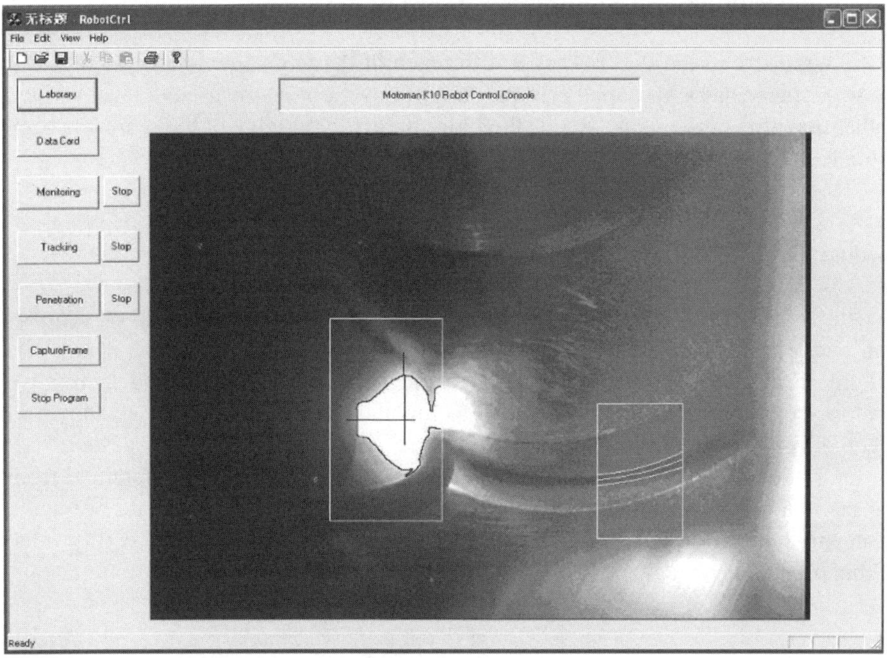

Fig. 6.41 The image of GTAW pool and seam

Firstly, recognition of the welding seam is presented. Since the size of the captured image is 768×576 pixels, most of which is useless and will cost plenty of CPU time, we selected an area, called window 1, shown in Fig. 6.41, where the seam center is extracted by digital image processing technology. Then the seam curve $\bar{s}$ is fitted with nonlinear least square method.

The second step is to calculate the tungsten electrode projection point on the work-piece in the image plane coordinate system. We selected the window 2 area with the same reason as shown in Fig. 6.41. Both the wire feeder and the visual sensor are in topside front of the torch. So the computer can't capture an entirely arc outline, shown in the window 2. The tungsten projection point, defined as point a, is on the orientation of tungsten electrode, defined as line $\overline{tv}$. Point e is defined as one of the widest edge points of the weld pool and line $\overline{ae}$ is normal to line $\overline{tv}$. According to the analysis, it is evident to be able to obtain the coordinates of point a if line $\overline{tv}$ and point e are both extracted.

At last, the offset of the tip of the torch to the seam center is accurately calculated. It is well known that the arc light is very intense and aluminum alloy has a good reflectivity, and the seam is obscure in the reflecting region in front of the weld pool. So the distant between window1 and point a is 10 mm at least in order to obtain the exact seam position. Then, the offset will be figured out after fitting the curve $\bar{s}$.

6.3.2.1 Recognition of Welding Seam Trajectory

In the window1 of the Fig. 6.41, the difference of the gay value is not too big. But the work-piece has a Y-shaped groove, and the gay value of the groove face is much higher than that of the seam. According this character, the edge of the seam has been extracted.

Median Filter

Generally, the noise maybe mixes into the image during the process of acquisition and transmission. It will play down the quality of the image, so it must be get rid of in advance. We used the efficient median filtering method to remove random noise mixed in the image and to maintain image sharpness, as shown in Fig. 6.42(b).

Set a window as 3×3, so the gray value of current pixel may be obtained from the median value of its eight-neighborhood. Suppose that the gray value of some pixel and its eight-neighborhood sort ascending as $\{p_1, p_2 \cdots p_8, p_9\}$, the gray value of this pixel is given as

$$p_0 = p_5 \tag{6.10}$$

where p_5 is the median value of the pixel and its eight-neighborhood.

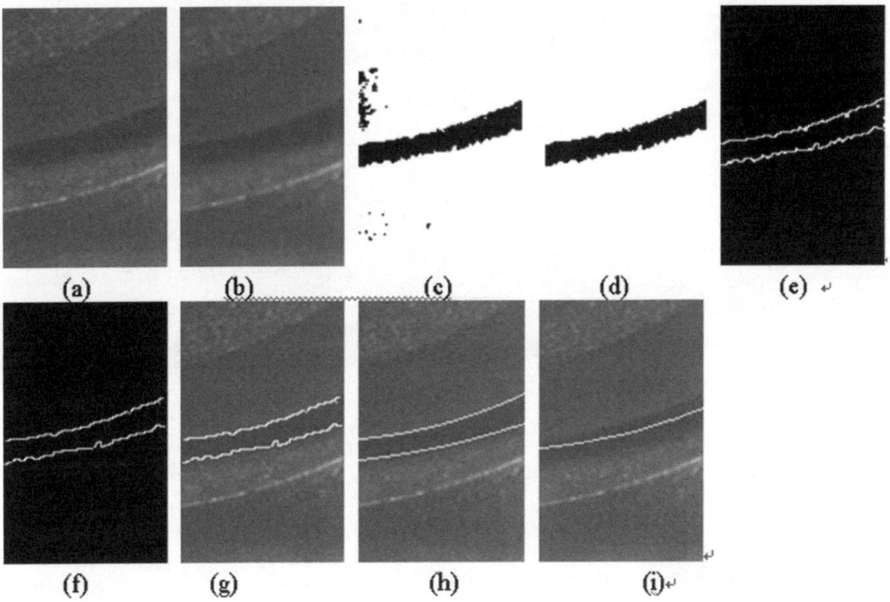

Fig. 6.42 Image processing of window 1 (**a**) original image (**b**) the filtered image by a median filter (**c**) the image with threshold value chosen to be 125 (**d**) the image after removing small area (**e**) the image detected using Roberts operator (**f**) the image after skeleton thinning (**g**) the welding seam points on original image (**h**) the welding seam edge points fitted by nonlinear least square method (**i**) the welding seam center

Thresholding

In the gray level image, the seam edge detected and its background own different gray level value. By choosing an appropriate threshold gray level value it is possible to separate the required seam edge from the background. Fig. 6.42(c) shows the result when the threshold value is chosen to be 125.

f is the gray level distribution function. The function is one of the mapping to T, which is a transfer function, where T_0 is the appropriate threshold value. The transfer function is as follows:

$$T(f) = \begin{cases} 0 & f < T_0 \\ 255 & f \geq T_0 \end{cases} \tag{6.11}$$

Removing Small Areas

Some false edge points exist in Fig. 6.42(c). So we take a 4-neighbors, as shown in Fig. 6.43, to remove them. In the thresholding image, the gray value of the characteristic points is zero. Suppose that $P[6.i][6.j] = 0$ and $P[i][j] \in R_i$ (a certain characteristic region), if $P[6.i - 1][6.j] =$ then $P[i - 1][j] \in R_i$, the same to

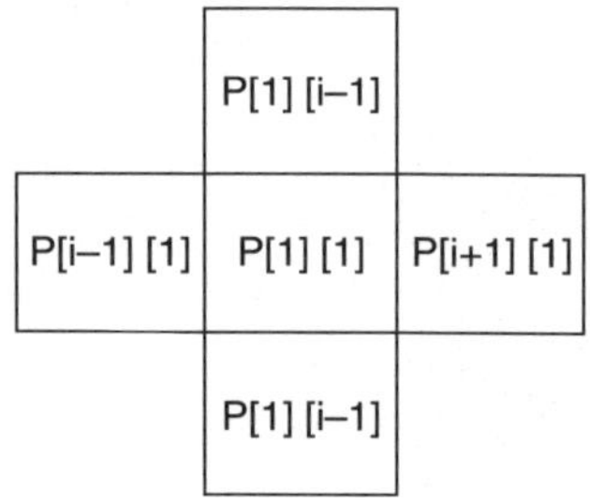

Fig. 6.43 4-neighbors of $P[i][j]$

$P[6.i][6.j-1]$, $P[6.i+1][6.j]$, $P[6.i][6.j+1]$. If the 4-neighbors $\notin Ri$, then count the total number (N_i) of the points in R_i. Then

$$\begin{cases} P[m][n]=0 & Ni \geq MinArea \\ P[m][n]=255 & Ni < MinArea \end{cases} \tag{6.12}$$

where *MinArea* is the area threshold value.

Figure 6.42(d) is the result after removing small area.

Edge Detection of the Seam

The gray level value of the image changes most dramatically when the gray level value moves from groove to seam, as shown in Fig. 6.42(a). Therefore, the gradient $G(x,y)$ at this point is the maximal. According to the image characters, Robert operator is chosen in this study. The Roberts operator is represented as follows:

$$G(x,y)=\left\{\left[\sqrt{f(x,y)}-\sqrt{f(x+1,y+1)}\right]^2+\left[\sqrt{f(x+1,y)}-\sqrt{f(x,y+1)}\right]^2\right\}^{1/2} \tag{6.13}$$

where $f(x,y)$ is the input image with integral coordinates.

The result using the Roberts operator is shown in Fig. 6.42(e).

Thinning of the Seam Edges

Thinning is an image-processing operation in which binary-valued seam image is reduced to lines that approximate their center lines. The purpose of thinning is to reduce the image components to their essential information so that further analysis and recognition are facilitated, as shown in Fig. 6.42(f).

A common thinning approach is to examine each pixel in the image within the context of its neighborhood region of 3×3 pixels and to peel the region boundaries, one pixel layer at a time, until the regions have been reduced to thin lines. The thinning results should approximate the medial lines and must be connected line structures.

Extracting Welding Seam Center

Figure 6.42(g) is the result after the digital image processing above. Then both seam edges are fitted by nonlinear least square method, as shown in Fig. 6.42(h). The edges curve is expressed as

$$\begin{cases} f_1(x) = a_1x^2 + b_1x + c_1 \\ f_2(x) = a_2x^2 + b_2x + c_2 \end{cases} \tag{6.14}$$

where $f_1(x)$ and $f_2(x)$ are the seam up-edge function and the seam down-edge function, respectively.

So the seam center function $f_3(x)$ is calculated easily, represented as follows:

$$f_3(x) = \left[f_1(x) + f_2(x)\right]/2 \tag{6.15}$$

Figure 6.42(i) is the comparison image of original and the welding seam center.

6.3.2.2 Tungsten Electrode Projection Point

There is a little difference can be seen between the window 1 and window 2 in the Fig. 6.41. The gray level value of the arc in the window 2 is much higher than that of the background. So after the median filter processing, the arc outline is obtained with appropriate threshold. Then the arc edge is extracted after edge detection and thinning. Figure 6.44 shows the image processing of the window 2.

In Fig. 6.44(g), we detect the center points of the tungsten electrode according to the edge points of the tungsten electrode. The orientation of the tungsten electrode is then expressed as

$$f(x) = kx + b \tag{6.16}$$

Then the rate of grade of line $\overline{pw}$ is $-\frac{1}{k}$ and point e is on line $\overline{pw}$. So line $\overline{pw}$ is as follow:

$$f'(x) = -\frac{1}{k}x + b' \tag{6.17}$$

By calculating Eqs. (6.16) and (6.17) the coordinates of point a is represented as

$$\begin{cases} a_x = \left[(b' - b)k\right] / \left[k^2 + 1\right] \\ a_y = \left[k^2b' + b\right] / \left[k^2 + 1\right] \end{cases} \tag{6.18}$$

6.3.2.3 Calculating the Offset

Figure 6.45 shows the projection of the tungsten electrode and seam center curve in image plane coordinate system xoy. Point $p(x_p, y_p)$ is the crossing point of line

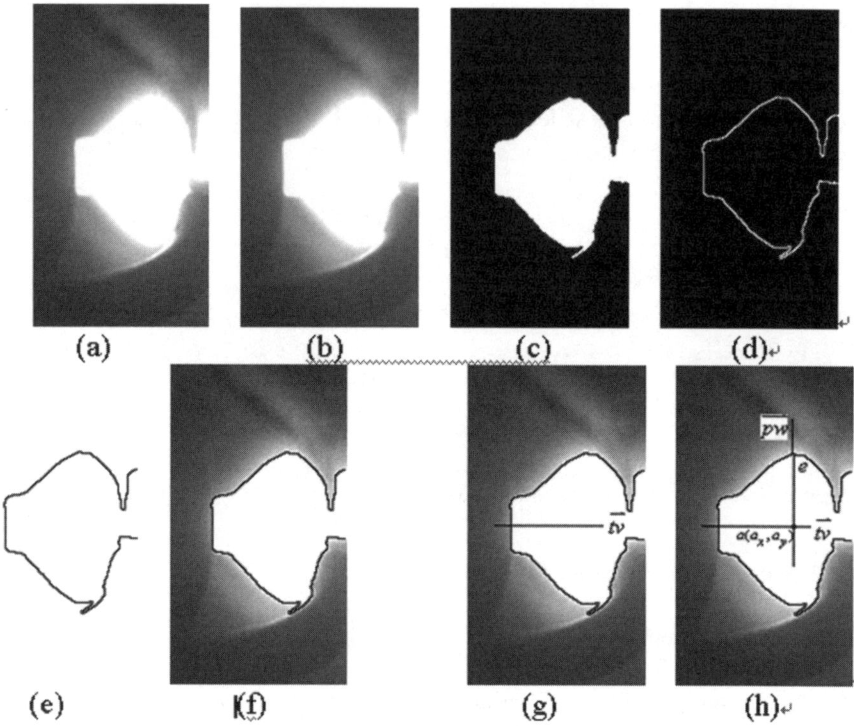

Fig. 6.44 Image processing of window 2 (**a**) original image (**b**) the filtered image by a median filter (**c**) the image with threshold value chosen to be 250 (**d**) the image detected using Roberts operator (**e**) the image after skeleton thinning (**f**) the arc outline on original image (**g**) the orientation of the tungsten electrode (**h**) the projection point of the tip of torch

$f'(x)$ and curve $f_3(x)$. Distance d is the offset of the torch to the seam in image plane coordinate system. According to Eqs. (6.9) and (6.18), the real offset d_{real} is given as follows:

$$d_{real} = \left\{ \left[X(a_x) - X(p_x)\right]^2 + \left[Y[a_y, X(a_x)] - Y[p_y, X(p_x)]\right]^2 \right\}^{1/2} \tag{6.19}$$

6.3.3 Seam Tracking Controller of the Welding Robot

A planar butt welding experiment is designed in order to build the controller. In Fig. 6.46, ae is the seam line and $abcde$ is the taught trajectory at the t_1 time. Suppose that the robot can weld the work-piece along afexactly, the robot will automatically adjust the taught trajectory to be $fd'e'$ at the t_2 time. For the same reason, ge'' is the taught trajectory at the t_3 time. So there is a negative offset trend in af and ge stages, and a positive offset trend in fg stage.

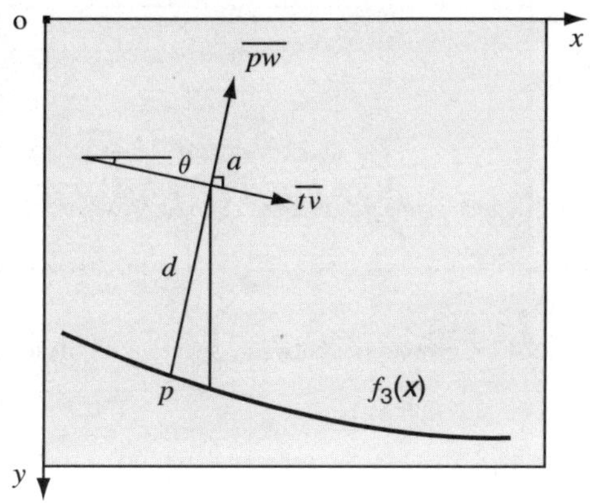

Fig. 6.45 The offset of the torch to the seam in the image plane coordinate system

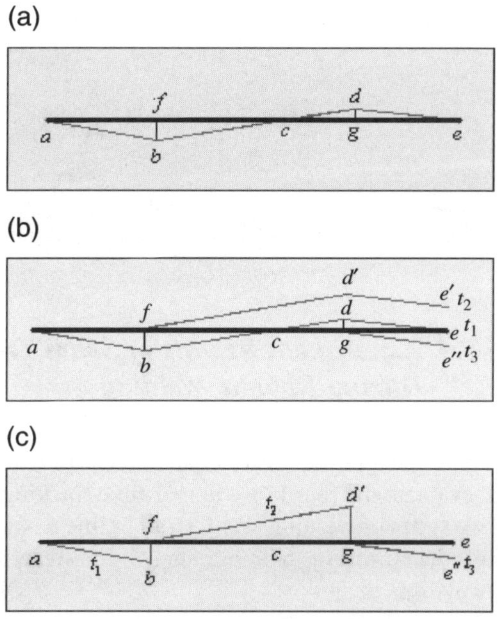

Fig. 6.46 The robot welding trajectory (**a**) the taught trajectory (**b**) robot trajectory at different time (**c**) the trend of offset at different stage

Then we respectively choose 1, 1.5, 2 and 3 v as the rectifying voltage for the work-piece designed in Fig. 6.46. Figure 6.47 shows the offset curve using the different voltage. When the rectifying voltage is 1 v, the rectifying speed is too small to weld along *ae*, especially in *fg* stage, the offset is so large that the seam is outside the scope of window 1 and the computer acquires a wrong result. When the rectifying voltage is 1.5 v, the offset curve has the same trend as Fig. 6.46(c). So the voltage is still evidently small. When it is 2 v, the offset curve fluctuates at the vicinity of zero. Then 2 v is an adaptive rectifying voltage for this seam. When it is 3 v, the fluctuation of the curve is too large and is worse than that of 2 v. So we

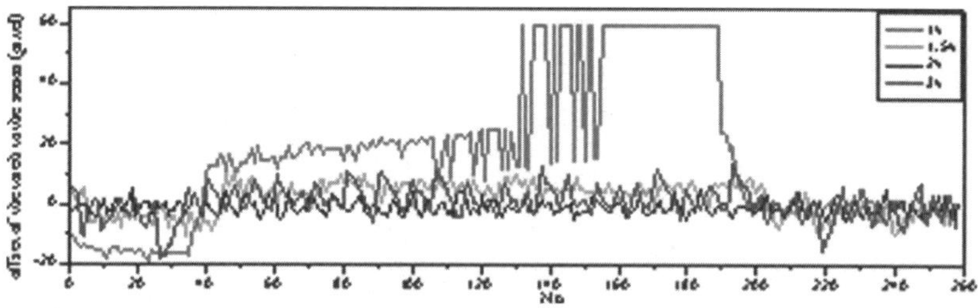

Fig. 6.47 Comparison between different rectifying voltage

came to a conclusion that the different offset must be corresponding with an adaptive rectifying voltage.

So a simple PID controller is researched after a lot of the experiments. The offset is the input signal and the rectifying voltage is the output signal. It is given in equation (6.20).

$$v(k) = k_p e(k) + k_i \sum_{j=0}^{k} e(j) + k_d \left[e(k) - e(k-1) \right] \tag{6.20}$$

where $v(k)$ is the rectifying voltage, $e(k)$ is the offset, k_p is the proportional gain, k_i is the integral gain and k_d is the derivative gain.

6.3.4 Experiment Results of Seam Tracking and Monitoring During Robotic Welding

Welding experiments are conducted with GTAW for the arc welding robot system to evaluate the feasibility in real-time tracking control of the backing weld process. Two types of welding seam, straight line designed in Fig. 6.46 and flange curve line introduced in Fig. 6.33 are chosen for seam tracking. Table 6.3 shows the experiments specifications.

Table 6.3 The experiments specifications

Work-piece	Straight line	Curve line
Material	aluminum alloy LD10	aluminum alloy LD10
Thickness/mm	6	6
Welding joint	Butt weld with Y-groove	Butt weld with Y-groove
Welding current/A	240	230–270
Welding speed/$(\mathrm{cm} \cdot \mathrm{min}^{-1})$	16	16
Wire feed rate/$(\mathrm{cm} \cdot \mathrm{min}^{-1})$	105	85–140
Welding wire diameter/mm	1.6	1.6

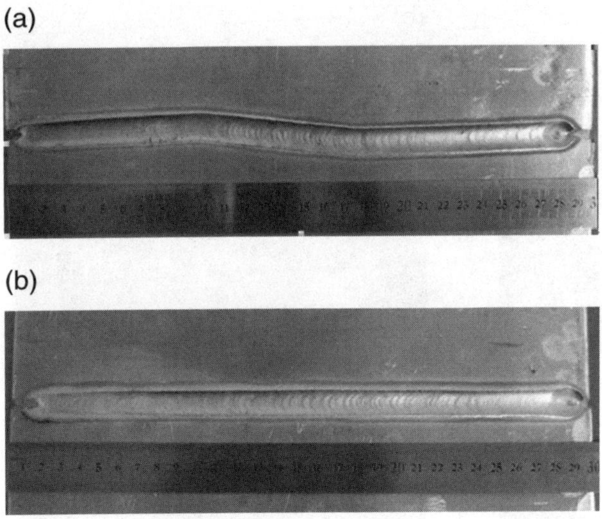

Fig. 6.48 Comparison picture of the backing weld with tracking control or without tracking control (**a**) with tracking control (**b**) without tracking control

Figure 6.48 shows two pictures illustrating the result of the welding experiment of the straight line seam with tracking control or without tracking control, respectively. Figure 6.48(a) shows the result of the backing weld for the preset trajectory without tracking control. And Fig. 6.48(b) shows the result for the same preset trajectory with tracking control. The offset error is in the range of $\pm 0.3\,mm$ during seam tracking process and the result is shown in Fig. 6.49.

An initial offset from 0 to 2 or -2 mm is preset along the circle seam by teaching the robot in the flange experiment. Figure 6.50 shows a favorable result in front side and back side of the flange and Fig. 6.51 shows the offset error is in the range of ± 0.5 mm.

According to these results, the real-time seam tracking system is feasible to control the offset for the different productions.

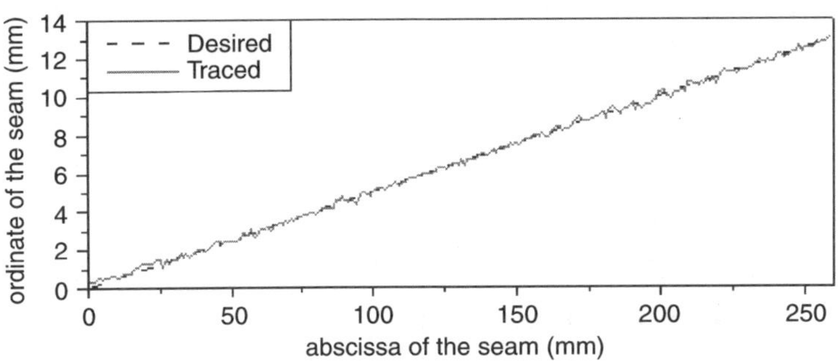

Fig. 6.49 The offset error of the straight line seam with tracking control

(a) (b)

Fig. 6.50 Flange with seam tracking (**a**) front side (**b**) back side

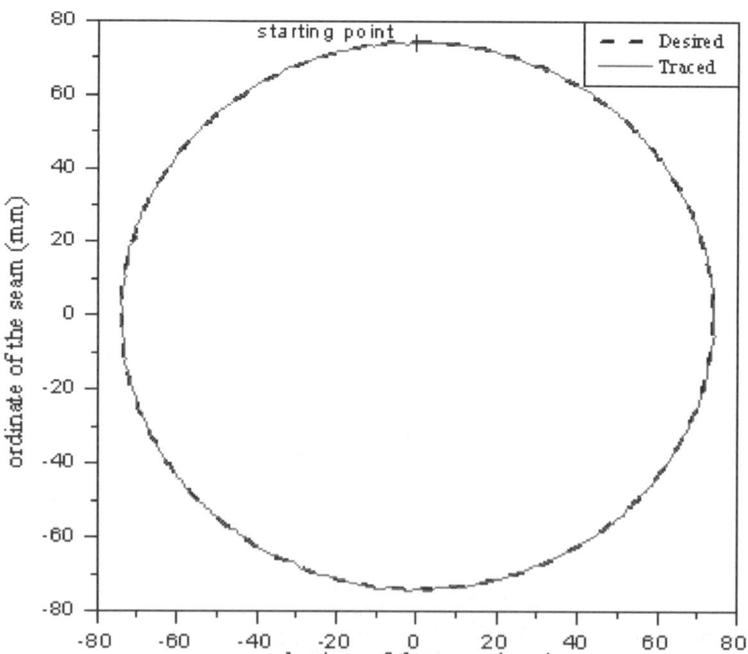

Fig. 6.51 The offset error of the flange seam with tracking control

6.4 Compound Intelligent Control of Weld Pool Dynamics with Visual Monitoring During Robotic Aluminium Alloy Pulsed GTAW

This section shows another robotic welding system with compound intelligent control scheme for full penetration monitoring in a practical welding process [41].

6.4.1 *The Robotic Welding Systems with Visual Monitoring During Pulsed GTAW*

The general structure for arc welding robotic system is shown in Fig.6.52. With the main control computer as its core, a vision sensor and the interface circuit box are designed to realize the autonomy and intelligence of the arc welding robot by

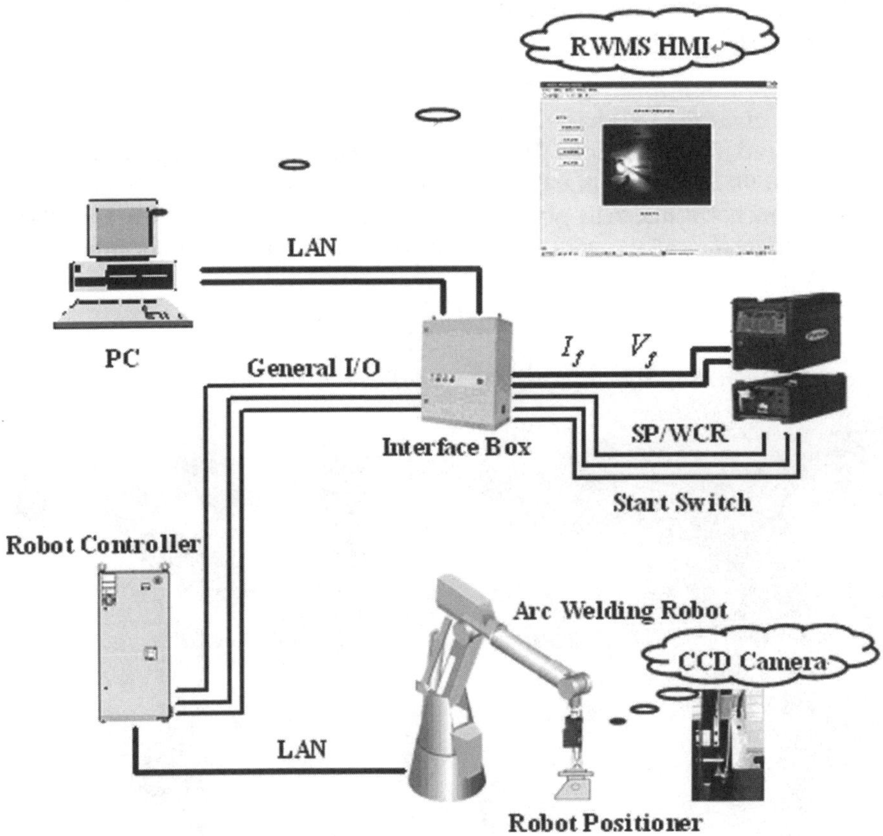

Fig. 6.52 Architecture of the robot arc welding system

existing INVERTER 500P dual inverter arc welding source power, two-axis positioner, and the correspondent software developed.

This is a local intelligentized welding robot system with online real-time control. The welding parameters and dynamic welding pool information are obtained by visual sensor and controlled by the PC controller. The main functions of this system are listed below:

1) The friendly human machine interface, which is used to control the digital welding power source and the robot controller as well as the status of the welding process.
2) Set up the arc start operation and send motion command to the robot controller.
3) Receive the welding pool information through the vision sensors and perform image display, storing and image processing.
4) Real time adjusting and controlling the main parameters of the arc welding robot based on the image analyzing results.

6.4.2 Image Obtaining and Processing for Weld Pool During Robotic Welding

Figure 6.53 gives the structure diagram of the vision sensor. The weld image is transferred into the computer by DH-CG400 PCI card, its transfer speed can reach 40 MB/s and the support with CPU is not needed. Additionally, the image is constructed from horizontal 640 pixels and vertical 480 pixels. In order to reduce the space limitation of the wire feeding nozzle, dual optical path construction is

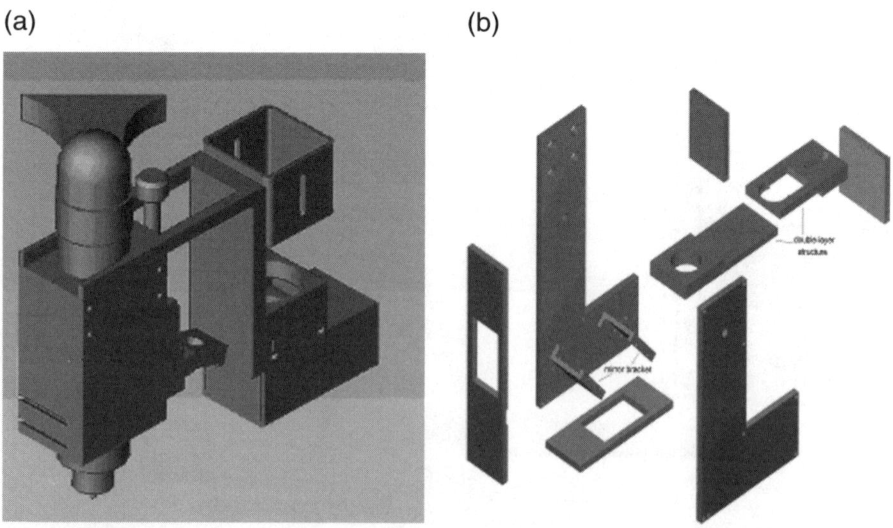

Fig. 6.53 Structure diagram of the robot vision sensor

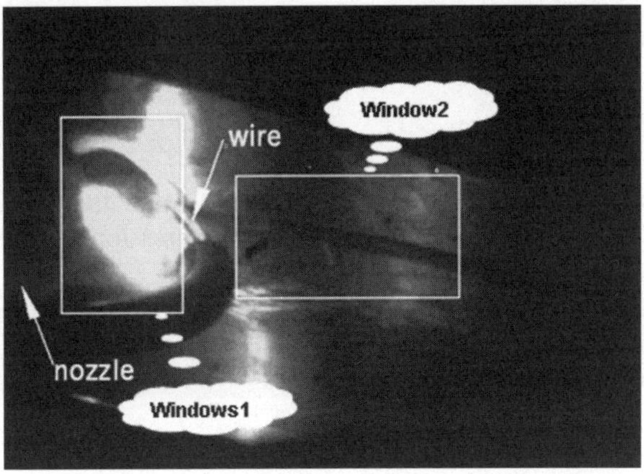

Fig. 6.54 Typical image of the weld pool and gap

employed in the vision sensing system. The CCD camera is attached to the welding torch. The center of the welding torch, plane mirror, camera lens and CCD target surface is nearly on the same plane. Thus the design structure would provide a reliable method to obtain the weld pool and gap.

The information of the weld pool and gap image is captured by the CCD camera with narrow-band composite light filter. From the analysis of the light spectrum of aluminum alloy, it is distributed from 580 nm to 720 nm.Based on the above analysis, the narrow-band filter centered at 630 nm and neutral glass slices of 10%+30%+50% are selected, which can reduce the arc noise to the weld pool. The typical image of the weld pool and gap is shown in Fig. 6.54.

In order to get a clear contour image of the weld pool, and then measure its size rapidly, a window, rather than the whole image is used to reduce the data which would be processed. Weld pool and the gap are defined as window1, window2 respectively.

Since the variation of the pixel gradient between the weld pool boundary and background is not significant, traditional edge detector should produce an edge indication localized to a single pixel located at the midpoint of the slope. Although this form of edge detection performs reasonably well, the detailed information is very poor in the weld pool image. Here, we selected improved canny operator, which has low signal-to-noise ratio and high detection precision [42]. According to the sharp discontinuities of the weld pool, the mask coefficient can be adaptively adjusted. More specifically, this process can be described as the following algorithm.

6.4.2.1 The Algorithm of Extracting Weld Pool Boundary from Window1

(1) Compute the edge gradient $G(x,y)$ in the discrete domain in terms of a row gradient $G_R(x,y)$ and a column gradient $G_C(x,y)$ according to the following functions,

$$G_R^{(m)}(x,y) = \frac{1}{2}\left[f^{(m)}(x+1,y) - f^{(m)}(x-1,y)\right] \tag{6.21}$$

$$G_C^{(m)}(x,y) = \frac{1}{2}\left[f^{(m)}(x,y+1) - f^{(m)}(x,y-1)\right] \tag{6.22}$$

(2) After the $m+1th$ iterations, the implementation for filtering with a weighted averaging filter is given by the expression ($\omega^{(m)}$ is a mask coefficient)

$$f^{(m+1)}(x,y) = \frac{\sum\limits_{s=-a}^{a}\sum\limits_{t=-b}^{b} f^{(m)}(x+s,y+t)\,\omega^{(m)}(x+s,y+t)}{\sum\limits_{s=-a}^{q}\sum\limits_{t=-b}^{b} \omega^{(m)}(x+s,y+t)} \tag{6.23}$$

(3) Do{ If $(m = M)$
    Then, end of iteration;
    Else, turn to step2;
    Go to next domain;
    } Until (end of image)

(4) At the end of procedure, a 8-connected region is determined according to a pixel, and therefore carrying out linking edge segment. At the same time, isolated false edges are deleted further by area filtering operator. In this case, connected area is 50.

Note that, the image of the weld pool using above processing algorithm is incomplete. However, each frame image is varied dynamically during welding process, so the spline fitting method is adopted. Figure 6.55 shows the processing flow for window1.

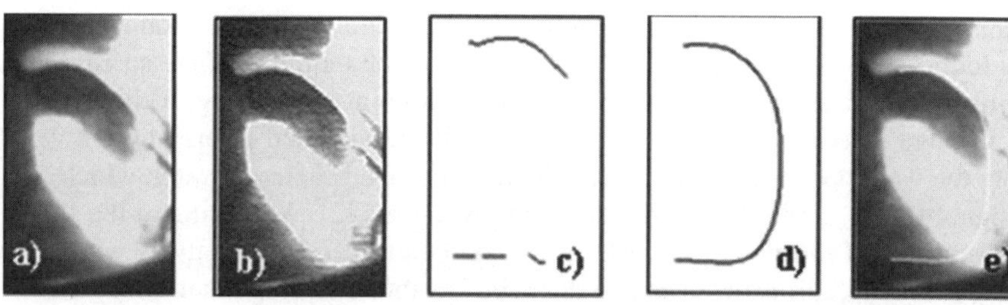

Fig. 6.55 Windows1 image processing **(a)** Original image **(b)** Laplacian filtered image **(c)** edge detection **(d)** spline curve fitting **(e)** validation

6.4.2.2 The Algorithm of Extracting Gap Boundary from Window 2

At the same time, the processing flow for window2 is similar to window1, and the result is shown in Fig. 6.56. The final image processing result indicate that the

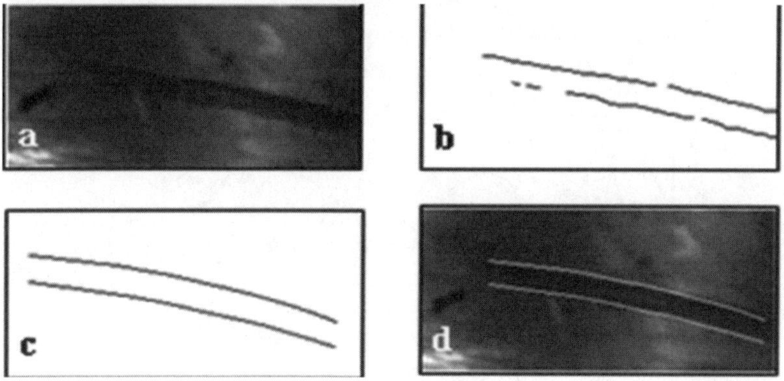

Fig. 6.56 Windows2 image processing (**a**) Original image (**b**) edge detection (**d**) spline curve fitting (**e**) validation

maximal absolute error of the top-side bead width and the gap are 0.92 and 0.1 mm respectively. Meanwhile, the total image processing time is 155 ms, which can meet the real-time control requirements.

6.4.3 Modeling and Control Scheme for Welding Robot System

6.4.3.1 Back-Side Bead Width Dynamic Neural Network Prediction Model

The back-side bead width is an important index in evaluating the weld penetration, However, since the welding backing is needed, the back-side bead information can not be get through the vision sensor in actual welding process of aluminum alloy. Moreover, since there are serious non-linear and uncertainties in the welding process, and the soft sensor model based on the traditional method hardly describe the dynamic state of the weld penetration depth. One of the advantages of the neural networks is that soft sensor model can be set up easily, which needn't understand the prior knowledge of the dynamic and steady welding process. By using error back propagation (BP) neural networks, a back-side bead width dynamic neural network prediction model is established, based on the welding parameters and weld pool geometry, i.e., the dynamical state of the penetration depth is described. The architecture of the each neural network, along with all the input and output variables, is shown in Fig. 6.57. Each neural network contains an input layer, a hidden layer, and an output layer. The input layer contains all the 17 input variables, which are connected to nodes in the hidden layer, represented by circles in Fig. 6.57.

As current and wire feed rate regulation exist time delay in GTAW welding dynamic process, the welding parameters and weld pool geometry information in historic moment should be introduced into the neural network model. Hence, the final welding parameters, namely, $C_p = \{I^{(0)}, I^{(1)}, I^{(2)}, I^{(3)}, V^{(0)}, V^{(1)}, V^{(2)}, V^{(3)}\}$, weld

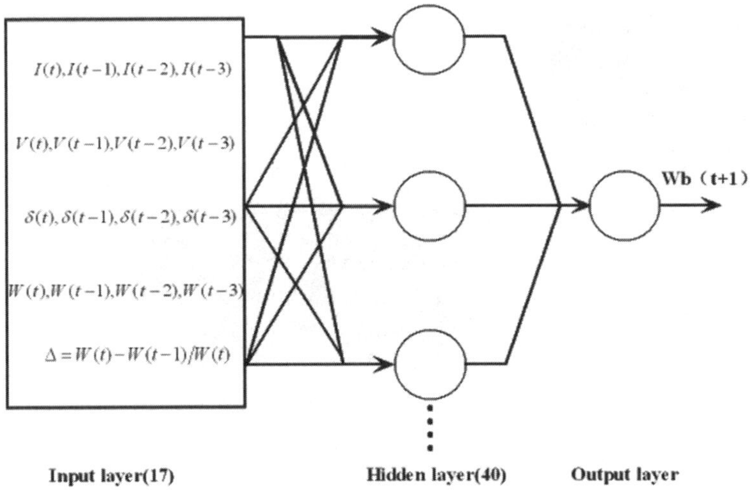

Fig. 6.57 Neural network architecture of the back-side bead width

pool geometry information, namely, $C_g = \{\delta^{(0)}, \delta^{(1)}, \delta^{(2)}, \delta^{(3)}, W^{(0)}, W^{(1)}, W^{(2)}, W^{(3)}\}$, and the variation ratio of the topside bead width, namely, $\Delta = W(t) - W(t-1)/W(t)$, are constructed for the input of the Mapper. And the back-side bead width is intended to be an approximation of the output vector $W(t+1)$.

Multilayer BP neural networks self-learning process is divided into two stages: The fist stage is calculated by using a nonlinear transfer function, which achieves the system nonlinear mapping capabilities; the second stage is to adjust the weights and the bias weight. BP network belongs to multilayer feed-forward networks, it adopts typical supervised learning algorithm, the cost function Least Mean Square (LMS) is used to approximate the target, and the optimized weights calculated by the gradient descent method are stored as one possible set of weights. Adjustment formula of the weight of each layer:

Weights of the output and hidden layer:

$$\Delta v_{ki}(n+1) = \eta \delta_k H_i \tag{6.24}$$

$$\delta_k = (O_k - T_k) O_k (1 - O_k) \tag{6.25}$$

Weights of the hidden and input layer:

$$\Delta w_{ji}(n+1) = \eta \delta_j I_i \tag{6.26}$$

$$\delta_j = H_j(1 - H_j) \sum_{k=0}^{m} \delta_k v_{kj} \tag{6.27}$$

where O_k is the output variable of the output layer, T_k is the desired output, I_i is the output of the input layer, H_j is the output of node j, η is called the learning rate, i is a node in the previous layer, and j is a node in the hidden layer. Traditional

BP algorithm only consider the adjustment the form of the Sigmoid function and learning rate η is equivalent with the influence of the entire network training speed, only by changing the value of η to achieve the adjustment of the learning speed, but this will increase iteration number, and finally lead to extending the learning time. The activation function is defined as the following:

$$O_k = \frac{1}{1+\exp(-\lambda * \sum_j v_{kj}H_i - \theta_j)} \tag{6.28}$$

where λ is the shape factor, v_{kj} is the weights from the previous input to the node j, θ_j is the bias weight of this neuron. The above formula (6.28), introducing the shape factor, makes learning process derivate very rapidly from "flat area" of the error curved surface (the derivative of the Sigmoid function is close to zero). Meanwhile, it can avoid any local optimal solution [43]. For the actual robotic arc welding process, the influence of welding process parameters on the welding pool shape include the pulse peak current, wire feed rate. The range of the welding process parameters and step are defined as: peak current $I_p = (145—175\,\text{A}, 2\,\text{A})$, wire feed rate $V_f = (6–16\,\text{mm/s}, 1\,\text{mm/s})$. Out of total 10 tests, each test contains 150 pulse sequences, 1000 datasets are chosen randomly and included in the training dataset, and the remaining 200 datasets formed the testing dataset for the validation of the neural network. The mean square error for back-side bead width is 0.207 mm in the testing dataset, where a typical value for GTAW of aluminum alloy is 4 mm. Thus, the error in back-side bead width is well within the error limits for process being considered. It can be seen that the neural networks can accurately predict the back-side width, and hence can be used for the prediction model of the feedback control.

6.4.3.2 Compound Adaptive and Fuzzy Controller for Robotic Welding Systems

Due to the influence of the variation of the heat dissipation and gap on the welding process, it is difficult to adjust the welding process parameters using conventional control strategy. Thus, in this paper, a peak current self adaptive regulating controller with weld gap compensation system is made to control the welding process. The block diagram of controlling the penetration depth is shown in Fig. 6.58. From this diagram, the vision system is employed as the feedback mechanism, and then the actual top-side bead width and the gap are input into the back-side bead width dynamic neural network model (BWDNNM). Where δ represents the state of the gap, I_p is the peak current, V_f is the wire feed rate, W_b is the target back-side bead width, and the $\overline{W_b}$ is the output variable from the BP. Meanwhile, in order to produce the corresponding compensation, the disturbance quantity is introduced to the feed forward link. Let the desired back-side bead width be W_{bset}.

The compound controller of the welding penetration is divide into the feed forward part and feedback part. On the one hand, the feedback part is the peak current

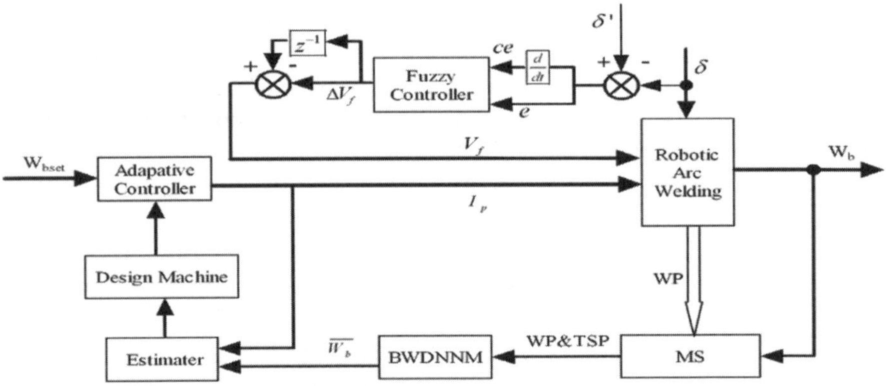

Fig. 6.58 Block diagram of compound adaptive and fuzzy controller

self adaptive regulating controller. It takes minimum variance strategy to control the welding penetration depth by adjusting the peak current [44], which bases on the time varying on-line identification model. The adaptive welding current adjuster based on the minimum variance theory is defined as,

$$\begin{aligned} I(k) = & -1/b_0(a)W_b(k) + a_2W_b(k-1) + a_3W_b(k-2) + a_4W_b(k-3) \\ & + b_1I(k-1) + b_2I(k-2) + b_3I(k-3) \end{aligned} \tag{6.29}$$

On the other hand, authors discuss the fuzzy controller in the feed forward part. Once the gap became wide, the penetration depth would be deeper. At the same time, the feed forward controller is conducted from the corresponding knowledge, i.e., the variation of the wire feed rate is determined by the disturbance variation of the gap. The deviation e and variation ce of the gap for the input variable are defined. The rule at this situation is described by the following if-then form.

$$\text{if } e \text{ is A and } ce \text{ is B, then } \Delta V_f \text{ is C}$$

where A, B, and C are the fuzzy variables. Let [6.-6 mm/s, 6 mm/s] is the universe of ΔV_k, and the membership functions ΔV_f of the fuzzy variables are adjusted based on the constant welding process parameters. If the deviation of the gap is beyond the universe [6.-2 mm, 2 mm], it is difficult to achieve the sound welding joint. The if-parts are determined from the control knowledge of experts in the situation. The control rule is described by using the ambiguousness negative big (NB), negative small (NS), zero (Z), positive small (PS) and positive big (PB) are shown in Table 6.4, which is constructed form 30 kinds of rules.

In general, we can regulate the peak current to the keep the top-side width constant using adaptive controller part and adjust the wire feed rate to resist the disturbance in the variation of the gap with fuzzy controller part.

Table 6.4 Control ruler of fuzzy controller

cee	NB	NS	Z	PS	PB
NB	NB	NB	NS	Z	PS
NS	NB	NS	Z	Z	PS
Z	NS	Z	Z	PS	PS
PS	Z	Z	PS	PS	PB
PB	PS	PS	PB	PB	PB

6.4.4 Penetration Control Procedure and Results by Robotic Welding

The practical experiment has been carried out in five-port connector for rocket motor system using above closed-loop control strategy. Here, the minimum adjustment step of the peak current I_p and wire feed rate are 2 A and 1 mm/s respectively. In this study, the desired value of the back-side width is set to 4 mm. Since the welding process become steady after 10 pulse cycles, the controller keep constant during the initial period. The basic experiments conditions of robotic GTAW for five-port connector are shown in Table 6.5.

The experiment is performed with the compound intelligent controller. Figure 6.59 is the curves of the parameters in closed-loop controller. The back-side bead width is controlled and kept steady regardless of the variation of the gap. In addition, we can see that the actual back-side bead width varied 5.9% around the average values (4.3 mm). In order to evaluate the welding penetration further, X-ray detection is examined. The uniform back-side bead width can be obtained by the control of the weld pool shape. Figure 6.60 is the photographs of the workpiece topside, backside width, and the X-ray results. A good quality of welding joint can be obtained, which can meet the requirements of the corresponding standard.

Table 6.5 Basic experiments conditions of robotic GTAW for five-port connector

Pulse frequency, Hz	2	Material	5456 Aluminum alloy
Welding current I_b/I_p, A	45/200	Specimen size, mm	Φ 180 × 3 (flange thickness:4)
Wire feed rate V_f, mm/s	14	Wire diameter, mm	Φ 1.2
Welding speed V, mm/s	3.2	Arc length, mm	4–5
Shielding gas flow rate, l/min	12.0	Tungsten diameter, mm	Φ 3.2

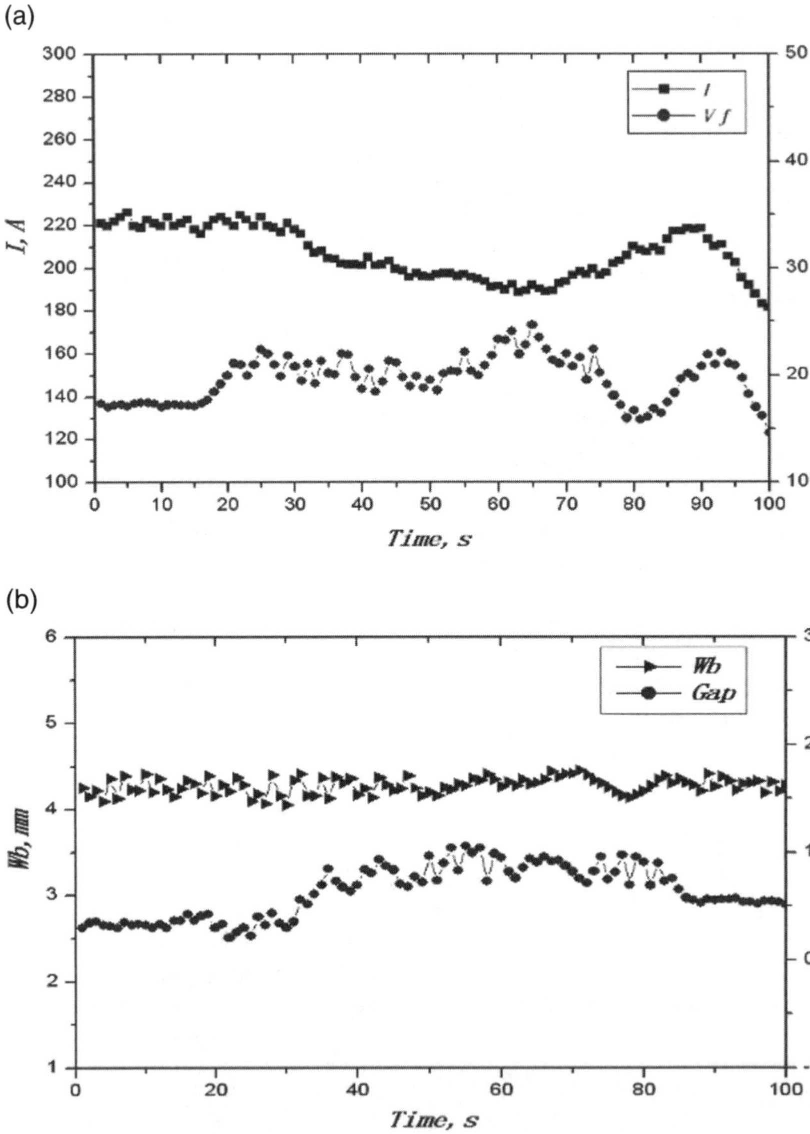

Fig. 6.59 The control process curve of five-port connector with compound controller (**a**) back-side bead width (**b**) welding parameters

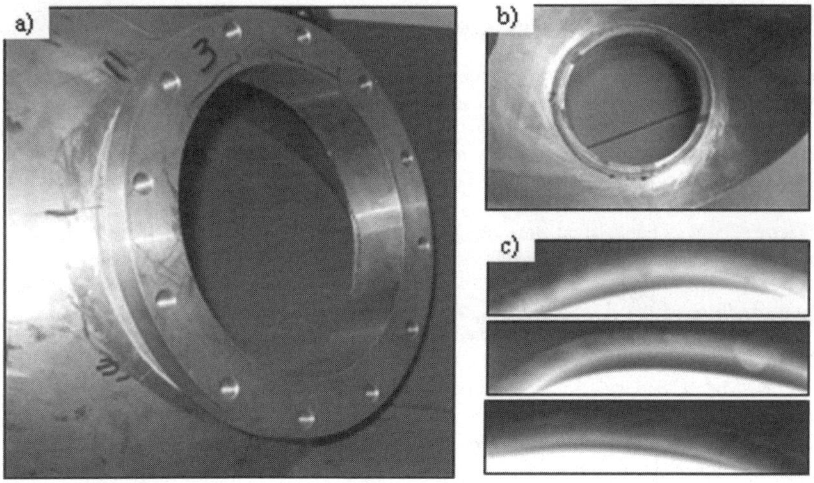

Fig. 6.60 A photo of five-port connector with compound control (**a**) Top-side bead (**b**) Back-side bead (**c**) X-ray inspection

6.5 The Chapter Conclusion Remarks

This chapter addresses the vision sensing and intelligentized control techniques for robotic arc welding. Current teaching play-back welding robot is not with this real-time function for sensing and control of weld process.

Using composed filtering technology, computer vision sensing systems are established and clear weld pool images are captured during robotic pulsed GTAW. Corresponding image processing algorithms are described to pick-up characteristic parameters of the weld pool in real time. Furthermore, intelligentized models and real time controller of weld pool dynamics during pulsed GTAW process have been developed in the robotic systems [19, 20]. Seam tracking is another key technology for welding robotic system. Seam tracking technology by computer vision sensing in real time without robot calibration is discussed. Image processing algorithms are presented to extract the seam trajectory and the offset of the torch to the seam in the weld pool images with grooves. The seam tracking controller is also analyzed and designed [45, 46].

References

1. P. Kopacek. Intelligent manufacturing: Present state and future trends. Journal of Intelligent and Robotic Systems: Theory and Applications. 1999, 26(3):217–229
2. C.S. Tzafestas, S.G. Tzafestas. Intelligent robotic assembly and disassembly: General architecture and implementation case studies, In: S. G. Tzafestas (ed.) Advances in Manufacturing: Decision, Control and Information Technology, Springer, Berlin/London, 1999, pp. 267–282

3. E.P. Vilkas. Automation of gas tungsten arc welding process. Welding Journal, 1966, 45(5):410s–416s
4. A. Suzuki, D.E. Hardt, L. Valavani. Application of adaptive control theory to on-ling GTA weld geometry regulation. ASME Journal of Dynamics System, Measurement, and Control. 1991, 113(1):93–103
5. S.B. Chen. "Visual Information Acquirement and Real-time Control Methodologies for Weld Pool Dynamics during Pulsed GTAW," Tthe invited paper of the THERMEC'2006, Journal Materials Science Forum. in March 2007, vols. 539–543
6. R. Kovacevic, Y.M. Zhang. Real-time image processing for monitoring the free weld pool surface. ASME Journal of Manufacturing Science and Engineering 119(5):161–169
7. K.A. Pietrzak, S.M. Packer. Vision-based weld pool width control. ASME Journal of Engineering for Industry. 1994, 116:86.92
8. D. Przakovic, D.T. Khani. Weld pool edge detection for automated control of weld. IEEE Transition on Robotics and Automation. 1991, 7(3):397–403
9. Y. Suga, M. Mukai, S. Usui, K. Ogawa. Estimation and adaptive control of penetration in GTAW by monitoring dimension of molten pool. Proceedings of the International Conference on Offshore Mechanics and Arctic Engineering – OMAE v 3 Apr 13-17, 1997, ASME 95–100
10. Y. Suga, M. Mukai, S. Usui. Measurement of molten pool shape and penetration control applying neural network in TIG weld of thin steel plate. ISIJ International. 1999, 39(10):1075–1080
11. S.B. Chen, Y.J. Lou, L. Wu, D.B. Zhao. Intelligent methodology for sensing, modeling and control of pulsed GTAW: Part 1–bead-on-plate welding Welding Journal. 2000, 79(6): 151s–163s
12. S.B. Chen, D.B. Zhao, L. Wu, Y.J. Lou. Intelligent methodology for sensing, modeling and control of pulsed GTAW: Part 2 – butt joint welding Welding Journal. 2000, 79(6):164-s–174-s
13. T.G. Lim, H.S. Cho. Estimation of weld pool sizes in GMA welding process using neural networks. The 3rd International Conference on Trends in Welding Research, Gaflinburg, Tennessee, 1993, 135–142
14. S.G. Tzafestas, E. Kyriannakis. Regulation of GMA welding thermal characteristics via a hierarchical MIMO predictive control scheme assuring stability. IEEE Transactions on Industrial Electron. 2000, 47(3):668–678
15. S.G. Tzafestas, S. Raptis, J. Pantazopoulos. A vision-based path planning algorithm for a robot-mounted welding gun. Images Processing and communications. 1996, 2(4):61–72
16. S. Murakami. Weld-line tracking control of arc welding robot using fuzzy logic controller. Fuzzy Sets and Systems. 1989, 32(2):31–36
17. Y. Suga, M. Naruse. Application of neural network to visual sensing of weld line and automatic tracking in robot welding. Welding in the World. 1994, 34:275–284
18. S.B. Chen, L. Wu, Q.L. Wang. Self-learning fuzzy neural networks for control of uncertain systems with time delays. IEEE Transactions on Systems, Man, and Cybernetics, Part B: Cybernetics, 1997, 27(1):142s–148s
19. S.B. Chen. On the key intelligentized technologies of welding robot. Lecture Notes in Control and Information Sciences. 2007, LNCIS 362:105–116
20. W.J. Chen, Research on local autonomous intelligent welding robot system and its remote control. Doctoral Dissertation Shanghai Jiao Tong University, 2004.
21. A. Bauchspiess, S.C. Absi Alfaro, L.A. Dobrzanski. Predictive sensor guided robotic manipulators in automated welding cells. Journal of Materials Processing Technology. 2001, 109: 13–19
22. S.B. Chen, Y. Zhang, T. Qiu, T. Lin. Welding robotic systems with vision sensing and self-learning neuron control of arc weld dynamic process. Journal of Intelligent and Robotic Systems. 2003, 36(2):191–208
23. S.B. Chen, Y. Zhang, T. Lin, T. Qiu, L. Wu. Welding robotic systems with vision sensing and real-time control of dynamic weld pool during Pulsed GTAW. International Journal of Robotic and Automation. 2004, 19(1):28–35
24. S.B. Chen, X.Z. Chen, J.Q. Li, T. Lin, Acquisition of welding seam space position information for arc welding robot based on vision. Journal of Intelligent & Robotic Systems. 2005, 43(1):77–97

25. Zh.Y. Zhu, T. Lin, Y.J. Piao, S.B.Chen. Recognition of the initial position of the weld based on the image pattern match technology. The International Journal of Advanced Manufacturing Technology. 2005, 26:784–788
26. S.B. Chen, X.Z. Chen, J.Q. Li, T. Lin. Acquisition of welding seam space position information for arc welding robot based on vision. Journal of Intelligent & Robotic Systems. 2005, 43: 77–97
27. Z.Y. Zhu. Research on welding robot recognizing and guiding of the initial welding position with visual method. Doctoral Dissertation Shanghai Jiao Tong University. 2004
28. L. Zhou. Vision-based autonomous programming methods of welding robotic path. Doctoral Dissertation Shanghai Jiao Tong University. 2006
29. J.-Y. Yu, S.-J. Na. A study on vision sensors for seam tracking of height-varying weldment. Part 1: Mathematical model. Mechatronics. 1997, 7(7):599–612
30. J.-Y. Yu, S.-J. Na. A study on vision sensors for seam tracking of height-varying weldment. Part 2: Applications. Mechatronics. 1998, 8(1):21–36
31. J.S. Smith, C. Balfour. Real-time top-face vision based control of weld pool size. Industrial Robot. 2005, 32(4): 334–340
32. L. Zhou, T. Lin, S.B. Chen. Autonomous acquisition of seam coordinates for arc welding robot based on visual servoing. Journal of Intelligent and Robotic Systems. 2006, 47:239–255
33. J.S. Kim, Y.T. Son, H.S. Cho, K. Il Koh A robust method for vision-based seam tracking in robotic arc welding. Mechatronics. 1996, 6(2):141–163
34. H.-C. Kuo, L.-J. Wu. An image tracking system for welded seams using fuzzy logic. Journal of Materials Processing Technology. 2000, 120(1):169–185
35. K.Y. Bae, T.H. Lee, K.C. Ahn. An optical sensing system for seam tracking and weld pool control in gas metal arc welding of steel pipe. Journal of Materials Processing Technology. 2002, 120(2):458–465
36. G. Jingguo, Z. Zhengqiang, H. Defu, C. Ligong. A vision-based algorithm for seam detection in a PAW process for large-diameter stainless steel pipes. International Journal of Advanced Manufacturing Technology. 2005, 26(10):1006–1011
37. S. Yamane, Y. Kaneko, N. Kitahara, K. Ohshima, M. Yamamoto. Neural network and fuzzy control of weld pool with welding robot. Ind. Appl. Soc. Annu. Meet.: Conf. Rec. 1993 IEEE. 1993, 3:2175–2180
38. D.B. Zhao, S.B. Chen, L. Wu, M. Dai, Q. Chen. Intelligent control for the shape of the weld pool in pulsed GTAW with filler metal. Weld Journal. 2001, 80(11):253–260
39. J.S. Smith, J. Lucas. Vision-based seam tracker for butt-plate TIG welding. Journal of Physics E: Scientific Instruments. 1989, 22(9):739–744
40. H. Luo. Robotic welding, intelligence and automation, laser visual sensing and process control in robotic arc welding of titanium alloys. 2004, LNCIS 299:110–122
41. H.B. Chen, T. Lin, S B. Chen. Compound intelligent control of robotic welding systems with visual monitoring during AI alloy pulsed GTAW. Submitted to Journal of Intelligent & Robotic Systems, 2008.
42. P. Meer, B. Georgescu. Edge detection with embedded confidence. IEEE Trans on PAMI. 2001, 23(120):1351–1365
43. P. Song, F. Zuxiang. The joint optimization of BP learning algorithm. Journal of Circuits and Systems. 2000, 5(3):26.30
44. J.J. Wang, T. Lin, S.B. Chen. Obtaining of weld pool vision information during aluminum alloy TIG Welding. International Journal of Advanced Manufacturing Technology. 2005, 26:219–227
45. H.Y. Shen, H.B. Ma, T. Lin, S.B. Chen. Research on weld pool control of welding robot with computer vision. Industrial robot. 2007, 34(6):467–475
46. H.Y. Shen, J. Wu, T. Lin, S.B. Chen, Arc welding robot system with seam tracking and weld pool control based on passive vision. The International Journal of Advanced Manufacturing, 2007.

Chapter 7
Conclusion Remarks

To overcome the bottleneck problems of effective control of weld quality during automatic and robotic arc welding process, this book presents our researching works on intelligentized methodology for arc welding dynamical process in the Intelligentized Robotic Welding Technology Laboratory (IRWTL), Shanghai Jiao Tong University, P. R. China. The content of the book involves visual information acquiring, knowledge modeling and intelligent control of arc welding dynamical process. The conclusions of the book are summarized as follows:

(1) Designed the appropriate visual sensors and systems based on arc spectrum and intensity of different welding materials to capture relevant clear images of welding pool during pulsed GTAW, e.g., the visual sensing systems and images for low carbon steel and aluminum alloy weld pools during pulsed GTAW.

(2) Developed the appropriate image processing algorithms for different weld pools to extract visual characteristic information of arc welding process, e.g. acquiring two and three dimensional characteristics from monocular image of weld pool during pulsed GTAW for low carbon steel, stainless steel and aluminum alloy weld pool images respectively.

(3) Developed the appropriate modeling methods for different welding dynamical processes to describe the dynamical characteristics of the weld pool during pulsed GTAW both by identification method and by intelligentized method, e.g., linear models and nonlinear transfer function models, artificial neural network models, fuzzy and knowledge models of weld pool dynamical process for predicting and control of weld pool dynamical characteristics.

(4) Completed various intelligent control strategies for arc welding process can realize real-time control of weld pool dynamics and seam formation during low carbon steel, stainless steel and aluminum alloy pulsed GTAW, such as developed the self-regulating PID controller, the fuzzy controller, the PSD controller, the neural network self-learning controllers, model free controller and the composite intelligent controller for dynamical weld pool during pulsed GTAW, and corresponding closed loop control systems for pulsed GTAW process.

(5) Integrated the visual sensing, intelligentized modeling and control strategies in a welding robot can realize real-time control of weld pool dynamics during robotic welding process, e.g., intelligentized welding robot systems with monitoring and real-time control of weld pool dynamics, which are the bottleneck technologies of intelligentized welding robot.

S.-B. Chen, J. Wu, *Intelligentized Methodology for Arc Welding Dynamical Processes*, Lecture Notes in Electrical Engineering 29,

As is well known, more and more modern welding technics are coming forth with welding multi-subject intersectant technology and with the coming of new welding materials. Up till now, there are about more than one hundred kinds of welding methods, and these welding technics have being transformed from traditional handcraft to a some extent of mechanization, half-automation and robotic welding. With the development of information science and technologies, such as computer, control theory, robotics and artificial intelligence, the dream of a few generation welders about real intelligent machine welding instead of manual work will come true in not far future. In intending intelligent welding robots, the key technologies on acquiring visual information, intelligetized modeling and real-time control strategies present in this book will be applied and further developed indispensably.

Index

Printing: Krips bv, Meppel, The Netherlands
Binding: Stürtz, Würzburg, Germany